Establishing Scientific Classroom Discourse Communities

Multiple Voices of Teaching and Learning Research

Establishing Scientific Classroom Discourse Communities

Multiple Voices of Teaching and Learning Research

Edited by

Randy Yerrick
San Diego State University

Wolff-Michael Roth
University of Victoria

LAWRENCE ERLBAUM ASSOCIATES, PUBLISHERS
2005   Mahwah, New Jersey   London

Copyright © 2005 by Lawrence Erlbaum Associates, Inc.
All rights reserved. No part of this book may be reproduced in
any form, by photostat, microform, retrieval system, or any other
means, without the prior written permission of the publisher.

Lawrence Erlbaum Associates, Inc., Publishers
10 Industrial Avenue
Mahwah, New Jersey 07430

Cover design by Randy Yerrick

Library of Congress Cataloging-in-Publication Data

Establishing scientific classroom discourse communities: multiple voices of teaching
   and learning research / edited by Randy Yerrick, Wolff-Michael Roth.
      p. cm.
Includes bibliographical references and indexes.
ISBN 0-8058-4434-1 (cloth : alk. paper)
   1. Science-Study and Teaching.   I. Yerrick, Randy, 1963-   II. Roth, Wolff-Michael,
1953-

Q181.E75   2004
507'.1'073—dc22   2004046926
                                                                                CIP

Books published by Lawrence Erlbaum Associates are printed on acid-free paper,
and their bindings are chosen for strength and durability.

Printed in the United States of America
10 9 8 7 6 5 4 3 2 1

Contents

Foreword   ix
Richard Duschl

Introduction: The Role of Language in Science
Learning and Teaching   1
Randy Yerrick and Wolff-Michael Roth

1   Language in the Science Classroom: Academic
Social Languages as the Heart of School-Based Literacy   19
James Paul Gee

METALOGUE: Situating Identity and Science Discourse   39
*James Paul Gee, Gregory J. Kelly, Wolff-Michael Roth,
and Randy Yerrick*

2   Telling in Purposeful Activity and the Emergence
of Scientific Language   45
Wolff-Michael Roth

METALOGUE: *Understanding*, of Science and Teaching   73
James Paul Gee, Wolff-Michael Roth, and Randy Yerrick

3 Discourse, Description, and Science Education 79
Gregory J. Kelly

METALOGUE: Contrasting Sociolinguistic and
Normative Approaches to Redesigning School 105
Gregory J. Kelly, Wolff-Michael Roth, and Randy Yerrick

4 Essential Similarities and Differences Between Classroom and Scientific Communities 109
G. Michael Bowen

METALOGUE: Authentic Science Education:
Contradictions and Possibilities 135
G. Michael Bowen, Wolff-Michael Roth, and Randy Yerrick

5 Dialogic Inquiry in an Urban Second-Grade Classroom: How Intertextuality Shapes and Is Shaped by Social Interactions and Scientific Understandings 139
Maria Varelas, Christine C. Pappas, and Amy Rife

METALOGUE: Lifeworld Language, Scientific Literacy,
and Intertextuality 169
G. Michael Bowen, Maria Varelas, Wolff-Michael Roth, and Randy Yerrick

6 Meaning and Context: Studying Words in Motion 175
Cynthia Ballenger

METALOGUE: Balancing Our View of Science With
Language Usage and Culture 193
Cynthia Ballenger, Randy Yerrick, and Wolff-Michael Roth

7 Science for All: A Discursive Analysis Examining Teacher Support of Student Thinking in Inclusive Classrooms 199
Kathleen M. Collins, Annemarie Sullivan Palincsar, and Shirley J. Magnusson

METALOGUE: A Question of Perspective in Supporting
the Learning of Students With Special Needs in
Inquiry-Based Science Instruction 225
*Margaret Gallego, Shirley J. Magnusson,
Kathleen M. Collins, Annemarie Sullivan Palincsar,
Wolff-Michael Roth, and Randy Yerrick*

CONTENTS

8 Playing the Game of Science: Overcoming Obstacles
in Re-negotiating Science Classroom Discourse   231
Randy Yerrick

METALOGUE: Emergent Nature of Classroom Talk
and the Game of Science   259
*Cynthia Ballenger, James Paul Gee, Randy Yerrick,
and Wolff-Michael Roth*

9 Teaching Science in Urban High Schools: When the
Rubber Hits the Road   265
Kenneth Tobin

METALOGUE: Expanding Agency and Changing
Social Structures   287
Kenneth Tobin, Randy Yerrick, and Wolff-Michael Roth

10 When the Classroom Isn't in School: The Construction
of Scientific Knowledge in an After-School Setting   293
Margaret A. Gallego and Noah D. Finkelstein

METALOGUE: Scientific Versus Spontaneous Concepts
at Work in Student Learning   315
*Maria Varelas, Margaret A. Gallego, Randy Yerrick,
and Wolff-Michael Roth*

Author Index   321

Subject Index   327

Foreword

Richard Duschl
King's College, London

Survey teachers about the meaning of the phrase "the language of the science classroom" and you will no doubt receive a plethora of perspectives that are rooted to personal experience, context, or educational commitments. Some teachers' points of view will be wedded to the language of content learning, knowing *what*, while some will be committed to that of process skill learning, knowing *how*. Others may well be grounded to problem-based learning for purposes of engendering the language of reasoning or inquiry learning for teaching the language of the nature of science. For some teachers, the language of the science classroom will be one associated with asking questions or designing and reporting experiments. Yet other teachers will embrace ideas that the language of the science classroom teach about the discourse of power and authority inherent in science rhetoric.

Survey a group of educational researchers about the "language of science teaching" and you could very well find yourself embedded in the throes of a "science war" debate. The contributions for this edited volume will certainly add fuel to the science wars debate. But what distinguishes the perspectives here from others is the grounding of voices in classroom and other learning environments. To be honest, I find it difficult to embrace or endorse several of the perspectives presented in this volume; perspectives that are at times against schools, against school science, against scientific inquiry, and against academic science. Thus, I was surprised to be asked to pen the foreword for this volume because editors Yerrick and

Roth certainly know where I stand on these matters. So, why invite me? And, why should I accept? The answer lies, I think, in the goal of this volume, a "volume devoted to examining the complexities of establishing scientific learning communities," a goal that I share and a topic that the editors are correct in saying needs to be debated among science education researchers.

Yerrick and Roth have brought together authors of different backgrounds and with varying perspectives to comment on ways for "Establishing Scientific Classroom Discourse Communities." As you will learn, particularly from reading the dialogues following each chapter, there is a great diversity of commitments among the contributing authors. Because they embrace sociocultural perspectives of learning, some important and significant differences among the contributors emerge. The chapters and the end-of-chapter dialogues expose the authors' competing points of view about the structure of science, the nature of learning, the purpose of schools, the nature of academic discourse, and, ultimately, the dynamics for establishing science discourse.

You, the reader, will learn about how academic social languages, phenomenological and linguistic frameworks and sociolinguistic and ethnographic components of science studies can function as theoretical frameworks for discourse communities. You will be guided to think about how classroom and scientific discourse communities are similar and different, how intertextuality shapes dialogical inquiry, and how we know what practices are best for teaching specific content knowledge and how these are best learned by diverse children. You will be asked to consider how out-of-school learning environments can inform classroom discourse communities and how classroom discourse communities might change to embrace learners with special needs and learners with histories of failure.

This volume presents many challenges to the reader, not the least of which is to get you thinking about what it is we mean by the language and discourse of classrooms. Establishing classroom discourse communities suggests to me a need to locate language and discourse somewhere along two ends of a continuum. At one end of the continuum are the strictly psychological explanations and biological cognitive mechanisms that help us understand the individual as a thinker. Here, language and discourse are perceived as a window into the mind. At the other end are the strictly anthropological explanations and cultural mechanisms that help us understand the individual in society. Here, language and discourse are perceived as tools for achieving, among other things, cultural capital and the construction, representation, and dissemination of knowledge claims. Personally, I don't believe it is an either/or debate; others may disagree. But that is the point of this volume, to engender the debate by exposing alternative perspectives.

FOREWORD

As I read this volume with all of its various theoretical and at times radical frameworks and perspectives about science discourse, I was reminded of an occasion when I heard Michael Apple speak in the mid-1980s at an annual meeting of the American Educational Research Association. The session was on critical theory, another radical perspective, and its role in shaping schools and the curriculum of schools. Presenters included such radical thinkers as Henry Giroux, Tom Popkewitz, among others I can't now recall who were making important contributions to the debate about schools reproducing society. What struck me then, and what motivates me to share this episode now, was that Michael Apple as the discussant for the session chose not to make any comments about the theoretical perspectives being put forth. Rather, and uncharacteristically for him I thought, he challenged the community of critical theorists to think about turning away from theory and toward practice, specifically to the design of materials and programs that instantiated critical theory. Otherwise, he argued, the radical ideas would never be taken seriously.

As you read this book, let me pose a question for you the reader to consider. Is the domain of language and discourse in science classrooms in need of more theorizing or in need of more fully developed materials and programs that implement theory? I think Apple's challenge is as apropos for critical theory then as it is for establishing science discourse communities now. What do you think? Along with the authors of the volume, I invite you to join the debate.

Introduction: The Role of Language in Science Learning and Teaching

Randy Yerrick
San Diego State University

Wolff-Michael Roth
University of Victoria

FRAMING THE SCIENCE EDUCATION REFORM ISSUES CONFRONTING TEACHERS, STUDENTS, EDUCATORS, AND RESEARCHERS

Once again, science learning in the United States has been put in the balance and found wanting by the National Commission on Mathematics and Science Teaching for the 21st Century (2000) and the Third International Mathematics and Science Study (TIMSS, 2001). Whether measured by student achievement-test scores or in class accounts of science teaching the message seems clear: As a nation the United States is sadly slipping away from our professed ideal for preparing our students. The consistency of these reports has spawned a variety of groups from legislators to scientists to educators who are drafting their plans for a new scientific literacy and rhetorical calls for new waves of reforms.

Professions outside of science and education have become involved in the quest for excellence in science learning through a variety of influences. Linguists have searched for ways that specific genres and norms of discourse inherent to culturally significant communities could be shared or appropriated to allow the entrance of "others" into foreign communities. Anthropologists and sociologists have turned their attention to scientific communities themselves, questioning the very nature of how scientists have been represented (or misrepresented) in their naturalistic settings. In an attempt to test popular constructions of science (based on scientists' own historical accounts), researchers have explored how scientists in research laboratories and other data-gathering contexts operate (Latour & Woolgar, 1986; Roth & Bowen, 1999; Traweek, 1988). The operation of science and the development of scientific theories in the laboratory discourse community are very different from the notion promoted prior to the 1970s that a standard scientific method was employed by scientists to make sense of the world (Millar, 1989).

Anthropological reports revealed that school characterizations of objective, rational scientific work drew on incomplete images of science in the making and theory development, which are far more subjective, messy human endeavors than had been represented in the past. The combination of sociological characterizations of science with the sociolinguistic revelation of the roles of culture and discourse stimulated new interests for educators to find ways to demystify science and thus open multiple ways of knowing (Roth & McGinn, 1997). Researchers even began to study ways of spawning communities of practice within classrooms that embraced certain discourse norms representative of those studied in scientific naturalistic settings (Roth, 1998). The hope was to broaden discussions of what counts as science for students in the process of constructing interesting alternatives (Aikenhead, 1997; Watson-Verran & Turnbull, 1995).

Our nation is confronted with a dilemma heretofore not experienced in the history of science education reform. While past school reforms focused upon the creation of exemplary curriculum, upgrading teacher knowledge, developing effective and transferable classroom management systems, and connecting teachers and students to both historically relevant and contemporary research findings, today's teachers find themselves in a tug-of-war between high-stakes standards-based education and the expectation that all students will succeed in science regardless of the sociocultural or socioeconomic conditions. Although student achievement has long been in the forefront of science education reform, historically, equity in science education had only remotely been addressed as an issue of knowledge access and gender. However, today's reform rhetoric calls upon science teachers in all contexts to connect science curriculum to children's lives in culturally, linguistically, sociologically significant ways. As science educators we have yet to address where current or future teachers will obtain this kind of knowledge, but education policy and legislation in many states falsely assume that we are well equipped as teachers to teach in such ways and that researchers and state assessments can accurately recognize scientific literacy among all students.

It is important to note that recommended reforms of the past differ from current science education reform recommendations. From the 1940s through the 1960s, reform efforts focused on specific content or pedagogy to correctly represent the discipline of science. Little attention was paid to the number or kinds of students who actually were succeeding in response to the massive investment of post-Sputnik science curriculum resources. Times changed. Once the space race had essentially been won, the attention of American educators turned to the disparity that school science had fostered—perhaps even propagated. Science classrooms were not representative of the democratic ideas we professed. For example, physics classes often consisted of small select homogenous communities of White, upper-middle-class males, free of gender or ethnic diversity. Although science reforms had been successful in replicating certain scientific attributes, the success achieved was not without its consequences including a sense of inequity toward specific groups of students.

The response of some researchers to such a revelation was to study the extent to which gender and ethnicity determined the predictive success of a student—a phenomenon that came to be known as *gap gazing* (Rodriguez, 2001). Others drafted new visions for scientific literacy with the expectation that all students could succeed and coined the phrase "Science for all Americans" (implying that Americans live within the boundaries of the U.S. borders as opposed to in South America or Central America). Some researchers turned their attention to teachers who had demonstrated success in science teaching and explored the nature of their pedagogy, content knowledge, and other components of teacher knowledge.

This book highlights the central issues of teaching and learning as they relate to an emerging body of cognition, ethnographic, sociolinguistic, and sociocultural research knowledge. As Kelly argues in this volume,

> ... analysis of the educational opportunities for students needs to consider the linguistic resources made available and how students are positioned to engage with such resources ... from a methodological point of view, ethnography and sociolinguistics provide ways of understanding the language forms of students, teachers, scientists, and community members, as they collectively construct school science. Sociolinguistics offers ways of understanding the differences between students' ways of talking and the discourse of more formalized epistemological communities. (p. 99)

Our chapters introduce the reader to some of the landmarks of past science education reform and highlight reasons for departure from past reform efforts. From this nonexhaustive overview the reader will come to appreciate that the growing body of sociocultural knowledge is part of the foundation required for assessing the current status of science teaching and learning of today's science education standards context. The issue of expanding teacher knowledge is reflected in the contemporary research examining issues of curriculum revision, teacher pedagogy, cultural sensitivity, teacher professional development, and classroom change. Some issues we address in this book include:

- What central tensions exist between today's literacy movements and the role science has played in school stratification and student achievement?
- Why is a sociolinguistic interpretation essential in examining science education reform?
- What similarities and differences exist between classroom communities and scientific communities?
- What central tensions are inherent to colonizing children with scientific ways of thinking, speaking, and acting, which need to be balanced when transforming classroom talk?
- What aspects of scientific communities are desirable in science classrooms, and can they be cleanly extracted from the whole?
- What special considerations must teachers heed when seeking to promote science discourse in language minority classrooms?
- What design implications should be considered before embarking upon the research, assessment, and evaluation of an evolving scientific classroom community?
- What other special needs of students could problematize our nation's goal of achieving "science for all Americans"?
- What obstacles confront teachers who attempt to renegotiate classroom knowledge?
- How does a teacher's central role of change agent affect the outcome of efforts to shift science classroom discourse?
- How are curricular issues intricately related to the process of transforming classroom talk and evaluating students' engagement and achievement?
- How might other educational contexts (e.g., after-school science programs) inform our efforts to change traditional science classroom teaching?

As editors we have selected leading authors in their respective fields to address the foregoing issues. We have purposefully summoned a variety of voices to enrich the discussions of classrooms from both internal and external perspectives. By weaving together a mélange of sociological, philosophical, educational, practical, and linguistic voices this volume explicates alternative frameworks through which to view successful teaching and learning, obstacles for change, and future directions for research and reform.

INTRODUCTION 5

We first offer chapters expounding upon the research and philosophical literature surrounding sociocultural views of science classroom communities. Our goal is to present in the first few chapters a solid theoretical framework for our interpretation of science and the uses of language in both classroom and scientific discourse communities. As each of the authors have adopted some aspect of sociocultural and sociolinguistic perspectives to examine their issue of science learning and teaching we have laid out this well-developed framework to subsequently examine classrooms themselves as active laboratories informing practice. We contend that if educators wish to discuss what counts as scientific activity, they must first agree upon individual and collective definitions for science manifest through social interaction and uses of language in particular ways for particular purposes. Finally, we explore less-than-ideal settings to inform our claims that all children can learn science and what circumstances most influence successful teaching. In doing so we expose the assumptions that underlie calls for equity and "science for all Americans." As authors of this volume, we agree that science is important for all children to learn and that all children can. However, such visions and subsequent pedagogical recommendations face challenges to their survival given increasing calls for standards-based teaching and assessment. As such, the examples presented in this volume demonstrate ways a sociolinguistic and sociocultural approach can contribute to normative considerations for educational reform.

CHAPTER ORGANIZATION AND STRUCTURE

Although many of the authors we have selected weave their analyses with strands common to sociolinguistic research, in several instances more anthropological, epistemological, and psychological perspectives are used to ground this work in a variety of educational arenas. We thank our contributors for using a broad base of literature (e.g., Derrida, Ehn, Holzkamp, Popper, Schwab, Vygotsky, Wertsch) that not only provides richer conversations and deeper connections but also extends the discussions beyond an American view of literacy and the challenges facing U.S. science teachers.

Each chapter is directed toward a specific issue that we feel is central to making change in science classroom discourse. These are identified in the following section and often returned to in subsequent chapters (e.g., Kelly's philosophical underpinnings expressed through Ballenger's design and Roth's analyses). In an attempt to further enrich the discussion and debate surrounding these issues, we invited the authors to respond summarily and critically to one another's chapters. We believe that modeling this kind of intellectual interchange will bring the issues and evidence closer together around our central themes. We believe neither that the current knowledge in these areas is complete nor that a necessarily complete consensus exists among researchers about where answers can be found. We offer these exchanges between authors at the conclusion of each chapter (credited to each author as a metalogue) because opportunities for insights and critiques may lead the reader to other interpretations in the field on these core issues. Such public exchanges in which jargon is frankly clarified and joint meanings are co-constructed among colleagues are rare at professional meetings and rarer still when captured in print. We hope that these exchanges will serve to broaden the voices, representations, and registers that currently contribute to the discussion of reform rhetoric and the process of formulating solutions to today's science teaching issues. Through the voices of researchers, ethnographers, teacher educators, linguists, teachers, philosophers, and reformers, the authors of this volume address core issues of science teaching and research—drawing upon unique perspectives that all play an integral role in defining the nature of meeting the challenges in an era of national standards for science education.

We hope that readers of this volume benefit as much from analyzing these varied perspectives as the authors did from one another during their compilation. Whether it improves the readers' science teaching, assists their research, or helps to better prepare tomorrow's science teachers, it is our goal as authors and editors that readers will join us after your reading of these perspectives and consider the challenges we all face as educators navigating the restless seas of reform as we strive to improve science for all in America.

CHAPTER OVERVIEWS

In chapter 1 James Gee sets the stage for our discussions by defining important components of current literacy rhetoric. Gee takes us through his interpretation of how students acquire formal discourses that differ from their home-based discourses, and he outlines the challenges science education reform faces in promoting science discourse in all classrooms. Gee outlines how academic social languages are at the heart of school-based literacy, arguing that reforms require students to experience immersion in the target discourse, experience multiple models, and be provided with adequate mentorship in order to understand not only language but also how language is at work within communities of practice.

If all students are to acquire school-based literacy, several preconditions—for which schools have yet to provide a venue—must be met, including the establishment of learning communities in which students can themselves become accepted and valued members who believe in their abilities and use their new social language to accomplish their own worthwhile goals. Gee is quick to remind us that establishing these preconditions has always been problematic in science because science makes demands on students to use language, orally and in print, as well as other sorts of symbol systems that are at the heart of higher levels of school success—and are often reserved for privileged students alone. Thus, Gee criticizes past educational reforms that encouraged students to embrace ethnic or economic vernaculars without situating science talk in children's lives, since it did not offer a way to bridge to and enhance the acquisition of one or more academic forms of language.

In considering why schools poorly train students to learn new social languages, Gee emphasizes that learners also require expanded texts that display the fuller forms of the social language they are to acquire, placed in the midst of practice and discussion, not just assigned as outside reading. While advocating for instruction for academic languages, Gee underscores the complexity of acquiring science language:

> Social languages can be understood in two different ways: either as largely verbal or as situated. I can understand a piece of a scientific social language largely as a set of verbal definitions or rather general meanings for words and phrases and, in turn, relate words and phrases to each other in terms of these definitions or general meanings and general knowledge about grammatical patterns in the language. However, such an understanding is not all that useful when one has to engage in any activity using a specialist language. (p. 24)

In addressing this complexity Gee recommends that students need reading lessons on such expanded texts, whereby teachers model how to read such texts and engage the students in overt discussions about the language and genre conventions of such texts and how these conventions arise out of history and relate to current practices.

Michael Roth extends this line of thinking in chapter 2 as he draws on phenomenological and linguistic frameworks to demonstrate that new discourses are neither homogenous nor learned in unambiguous ways. Rather, Roth discourages the practice of guiding students to specific discourses, which has a lot to do with cultural reproduction and the associated cultural domination by those who have most of the cultural capital (Foucault, 1979). Discourse acquisition has more to do with allowing students to find their own ways of participating in culture and create their own forms of talking. As such the problem of science as language must be decentered to be solved. In his chapter Roth does this by focusing on phenomenological and pragmatic aspects of science conceived as language. Roth argues that the idea of science as language has not been theorized sufficiently; a full articulation of the idea brings to the foreground major issues such as learning as by-product and identity that encourage a radical rethinking of current instructional practices.

Roth's chapter unfolds in four stages. First, he provides a phenomenological and pragmatic perspective of language-in-use, and second the exemplification of a best-case scenario in which upper-middle-class students are engaged in collaboratively building concept maps and in the process evolving proto-scientific languages. Third, Roth offers vignettes that illustrate the problems immersion students have in science conversations despite eight years of French as school language. Finally, he broadens the discussion of obstacles in appropriating science discourse as incorporating larger constructs like language interaction, self, and the purposes of everyday activities within communities.

Roth concludes with a challenge that changing the standard classroom science discourse will require nothing less than rethinking the colonization approach to school science. He argues that under current classroom conditions students are never really immersed, and they subsequently hardly ever learn situated meaning and phrase in practice. Roth surmises,

> Language is more than a code; it is part of a form of life and therefore is a central aspect of our identity, a means of establishing who we are in relation to others and how we understand ourselves. Science educators rarely consider the learner as someone whose identity development is profoundly bound up with language and culture. (p. 63)

Even when a teacher models a better phrasing for one individual, many other students in the class never benefit because, for one reason or another, they have not been attuned to the current conversation. To make any progress educators must resist the urge to hide behind the jargon of standards or dodge the paradox central to his chapter that acquiring understanding of a social language requires access also to a specific set of cultural resources that are deemed more central than others. Roth argues that still more needs to be said about classrooms if we are to speak to salient equity issues (e.g., children must, unless we substantively change schools, come to control forms of academic registers, at least the sorts of social languages that are used in their textbooks and on their tests, if they are to get out of high school, go to college, and compete on equal terms in society).

In chapter 3 by Gregory Kelly, we offer the philosophical bases for our belief that a sociolinguistic approach is a reasonable lens through which to interpret past failures of reform and future directions for reaching the national vision that all students can have access to scientific discourse. Kelly has devoted his chapter to explicating the philosophical underpinnings associated with this particular line of inquiry—offering a three-pronged argument drawn from evidence diverse enough to be compelling from a variety of perspectives—even those outside of sociolinguistic research. For example, Kelly argues that the kind of science studies described in chapter 3 offer educators more detailed information about scientific practices and expand the possible repertoire of what counts as science education. The following quote extracted from his chapter conveys Kelly's balance of theoretical integrity with pragmatic methodology.

> Science studies can serve as analogous cases for similar investigations in educational settings. Simply put, a central lesson of science studies is not conclusive evidence about the true nature of science, but rather the methodological orientation to investigate science education from an empirical, descriptive point of view as it is created through social activity. (p. 86)

Kelly's argument draws on descriptions of the scientific discipline like Schwab's (1978) syntactic structures, which are discussed more centrally in later chapters (e.g., chap. 7).

Kelly makes the case that the methodological perspective used in science studies, with its sociolinguistic and ethnographic components, helps educators to learn about how science in action develops in the classroom. Thus, the extent to which we achieve reform visions for developing scientific classroom discourse communities will likely be viewed by our clarity of vision of how researchers see school-science-in-the-making through the concerted activities of members of a relevant community (Kelly, Chen, & Crawford, 1998). Kelly's chapter provides a rationale for ethnographic and sociolinguistic descriptions of everyday life in science classrooms and helps his audience to consider the value of social studies of science for framing the issues of what counts as science in schools.

While acknowledging the substantive critique of certain descriptive studies of scientific practice, Kelly argues that, despite interpretative problems, attention to the concerted actions of people in educational settings offers insight into epistemological issues in science education. By distinguishing descriptive study from normative considerations, Kelly proposes ways of using empirical studies of scientific practice for applications in educational research.

The value of such ethnographically oriented studies in science and science education is supported in the form of three reasons for description: Description specifies uses of language in social action, description offers intellectual puzzles for continued deliberation regarding educational issues, and description provides ways for increasing understanding of people in different life situations. Kelly warns us, however, not to translate descriptive accounts of science directly into normative considerations for education situations (the naturalistic fallacy). If we do, we are likely to be sorely disappointed inasmuch as students are not scientists (Driver, 1990) and classroom communities are not scientific communities.

In chapter, 4 G. Michael Bowen picks up the development of a sociolinguistic perspective for scientific classroom discourse. Bowen demarcates important similarities and differences in classroom and scientific communities. Standing on Kelly's premise that scientific communities are places in which members generate, debate, evaluate, accept, and reject proposed knowledge claims within a relevant discourse community, Bowen analyzes middle-school classrooms and compares them with scientists operating in naturalistic settings. Further, Bowen uses this sociolinguistic and ethnographic description to distill the nature of scientific and classroom communities by identifying substantial differences in the two cultures—questioning ultimately whether it is actually desirable to replace the traditional classroom culture with a truly authentic science one. For example, although science curriculum reform documents often assume that the culture(s) of scientific communities can be mapped, more or less directly, onto classroom cultures, Bowen draws our attention to the development of competence in both communities and examines the contradictions people face between what is learned "about science" and what scientists do in their own practices.

Bowen challenges us to consider the construction of science knowledge in science communities where the complexities of what is known are further elaborated so as to generate more precise, complex, predictive, and explanatory models. The contrasts with classroom practices illuminate why schools so poorly develop student interest in science and understanding of the complex relationships in the natural world. Bowen emphasizes how philosophical and practical tenets of science must be considered before making the case for good science teaching explaining,

> An enculturation into science practices—a developed competency in tool use, language use, and interpretive stances—needs to be balanced with an enculturation into a critical reflection towards those same scientific tools. Teachers therefore need to engage students in communities of practice which both develop an understanding of the tools and concepts as well as a meta-awareness of the shortcomings of those tools to explain the surrounding world. (pp. 131–132)

Bowen warns that science educators need to understand in far greater detail the trajectories of competencies and practices that are nurtured within scientific communities before we can offer sound pedagogical advice for science teachers and teacher educators. He then provides examples of how classroom communities can better model knowledge-generating communities and challenges the reader to involve students in self-sustaining scientific communities that address questions central to students' interests.

In chapter 5 Maria Varelas and her colleagues Christine Pappas and Amy Rife explore the nature of dialogic inquiry that takes place in an elementary classroom where a teacher and second-grade students engaged in integrated science–literacy experiences associated with the water cycle. Speaking from a sociocultural constructivist perspective, they claim that scientific understanding does not reside within individuals but in transactions between individuals such as in the voices of the teacher and the students that come together as they co-construct understandings. Varelas et al. provide educators with a glimpse of the collaborative microinteractions that allow collective knowledge to be corporate, tentative, and open to alternative interpretations and revision of the community.

> Examining intertextuality allows us to appreciate the funds of knowledge (Moll, 1992) that young urban children bring to the class along with the teacher's role in legitimizing and using these funds to facilitate the building of new understandings or elaborate prior understandings. In dialogically oriented instruction, intertextuality takes place as a negotiated dance among teacher, children, and texts in the construction of knowledge (Oyler, 1996). It is a complicated endeavor. (p. 142)

Using their notions of intertextuality we see how scientific literacy emerged in Rife's class reciprocally as children used their own reasoning to make sense of sociocultural achievements, while at the same time being influenced by them to reorganize their understandings.

Varelas et al. used the construct of *intertextuality*—the juxtaposition or reference to other texts—to support children's scientific understandings (Pappas, Varelas, Barry, & Rife, 2003). This process entails learning by encountering concepts, procedures, and strategies that have been established by others over the course of time. In this way, the intertextuality of classroom talk, as dialogic inquiry, enabled children to appropriate scientific ideas, further extend them, question their validity, and even generate new ones. In this chapter, Varelas focuses on intertextuality as a way to examine the ways in which specific aspects of classroom discourse contributes to the acquisition of science and literacy. Thus, creating scientific understandings entails constructing new conceptual entities and the wordings to express those entities.

Like Collins et al. in chapter 7, Varelas and her colleagues directly challenge the deficit orientation of science standards prevalent within the general public and among schoolteachers. Their challenge to science educators of all levels is to share authority over scientific ideas and to mutually co-construct scientific identities with their students. Teachers need to view students' prior experiences and understandings as the capital in which to invest in order to achieve the views of scientific literacy described by current reform rhetoric.

In chapter 6, Cynthia Ballenger puts methodologies of sociolinguistic research to the test as she explores how two students in a Haitian bilingual class learn about motion. Ballenger's work is based on a long, fruitful, and rewarding project (Chèche Konnen) that has helped educators understand the role of culture and language in learning science and the intricacies of making science accessible to all. Ballenger models the kind of methodological research choices required to uncover children's knowledge as she describes how a teacher gives children opportunities to express their ideas while she learns from children different aspects of the content knowledge herself.

Ballenger explores the thinking of two children and uses their responses to bring forth the relevance of the children's culture as only an insider can. Her deep knowledge of the Haitian culture and knowledge of successful teaching strategies allow her to see connections that the children are attempting to make. Ballenger's strategies give centrality to children's own ideas and questions with their everyday language and experience; she hopes that like-minded analyses of other contexts will build a knowledge base to assist teachers in connecting science concepts with children's experiences.

Through the examples of these two students Ballenger challenges the assumptions that we know what practices are best for teaching specific content knowledge and how these areas are best learned by diverse children. She reminds us that teachers are faced with many self-imposed obstacles when they use what they think are the best practices. When standard or innovative practices fail to work for all children, teachers (sometimes unconsciously) generally blame the child's background. Ballenger stretches all of us as teachers and researchers to consider the question, "What children do we choose to learn from?" when our teaching does not seem to accomplish what we intend.

> The question . . . is not often asked in teacher preparation. And yet I think it might be a question that, as educators, we ought to ask routinely, picking out our puzzling children in various ways and at different points for interview, for further probing and study. When we're using what we think are the best practices and they don't work for all children, we generally blame the child's background, sometimes without even realizing it. . . . It takes our attention away from the process of reflective practice. It takes us away from a stance of inquiry towards the child's ideas and towards what we are teaching. It prevents us from learning from these children. (p. 188)

Through her teaching, research, and challenging questions to teacher educators, Ballenger continually raises the bar for science educators for keeping children's voices in the forefront of our science education efforts.

INTRODUCTION 13

In chapter 7, Kathleen Collins, Annemarie Palincsar, and Shirley Magnusson remind us that in a volume devoted to examining the complexities of establishing scientific learning communities in classrooms, we must also examine how students with special needs navigate the intellectual and discursive demands of such communities. This issue is particularly important in an era of standards-based reform that has left the participation and contributions of students with special needs largely unexamined. Collins et al. emphasize that holding all students to high content-area learning standards in science does not necessarily require standardizing ways of knowing, learning, and teaching.

To make their case Collins et al. begin by describing a guided inquiry orientation to teaching and the research literature supporting instructional conversations in school science contexts that shaped their teaching and current study. To assist in the interpretation of what counts as scientific activity Collins et al. employ the relevant theoretical works of Lev Vygotsky and Joseph Schwab. Their analysis reveals that students' appropriate ways of speaking and thinking science parallel cited substantive and syntactic structures of the scientific discipline in ways that surprise even the teachers of special-needs students—thus demonstrating the impact of moving away from viewing student progress solely through standardized structures or frameworks. Collins et al. expose the pervasive nature of deficit approaches to teaching, noting that,

> In various forms, deficit discourses have been used to explain the low school achievement of students of color, poor and working-class students, students whose first language is one other than English, and students identified as having learning disabilities. . . . One common form that deficit discourses take is the assertion by (often well-meaning) teachers that some students . . . bring less to school learning contexts in terms of background knowledge and experience than their peers. (p. 200)

In contrast to the commonly used deficit, Collins et al. examine the intersection of environment and individual to understand how they mutually construct each other. Through their close analysis of the participation of students with special needs in a particular form of inquiry science instruction they refer to as Guided Inquiry supporting Multiple Literacies, Collins et al. reveal the ways in which specific aspects of instructional conversations influenced the appearance of perceived ability and disability in this classroom. Their findings bear on the design of scientific learning communities that are inclusive of all learners rather than limited to those who have historically been successful in school science contexts.

Although there is widespread agreement among reformers, legislators, and researchers that changes in science teaching practices are necessary, there is little direction offered in the rhetoric of national or state standards for how teachers can make initial steps toward establishing scientific classroom discourse communities. Teachers attempting to change classroom discourse are confronted with a number of obstacles, which include the recognition and negotiation of the insertion of home-based discourse into attempts to promote scientific inquiry. What are the major obstacles that teachers face when they attempt to establish scientific classroom communities?

Randy Yerrick, in chapter 8, focuses on important obstacles confronting the creation of scientific classroom discourse communities and potential strategies for renegotiating classroom discourse given common origins of resistance and socialization. Yerrick provides insights into the problematic nature of these pedagogical dilemmas as he speaks from research in culturally diverse classrooms. Like Ballenger, Yerrick speaks from the perspective of teacher-researchers as he draws on his own teaching of lower track high school science students. Rather than draw on the framework described by Collins et al., Yerrick employs one of Gee's metaphors for appropriating foreign discourses. Through the analogy of games Yerrick helps researchers interpret initial shifts in discourse among marginalized science students in several different contexts.

Yerrick reminds teachers and researchers that embracing Vygotskian theories concerning discourse and orchestrating classroom settings to reflect Schwab's syntactic and substantive structures provide no guarantee of success. In fact, when presented with alternative ways to speak, think, and act in science classrooms, students will sometimes not immediately embrace scientific discourse. Instead, marginalized students often insert other ways of speaking, thinking, and acting that run cross-purpose to the goals of scientific inquiry. Yerrick explicates and analyzes obstacles encountered by teachers in making shifts in classroom discourse and offers hope for change amidst even the toughest of science students. At the same time, Yerrick offers hope for change amidst even the most difficult of contexts. His suggestions for teachers include the attention to students' agendas and interests as well as a continued commitment to reinventing classroom discourse through scaffolding and collective reflection.

> Sustaining regular metadiscourse surrounding commonly shared learning events allows the teacher opportunity to provide students assistance for interpreting cues, finding alternative ways to transcend intellectual obstacles, and better orchestrate discussions where resolving a diverse set of students' opinions and observations can indeed be a challenge. (p. 249)

On the heels of Yerrick's examination of obstacles to changing science classroom discourse among historically unsuccessful students, in chapter 9 Kenneth Tobin explores gaps in our knowledge of teaching required to make profound changes. Tobin guides us through his esteemed career of learning to teach science expertly and shows us how, despite the expectations of others, he struggled to connect science with the lives of students he desired to touch. His autobiographical account stands as a stark reminder that content knowledge, teaching experience, and a desire to make a difference do not assure any teacher's success with urban science students in today's schools. Rather, a socioculturally specific kind of knowledge is required of all teachers of their students and science in order to turn around histories of failure in today's U.S. science classrooms.

Tobin exposes the current context of science education reform rhetoric that paints pristine visions of classroom contexts. He is among the few researchers and practitioners who turn a critical lens upon informed practices—subjecting teaching and self to the risks associated with bringing professed beliefs and classroom reality in conjunction. Such stories promote extended discourse surrounding evolved constructivist classrooms of the next school era as they more fully develop. These useful stories are neither cleared of rough terrain nor oversimplified for the purpose of speaking to particular aspects of classroom teaching and learning. They are real stories with real students and can promote more grounded reform discussions.

Tobin's primary purpose for telling his story is to address the challenges of teaching and learning science in an inner-city neighborhood high school. Given the current backdrop of policies and practices being adopted within a framework of applying national, state, and district standards and holding schools and teachers accountable for performance, Tobin warns us of the danger of such a hegemonic approach and argues that standards for scientific literacy must take into account who, where, and when as well as with what. His chapter challenges the narrow focus on scientific literacy that pays undue attention to the endpoints and relatively insufficient attention to the pathways and what can be attained along the way through active participation. Tobin's commitment to his ideals placed him in a struggle to test the limits of developing scientific habits "at the elbows of another."

> The act of doing science creates habitus and the habitus then supports particular patterns of action that accord with the structure of the field of which the practices are part. . . . Just as habitus is formed by being with others in particular cultural fields, so too a scientific habitus is built by coparticipating with others (i.e., teachers and students) as they do science. (p. 283)

Through Tobin's peaks and valleys he demonstrates that a standards-based approach fails to acknowledge that what is done in the process of becoming scientifically literate might be beyond the awareness of teachers and students—which might constitute the greatest part of what it means to be scientifically literate.

Of course, no discussion of science learning would be complete without the consideration of contexts outside the classroom. In chapter 10, the ways that Vygotsky's theories and related sociocultural frameworks can be used to interpret the success of certain after-school programs are explored as well as possible implications that would naturally follow for developing science discourse in regular classroom settings. While educators entertain discussions of connecting everyday experiences with science learning as a means for supporting the end (namely, scientific literacy), Gallego and Finkelstein contend that the relationship between everyday and scientific concepts is not unidirectional. Rather, everyday and scientific concepts are interdependent, both mediated through the use of root languages, as demonstrated by the illustrations of children's development of academic discourse and science knowledge while participating in a non-school learning setting.

Gallego and Finkelstein point to the shortcomings of such a dichotomization that ignores Vygotsky's view of scientific and everyday concepts as aspects of a unified process of concept formation. In their opinion, separating the experiences from concepts limits teachers' and researchers' interpretations of moment-to-moment discoveries made by each child as they experience the process of reformulation, or re-cognition, of an academic discipline like science. The authors explore how certain characteristics of school setting (e.g., compulsory attendance, standardized curriculum, mandatory assessment) influence content and how teachers and students participate in these settings. Through their analysis, they demonstrate how the process of exploring children's concept formations can inform our understanding of teaching and learning more generally. In their chapter, a sociocultural framework (Cole, 1996; Wertsch, 1991) was again employed to illustrate children's development of academic discourse and scientific knowledge while participating in a non-school learning setting. Within this approach, learning in context, concept formation (everyday and scientific), and the role of language as a tool for learning are highlighted. Analysis of data derived from four excerpts from videotaped recordings of participants of Science and Tech Club, an after-school activity, illustrates children's concept formation (everyday and scientific) and construction of academic discourse.

Finally, Gallego and Finkelstein address a recurring issue that has been a strand throughout each chapter: They challenge educators to reconsider how and what counts for school science:

> The challenge is not only to arrange learning contexts that provide student opportunities to use language as a tool for communication (overt and intended for others) as well as the use of language as instrument of thought and reason (intended for self, though at times overtly stated). Vygotsky's

(1934/1998) central question, "How children's learning changes upon entering school?" presupposes that it would indeed (and necessarily) change. As a unified process, concept formation, everyday and scientific are mutually influenced by the other. While the physical differences of non-school settings may provide for variable social organization, schools and classroom can also be modified to support the academic utility of everyday and scientific concepts. (pp. 310–311)

Harboring a strong belief in the ability to import, transfer, and replicate the patterns of interaction and the collaborative relationships found in some out-of-school programs into more standardized settings, Gallego and Finkelstein encourage us to consider how educators can organize social activity to maximize student science learning as they address how research conducted in non-school settings can support school-based learning.

REFERENCES

Aikenhead, G. S. (1997). Toward a First Nations cross-cultural science and technology curriculum. *Science Education, 81*, 217–238.
Cole, M. (1996). *Cultural psychology: A once and future discipline.* Cambridge, MA: Belknap Press.
Driver, R. (1990). *Children's ideas in science?* Philadelphia: Open University Press.
Foucault, M. (1979). *Discipline and punish: The birth of the prison.* New York: Vintage Books.
Kelly, G. J., Chen, C., & Crawford, T. (1998). Methodological considerations for studying science-in-the-making in educational settings. *Research in Science Education, 28*, 23–49.
Latour, B., & Woolgar, S. (1986). *Laboratory life: The construction of scientific facts.* Princeton, NJ: Princeton University Press.
Millar, R. (1989). *Doing science: Images of science in science education.* London: Falmer Press.
National Commission on Mathematics and Science Teaching for the 21st Century. (2000). *Before it's too late.* http://www.ed.gov/americacounts/glenn/toc.html
Oyler, C. (1996). Sharing authority: Student initiations during teacher-led read-alouds of information books. *Teaching & Teacher Education, 12,* 149–160.
Pappas, C. C., Varelas, M., Barry, A., & Rife, A. (2003). Dialogic inquiry around information texts: The role of intertextuality in constructing scientific understandings in urban primary classrooms. *Linguistics and Education, 13,* 435–482.
Rodriguez, A. J. (2001). From gap gazing to promising cases: Moving toward equity in urban education reform. *Journal of Research in Science Teaching, 38,* 1115–1129.
Roth, W. -M. (1998). *Designing communities.* Dordrecht, Netherlands: Kluwer Academic Publishing.
Roth, W. -M., & Bowen, G. M. (1999). Digitizing lizards or the topology of vision in ecological fieldwork. *Social Studies of Science, 29,* 719–764.
Roth, W. -M., & McGinn, M. K. (1997). Science in schools and everywhere else: What science educators should know about science and technology studies. *Studies in Science Education, 29,* 1–44.

Schwab, J. J. (1978). The nature of scientific knowledge as related to liberal education. In I. Westbury & N. J. Wilkof (Eds.), *Science, curriculum, and liberal education: Selected essays* (pp. 68–104). Chicago: University of Chicago Press.

Third International Mathematics and Science Study. (2001). *The condition of education.* http://nces.ed.gov/pubsearch/

Traweek, S. (1988). *Beamtimes and lifetimes: The world of high energy physicists.* Cambridge, MA: Harvard University Press.

Vygotsky, L. S. (1987). Thinking and speech. In R. W. Rieber & A. S. Carton (Eds.), *The collected works of L. S. Vygotsky (Vol. 1): Problems of general psychology* (N. Minick, Trans., pp. 39–285). New York: Plenum Press. (Original work published 1934)

Watson-Verran, H., & Turnbull, D. (1995). Science and other indigenous knowledge systems. In S. Jasanoff, G. E. Markle, J. C. Peterson, & T. Pinch (Eds.), *Handbook of science and technology studies* (pp. 115–139). Thousand Oaks, CA: Sage.

Wertsch, J. V. (1991). *Voices of the mind: A socio-historical approach to mediated action.* Cambridge, MA: Harvard University Press.

CHAPTER ONE

Language in the Science Classroom: Academic Social Languages as the Heart of School-Based Literacy

James Paul Gee
University of Wisconsin–Madison

There are different "ways with printed words" embedded in different social and cultural groups, in different institutions, and in different social practices. Literacy is, in that sense, multiple. Furthermore, literacy viewed in terms of the different sorts of social practices in which it is embedded, is almost always integrally involved with oral language and with ways of acting, interacting, and thinking, and not just reading and writing. In regard to schooling we need to focus on the acquisition of academic sorts of language within specific social practices and not on literacy as a general thing or as only reading and writing. In school, especially as one moves beyond the first couple of years, reading and writing become (or most certainly should be) fully embedded in and integrated with learning, using, and talking about specific content.

No domain represents academic sorts of language better than science. Science makes demands on students to use language, orally and in print, as well as other sorts of symbol systems, that epitomize the sorts of representational systems and practices that are at the heart of higher levels of school success. They are at the heart, too, of living in and thinking critically about modern societies.

In this chapter, I make six claims about the connections between language and learning science, although I construe language in a particular way, that is, as integrally involved with identities and social practices. The connections, however, are the same for all school subjects. In much of science education, language is moved to the background or ignored, while thinking or doing are foregrounded as if these had little to do with language. I do not think this is true. Each of my claims implies a particular property that science instruction ought to have, though little work on science instruction that I am aware of speaks directly to these properties.

ACADEMIC LANGUAGE AND THE SOCIAL LANGUAGE CODE

Claim 1: Success in school is primarily contingent on willingness and ability to cope with academic language.

In the recent debates over early reading, the alphabetic code has been given pride of place (Gee, 1999a; National Reading Panel, 2000; Snow, Burns, & Griffin, 1998). However, more children fail in school, in the long run, because they cannot cope with "academic language" than because they cannot decode print. Academic language is a related family of what I will call "social languages." English, like any human language is not a single language, but is composed of a myriad of different "social languages," some of which are forms of academic language (Gee, 1996, 1999b).

A social language is a way of using language so as to enact a particular socially situated identity and carry out a particular socially situated activity. For example, there are ways of speaking and acting like a (specific type of) doctor, street-gang member, postmodern literary critic, football fanatic, neoliberal economist, working-class male, adaptationist biologist, and so on, through an endless array of identities. Often, of course, we can recognize a particular socially situated "kind of person" engaged in a particularly characteristic sort of activity through his or her use of a given social language without ourselves actually being able to enact that kind of person or actually being able to carry out that activity. We can be producers or consumers of particular social languages, or we can be both.

Recognizing a particular social language is also a matter of a code. Let us call it the "social language code." Different patterns or co-relations of grammatical elements (at the levels of phonology/graphology, morphology, words, syntax, discourse relations, and pragmatics) are associated with or map to particular social languages (specific styles of oral and/or written language) associated with specific socially situated identities and activities. So here the code is not between sounds and letters, but between grammatical patterns and styles of language (and their associated identities and activities). For example, the following text is taken from a school science textbook:

The destruction of a land surface by the combined effects of abrasion and removal of weathered material by transporting agents is called erosion. . . . The production of rock waste by mechanical processes and chemical changes is called weathering. (Martin, 1990, p. 93)

A whole bevy of co-related grammatical features (i.e., features that hang together or pattern together in oral and/or written texts) mark these sentences as part of a distinctive social language or style of (in this case, academic) language. Some of these features are: "heavy subjects" (e.g., "the production of rock waste by mechanical processes and chemical changes"); processes and actions named by nouns or nominalizations, rather than verbs (e.g., "production"); passive main verbs ("is called") and passives inside nominalizations (e.g., "production by mechanical means"); modifiers which are more contentful than the nouns they modify (e.g., "transporting agents"); and complex embedding (e.g., "weathered material by transporting agents" is a nominalization embedded inside "the combined effects of . . ." and this more complex nominalization is embedded inside a yet larger nominalization, "the destruction of . . .").

This style of language also incorporates a great many distinctive discourse markers, that is, linguistic features that characterize larger stretches of text and give them unity and coherence as a certain type of text or "genre." For example, the genre here is *explanatory definition* and it is characterized by *classificatory language* of a certain sort. Such language leads adept readers to form a *classificatory scheme* in their heads something like this: There are two kinds of change—erosion and weathering—and two kinds of weathering—mechanical and chemical.

Academic social languages constitute a large family of related (but, of course, different) social languages. Academic social languages are not primarily associated with things as big as whole disciplines, but with particular ways of being-doing intellectual inquiry, ways that are sometimes subdomains of a discipline, but, especially today, often cross-disciplinary boundaries. For example, different sorts of biologists talk, write, and act in different ways, and in some cases they share these ways more closely with some people outside biology (e.g., certain sorts of physicists) than they do with certain other biologists. The same can be said of almost any other discipline.

ACQUISITION AND IDENTITY: LIFEWORLD VERSUS SPECIALIST WORLDS

Claim 2: To acquire an academic social language, students must be willing to accept certain losses and see the acquisition of the academic social language as a gain.

When one comes to the acquisition of an academic social language—most especially one associated with science—there are both significant *losses* and *gains* (Halliday & Martin, 1993). To see this, consider the two sentences below. The first is in the social language of the "lifeworld" and the second is in an academic social language. By the "lifeworld" I mean that domain where we speak and act as ordinary, everyday, nonspecialist people. Of course, different social and cultural groups have different lifeworlds and different lifeworld social languages, though such lifeworld languages do share a good many features that cause linguists to refer to them as "vernacular language."

1. Hornworms sure vary a lot in how well they grow.
2. Hornworm growth exhibits a significant amount of variation.

Subjects of sentences name what a sentence is about (its "topic") and (when they are initial in the sentence) the perspective from which we are viewing the claims we want to make (the sentence's "theme"). The lifeworld sentence (1) is about hornworms (cute little green worms) and launches off from the perspective of the hornworm. The presence of "sure" helps to cause the subject here (hornworms) also to be taken as naming a thing with which we are empathizing. The specialist sentence (2) is not about hornworms, but about a trait or feature of hornworms (in particular one that can be quantified) and launches off from this perspective. The hornworm disappears.

The lifeworld sentence involves dynamic processes (changes) named by verbs ("vary," "grow"). People tend to care a good deal about changes and transformations, especially in things with which they empathize. The specialist sentence turns these dynamic processes into abstract things ("variation," "growth"). The dynamic changes disappear. We can also mention that the lifeworld sentence has a contentful verb ("vary"), while the specialist one has a verb of appearance ("exhibit"), a class of verbs that is similar to copulas and are not as deeply or richly contentful as verbs like *vary*. Such verbs are basically just ways to relate things (in this case, abstract things, to boot).

The lifeworld sentence has a quantity term (*how well*) that is not just about amount or degree, but is also "telically evaluative," if I may be allowed to coin a term. "How well" is about both a quantity and evaluates this amount in terms of an end-point or "telos" germane to the hornworm, that is, in terms of a point of good or proper or full growth toward which hornworms are *meant* to move. Some hornworms reach the telos and others fall short. The specialist sentence replaces this "telically evaluative" term with a more precise measurement term that is "Discourse evaluative" (significant amount). *Significant amount* is about an amount that is evaluated in terms of the goals, procedures (even telos, if you like) of a Discourse (here a type of biology), not a hornworm. It is a particular area of biology that determines what amounts to significant and what does not. All our hornworms could be stunted or untypical of well-grown hornworms ("well grown" from a lifeworld, nonspecialist perspective) and still display a significant amount of variation in their sizes.

This last difference is related to another one: The lifeworld sentence contains an appreciative marker ("sure"), while the specialist sentence leaves out such markers. The appreciative marker communicates the attitude, interest, and even values of the speaker/writer. Attitude, interest, and values, in this sense, are left out of the specialist sentence.

So, when one has to leave the lifeworld to acquire and then use the specialist language above, these are some of the things that are lost: concrete things like hornworms and empathy for them; changes and transformations as dynamic on-going processes; and telos and appreciation. What is gained are: abstract things and relations among them; traits and quantification and categorization of traits; evaluation from within a specialized domain. The crucial question, then, is this: *Why would* anyone—*most especially a child in school—accept this loss?*

My view is that people will accept this loss only if they see the gain *as a gain*. So a crucial question in science education ought to be: *What would make someone see acquiring a scientific social language as a gain?* Social languages are tied to socially situated identities and activities, as we have seen earlier. People can only see a new social language as a gain if they recognize and understand the sorts of socially situated identities and activities that recruit the social language; if they value them or, at least, understand why they are valued; and if they believe they (will) have real access to them or, at least (will) have access to meaningful versions of them.

Thus, acquisition is heavily tied at the outset to identity issues. It is tied to the learner's willingness and trust to leave (for a time and place) the lifeworld and participate in another identity, one that, for everyone, represents a certain loss. For some people, as we all know, it represents a more significant loss in terms of a disassociation from, and even opposition to, their lifeworlds, because their lifeworlds are not rooted in the sort of middle-class lifeworlds that have historically built up some sense of shared interests and values with some academic specialist domains.

SITUATED MEANING

Claim 3: One does not know what a social language means in any sense useful for action unless one can situate the meanings of the social language's words and phrases in terms of embodied experiences.

Recent debates over reading have pointed out that although one's first ("native") oral language is acquired by immersion in practice (socialization as part of a social group), literacy often requires some degree of overt instruction (e.g., Gee, 1994). So, then, what can we say about the acquisition of an academic social language? The language of adaptationist biology, for example, is no one's "native language." Are these acquired largely through immersion in practice or do they require a good deal of overt instruction, as well? In reality—given that scientists tend to ignore language as a relevant factor—many a scientist has picked up a scientific social language through immersion in practice, with little or no overt instruction in regard to the language.

Although I advocate overt instruction in regard to academic social languages, both in the midst of practice and outside it, what I want to point out here is this: Social languages can be understood in two different ways, either as largely verbal or as situated. I can understand a piece of a scientific social language largely as a set of verbal definitions or rather general meanings for words and phrases and, in turn, relate words and phrases to each other in terms of these definitions or general meanings and general knowledge about grammatical patterns in the language. However, such an understanding is not all that useful when one has to engage in any activity using a specialist language.

Words and phrases in use in any social language, including lifeworld social languages, have not only relatively general meanings (which basically define their meaning potential), but *situated meanings* as well. Words and phrases are associated with different situated meanings in different contexts, in different social languages, and in different Discourses. For example, consider what happens in your head when I say, "The coffee spilled, get a mop" versus "The coffee spilled, get a broom" versus "The coffee spilled, can you restack it." In each case, you actively assemble a different situated meaning for "coffee" (as liquid, grains, or containers).

Let me develop a bit more what I mean by saying that "meanings" are assembled on the spot (and for the spot, so to speak). In the context of the lifeworld, if we are asked a question like "How far does the light go?" while staring at a lamp, we are liable to answer: "Not far, only as far as I can see the light illuminate an area near the light source." In this case, we assemble a meaning for "light" that is something like "illuminated region."

In the context of physical science, on the other hand, we would answer the question quite differently and assemble a different meaning. In the case of physics, there are a number of different assemblies that could be associated with "light," one of which is "rays" ("lines of light") that travel indefinitely far and which can reflect off of surfaces. In terms of this assembly, we would answer to our question: "The light travels forever unless it reflects off a surface."

Of course, there are other assemblies for light in yet other Discourses. For example, in the context of theater, light is associated with assemblies dealing with things like "lighting effects." Further, new assemblies do and can arise as Discourses change and interact or as new situations arise.

Situated meanings are, crucially, rooted in embodied experience—one has to "see" the meaning as a pattern extracted from the concrete data of experience (Barsalou, 1999a). To do so one must have had lots of practice with experiences that trigger the pattern. Situated meanings are not "definitions." If one cannot situate the meaning of a word or phrase, then one has only a verbal or definitional meaning for the word or phrase. Such verbal meanings, while they can facilitate passing certain sorts of tests, are of limited value when language has to put to use within activities.

Recent work in cognitive psychology has developed this notion of situatedness (long taken for granted in work on pragmatics) further. Consider, for instance, the two statements, "comprehension is grounded in perceptual simulations that prepare agents for situated action" (Barsalou, 1999b, p. 77) and "to a particular person, the meaning of an object, event, or sentence is what that person can do with the object, event, or sentence" (Glenberg, 1997, p. 3). These two quotes are from work that is part of a "family" of related viewpoints (e.g., Barsalou, 1999a, 1999b; Brown, Collins, & Duguid, 1989; Engeström, Miettinen, & Punamaki, 1999; Gee, 1992; Glenberg, 1997; Glenberg & Robertson, 1999; Hutchins, 1995; Latour, 1999; Lave, 1996; Lave & Wenger, 1991; Wenger, 1998). Although there are differences among the different members of the family (alternative theories about situated cognition), they share the viewpoint that situated meaning in language is not some abstract propositional representation that resembles a verbal language. Rather, meaning in language is tied to people's experiences of situated action in the material and social world. Furthermore, these experiences (perceptions, feelings, actions, and interactions) are stored in the mind/brain not in terms of propositions or language, but in something like dynamic images tied to perception both of the world and of our own bodies, internal states, and feelings: "Increasing evidence suggests that perceptual simulation is indeed central to comprehension" (Barsalou, 1999a, p. 74).

It is almost as if we "videotape" our experiences as we are having them, create a library of such videotapes, edit them to make some "prototypical tapes" or a set of typical instances, but stand ever ready to add new tapes to our library, reedit the tapes based on new experiences, or draw out of the library less typical tapes when the need arises. As we face new situations or new texts we run our tapes, perhaps a prototypical one, or a set of typical ones, or a set of contrasting ones, or a less typical one, whatever the case may be, in order to apply our old experiences to our new experience and to aid us in making, editing, and storing the videotape that will capture this new experience, integrate it into our library, and allow us to make sense of it (both while we are having it and afterwards).

These videotapes are what we think with and through. They are what we use to give meaning to our experiences in the world. They are what we use to give meaning to words and sentences. But they are not language or "in language" (not even in propositions). Furthermore, because they are representations of experience (including feelings, attitudes, embodied positions, and various sorts of figure and ground of attention), they are not just "information" or "facts." Rather, they are value-laden, perspective-taking "movies in the mind." Of course, talking about videotapes in the mind is a metaphor that, like all metaphors, is incorrect if pushed too far (see Barsalou, 1999b, for how the metaphor can be cashed out and corrected by a consideration of a more neurally realistic framework for "perception in the mind").

On this account, too, the meaning of word (the way in which we give it meaning in a particular context) is not different than the meaning of an experience, object, or tool in the world (i.e., in terms of the way in which we give the experience, object, or tool meaning):

> The meaning of the glass to you, at that particular moment, is in terms of the actions available. The meaning of the glass changes when different constraints on action are combined. For example, in a noisy room, the glass may become a mechanism for capturing attention (by tapping it with a spoon), rather than a mechanism for quenching thirst. (Glenberg, 1997, p. 41)

While Glenberg here is talking about the meaning of the glass as an object in one's specific experience of the world at a given time and place, he could just as well be talking about the meaning of the word "glass" in one's specific experience of a piece of talk or written text at a given time and place. The meaning of the word "glass" in a given piece of talk or text would be given by running a simulation (a videotape) of how the glass fits into courses of action being built up in the "theater" of our minds. These courses of action are based on how we are understanding all the other words and other goings on in the world that surrounds the word "glass" as we read it, "the embodied models constructed to understand language are the same as those that underlie comprehension of the natural environment" (Glenberg, 1997, p. 17).

However, I want to state one important caution here. Much of the work in cognitive psychology stresses that the experiences in which situated meanings are rooted are ones we have had as embodied perceivers of and actors in the material world. However, situated meanings are also importantly rooted in experiences we have had as participants in specific rhetorical practices and activities, which themselves are forms of embodied engagement with the world. Consider, for example, the following case of a high school student's first and second draft of a paper on "Albinism" (the two drafts were separated by a good deal of work on the part of the teacher):

First Draft: Then to let people know there are different types of Albinism, I will tell and explain all this.

Second Draft: Finally, to let people know there are different types of Albinism, I will name and describe several.

In the first case, the student appears to have formed situated meanings for "tell" and "explain" that have to do with telling a story and explicating the "big picture" ("all this") through that story. This is, of course, an activity the student has experienced a great many times, including in his lifeworld. Unfortunately, in the sort of academic writing in which the student was engaged, the phrase "different types of Albinism" requires its meaning to be situated in a quite different way, one inconsistent with how the student has situated the meanings of *tell* and *explain*.

If one has experienced *the activity of classifying in academic Discourses* of certain types, one would not be tempted to use *tell* and *explain* as this student has done (note that *tell* and *explain* have other situated meanings in terms of which they could occur with the phrase "different types of Albinism"—for example, a teacher could write on a student essay: "You first need to tell your readers what the different types of Albinism are and then to explain how these types are distinguished").

On the other hand, one can readily situate more appropriate meanings for "name" and "describe" consistent with the appropriate situated meaning for "different types of Albinism" (though one can situate wrong ones for *name* and *describe,* as well, of course, as in "Scientists name different types of Albinism differently"). To do this one needs *to have experienced certain sorts of acts of classification within certain sorts of Discourses*.

I should note that even the student's second version is not quite right in terms of many academic Discourses' ways with classifying ("people" is wrong, and "name" is just a bit off—something like, "There are different types of Albinism. Below I list several of these and describe them" would have been yet better). This example, simple as it is, tells us how subtle a process of situating meaning can be.

When anyone is trying to speak/write or listen/read within a given social language within a given Discourse, the crucial question becomes, *what sorts of experiences (if any)—in terms of embodied practices and activities, including textual, conversational, and rhetorical ones—has this person had that can anchor the situated meanings of words and phrases of this social language?* Otherwise, one is stuck with a merely general and verbal understanding (the sort that, alas, is often nonetheless rewarded in school).

LANGUAGE ACQUISITION AND PERSPECTIVE TAKING

Claim 4: Language acquisition crucially involves access to and simulations of the perspectives of more advanced users of the language in the midst of practice.

In my remarks on situated meaning, I argue that embodied experiences in the world and the active assembly of meaning based on these experiences are the foundation of meaning-in-use in language. However, I believe a strong case can be made that social interaction and dialogue of a certain sort are also crucial to the acquisition of any social language whatsoever.

Consider, in this regard, the following quote from Michael Tomasello's (1999) book *The Cultural Origins of Human Cognition*:

> [T]he perspectivial nature of linguistic symbols, and the use of linguistic symbols in discourse interaction in which different perspectives are explicitly contrasted and shared, provide the raw material out of which the children of all cultures construct the flexible and multi-perspectival—perhaps even dialogical—cognitive representations that give human cognition much of its awesome and unique power. (p. 163)

Let us briefly unpack what this means. From the point of view of the model Tomasello is developing, the words and grammar of a human language exist to allow people to take and communicate alternative perspectives on experience (see also Hanks, 1996). That is, words and grammar exist to give people alternative ways to view one and the same state of affairs. Language is not about conveying neutral or "objective" information; rather, it is about communicating perspectives on experience and action in the world, often in contrast to alternative and competing perspectives: "We may then say that linguistic symbols are social conventions for inducing others to construe, or take a perspective on, some experiential situation" (Tomasello, 1999, p. 118).

Let me give some examples of what it means to say that words and grammar are not primarily about giving and getting information, but, rather, about giving and getting different perspectives on experience. I open Microsoft's web site: Is it "selling" its products, "marketing" them, or "underpricing" them against the competition? Are products I can download from the site without paying a price for them "free," or are they being "exchanged" for having bought other Microsoft products (e.g., Windows), or are there "strings attached"—and note how metaphors (like "strings attached") add greatly to, and are a central part of, the perspective taking we can do. If I use the grammatical construction "Microsoft's new operating system is loaded with bugs" I take a perspective in which Microsoft is less agentive and responsible than if I use the grammatical construction "Microsoft has loaded its new operating system with bugs."

Another example: Do I say that a child who is using multiple cues to give meaning to a written text (i.e., using some decoding along with picture and context cues) is "reading" or (as some of the pro-phonics people do) do I say that she is "not really reading, but engaged in emergent literacy" (for these latter people, the child is only "really reading" when she is decoding all the words in the text and not using nondecoding cues for word recognition). In this case, contending camps actually fight over what perspective on experience the term *reading* or *really reading* ought to name. In the end, the point is that no wording is ever neutral or just "the facts." All wordings—given the very nature of language—are perspectives on experience that comport with competing perspectives in the grammar of the language and in actual social interactions.

How do children learn how words and grammar line up to express particular perspectives on experience? Here, interactive, intersubjective dialogue with more advanced peers and adults appears to be crucial. In such dialogue, children come to see, from time to time, that others have taken a different perspective on what is being talked about than they themselves have. At a certain developmental level, children have the capacity to distance themselves from their own perspectives and (internally) simulate the perspectives the other person is taking, thereby coming to see how words and grammar come to express those perspectives (in contrast to the way in which different words and grammatical constructions express competing perspectives).

Later, in other interactions, or in thinking to oneself, the child can rerun such simulations and imitate the perspective-taking the more advanced peer or adult has done by using certain sorts of words and grammar. Through such simulations and imitative learning, children learn to use the symbolic means that other persons have used to share attention with them. Thus, "in imitatively learning a linguistic symbol from other persons in this way, I internalize not only their communicative intention (their intention to get me to share their attention) but also the specific perspective they have taken" (Tomasello, 1999, p. 128).

Tomasello (1999, pp. 129–130) also points out—in line with our previous discussion that the world and texts are assigned meanings in the same way—that children come to use objects in the world as symbols at the same time (or with just a bit of a time lag) as they come to use linguistic symbols as perspective taking devices on the world. Furthermore, they learn to use objects as symbols (to assign them different meanings encoding specific perspectives in different contexts) in the same way they learn to use linguistic symbols. In both cases, the child simulates in her head and later imitates in her words and deeds the perspectives her interlocutor must be taking on a given situation by using certain words and certain forms of grammar or by treating certain objects in certain ways. Thus, meaning for words, grammar, and objects comes out of intersubjective dialogue and interaction: "human symbols [are] inherently social, intersubjective, and perspectival" (Tomasello, 1999, p. 131).

Now one might argue that Tomasello's work is germane only to the acquisition of one's "native" language early in life. But this would be to miss one of the important points of work like that of Tomasello and Hanks. A crucial point they are making, to my mind, at least, is that grammar by its very nature, regardless of what social language it is used within, is a perspective-taking device, learned in part by the human capacity to run simulations of experience (even from someone else's perspective) in our heads.

SCIENCE AND LIFEWORLD LANGUAGE

Claim 5: Lifeworld language is problematic for science.

There are many who argue that the way to solve the identity issues with which I started this chapter (how to get children to "buy into" the situated identity associated with a given academic Discourse) is to allow and encourage them to use their "everyday" lifeworld (vernacular) languages for "science talk." In my view, this idea comports poorly with everything I have said before. However, the issue remains: Why can't we marry science to lifeworld languages?

I believe there are good reasons to encourage children, even early on, to marry scientific activities with scientific ways with words, and not lifeworld languages, though lifeworld languages are obviously the starting point for the acquisition of any later social language, as Vygotsky pointed out. To make this argument, I want to work through one example (the following data are from a 1994 AERA session put together by Cindy Ballenger, Ann Rosebery, and Beth Warren of TERC). I draw on the genre work of Bakhtin (1986) and Halliday (1994) to build the case for blending science and lifeworld language found in daily classroom events.

1. LANGUAGE IN THE SCIENCE CLASSROOM

Students in an elementary classroom have been discussing the question, "What makes rust?" They go on to develop a set of questions about rust including, Does ordinary water cause rust? Does rain water cause rust? Does paint protect rust? Do other things besides metal get rusty? To help answer these questions, the children place a number of metal and nonmetal objects in tap water and rain water to examine whether and what "gets rusty," as well as a variety of other questions. In the following segment, the children have taken a metal bottle cap and plastic plate out of the water. The cap had been sitting on the plate. Below they are discussing the cap and plate. Since, like all interactive language, their language is replete with deictics, I have placed material in brackets to make clear what they are referring to.

Elizabeth:	Do you think plastic would get rusty?
Philip:	No 'cause we've got this [plastic plate here and it's not rusty]—maybe [it would get rusty] if something metal was on this [plastic plate].
Jill:	It [the plastic plate] will get old [and old things get rusty].
Philip:	[If it was old and got rusty, then] Metal got on it like [it got on] this plastic plate.
Jill:	Cuz like the bottle cap where the rust is from, the rust got off the metal [bottle cap] and went on to the plate.
Philip:	But without the metal on top of the plastic [plate], you would never get, it [the plate] probably wouldn't ever get rusty.
Jill:	You wouldn't get marks.
Elizabeth:	Why these marks here [on the plastic plate]?
Jill:	Because the rust comes off the things and it goes on to there and it stays.
Philip:	And the rust gots nowhere to go so it goes on the plate.
Jill:	But if we didn't put the metal things on there [on the plate] it wouldn't be all rusty.
Philip:	And if we didn't put the water on there [on the metal bottle cap/on the plastic cap plate combination/on the plate?] it wouldn't be all rusty.

Jill points out that rust comes off rusty metal things like the bottle cap and leaves marks on other things, like the plastic plate the bottle cap was sitting on. She later points out that if the rusty metal bottle cap had not been placed on the plastic plate, then the plate would not be "all rusty." Philip formulates his last contribution above as a direct "copy" of Jill's immediately preceding contribution, pointing out that if water had not been placed on the metal bottle cap (and the plate it's on), then it would not be "all rusty":

Jill: But if we didn't put the metal things on there it wouldn't be all rusty

Philip: But if we didn't put the water on there it wouldn't be all rusty

This whole segment is typical of everyday language, and, in fact, reflects both the strength and weakness of such language. In everyday language, when we are trying to make sense of a problematic situation, we use patterns and associations, repetitions and parallelism, what might loosely be called "poetic" devices, to construct (or, as here, to co-construct) a senseful (sense making) design. This is a feature that much everyday talk shares with poetry, myth, and storytelling, regardless of what "genre" it is in formally. Far from meaning to denigrate this approach to sense making, I have elsewhere celebrated it (e.g., Gee, 1991) and have no doubt that it has given rise to some of human beings' "deepest" insights into the human condition. But it is not how the Discourses of the sciences operate; in fact, historically, while these Discourses most certainly grew out of this method of sense making, they developed partly in overt opposition to it (Bazerman, 1988).

Everyday language allows for juxtapositions of images and themes in the creation of patterns (like "it wouldn't be all rusty"), and, as I have just said, this is very often an extremely powerful device in its own right. But, from the perspective of scientific Discourse, it can create a symmetry that is misleading and which obscures important "underlying" differences (e.g., the underlying "reality" behind "a plastic plate that is all rusty" and the one behind "a metal cap that is all rusty"). Unfortunately, in science it is often this "underlying" level that is crucial.

The children's everyday language obliterates what we will see is a crucial distinction and it obliterates the "underlying mechanisms" (here cause and effect) that are the heart and soul of physical science. Jill and Philip's parallel constructions above—in particular, their uses of "all rusty" and "if we didn't put . . . on, it wouldn't be . . ." obscure the fact that these two linguistic devices here mean (or could mean) two very different things. Rusty metal things "cause" things like plastic plates to "be all rusty" (namely, by physical contact) in a quite different way that water "causes" metal things to "be all rusty" (namely, by a chemical reaction).

Furthermore, the plastic plate and the metal bottle cap are "all rusty" in two crucially different senses—that is, crucial for scientific Discourse, though not necessarily for everyday Discourse, which is content to pattern them together through the phrase "all rusty." In Jill's statement, "all rusty" means (or could mean) "covered in rust," while in Philip's statement it means (or could mean) "a surface which has become rusted." In other words, the distinction between "having rust" (a state) and "having rusted" (a process) is obliterated.

It is typical of everyday language that it tends to obscure the details of causal, or other systematic, relationships among things in favor of rather general and vague relations, like "all rusty" or "put on." Everyday language, in creating patterns and associations, is less careful about differences and underlying systematic relations, though these are crucial to science. Again, I do not intend to denigrate everyday language. The very weaknesses I am pointing to here are, in other contexts, sources of great power and strength. Everyday language is much better, in fact, than the language of science in making integrative connections across domains (e.g., light as a psycho-social-physical element in which we bathe).

CONVERSATION AND EXPANDED LANGUAGE

Claim 6: A face-to-face conversational framework is problematic for the acquisition of scientific academic languages.

Very often even when classrooms import a form of a scientific social language into their discussions, they keep this talk in the typical style of interactive conversation among peers. In face-to-face conversations, both in the lifeworld and in science, people tend to used "truncated" language. That is, they use deictics, vague references, and ambiguous structures that are resolved by the shared knowledge the interlocutors have of what they are talking about and of the context they share in their face-to-face encounter. This is fine for people who share a good deal of knowledge, know a good deal about what they are talking about, and have mastered the social language in which they are communicating.

But such a framework is paradoxical for people trying to acquire a new social language. Such beginners cannot see the expanded forms of the social language used in overt ways that make public what they mean. The deictics, vague references, and ambiguous structures typical of face-to-face conversations hide meanings from these beginners and also hide the beginners' own meanings from more advanced peers and teachers who could otherwise scaffold them in their acquisition of the social language. Yet it is the expanded form of the language that ultimately underlies the meanings people are expressing and that appears in the sorts of writing and monologues that are very often the basis of evaluation in the schools.

Let me give one example (data from Rosebery, Puttick, & Bodwell, 1996). The following discussion occurred in a second grade. Three girls in the class had designed an experiment to test how much light plants needed to grow well (they had raised fast growing plants under different light conditions, including plants grown in a dark closet—these plants had grown tall, but were a sickly white color and very droopy). The girls have given a report to their class and are now engaged in a whole-class discussion. I place material in brackets to indicate what certain deictic and vague expressions mean (". . ." represents material—usually the teacher or other students designating next speaker—left out):

Teacher: . . . does anybody have any idea about why those [pale plants grown in the closet, JPG] might bae that color [i.e., not green, JPG]?

. . .

Karen: Because, um, that's in the dark and it doesn't get any light maybe.
Girl: It does get a little light.
Girl: It gets the teeniest bit.

. . .

Aleisha: I think it's that color because it doesn't get that much light, and, it—it has—and plants grow with light, so.

. . .

Michael: Well, I think these are—there are these special rays in light that make it turn green and it's not getting those rays, so it won't turn green.

. . .

Michael: Like a laser and a light beam are almost the sa—are almost different—I mean they are different kinds of light. So, maybe there's this kind of light in the air that maybe we can't see, but maybe the plants need it maybe to turn green.

. . .

Anna: I think, um, the rays, um, gives the plant food, and um, they like store the food in the leaves and cotyledon, and the food like makes it turn green? And stuff.

. . .

Michael: Yeah, that sounds like an idea behind my idea.
Will: Um, maybe it's not the light. Maybe it's heat.

. . .

Go: Maybe, um, light has something to make plants green, but in the closet it gets—it gets just a little air—little light, so it has, it doesn't has enough—enough things to make green. So hmm, that stem can't get green.

The teacher's question, "Does anybody have any idea about why those might be that color?" requires the children to know which plants he is indicating by "those plants." While this is not a problem, since he is gesturing to them, the deictic reference coupled with the vague reference "that color" makes this sentence radically ambiguous. Does he mean "Does anybody have any idea why the tall, droopy, sickly white plants are not green?" or does he mean, "Does anybody have any idea why the tall, droopy, sickly white plants are white?" These are different questions.

The girls' experiment has answered one of these questions ("Why are these plants not green?")—the first and most probable interpretation. Their answer would be something like: Because plants require light to grow green (and greenness is the criterion the girls had agreed upon earlier as the test of healthy plants), these plants are not green because they did not get enough light. Though this ("Why are these plants not green?") is the most probable interpretation, since the girls have just given a report that answers the question on this interpretation, it is rather paradoxical to ask this question from a pragmatic point of view. Why would someone ask, "Does anybody have an idea?" when three people have just presented such an idea? Of course, in a typical sort of teacher style, the teacher is trying to open up a whole class "sense making" discussion, in which all viewpoints will be heard. The second question, "Why are these plants white?," has not been answered by the girls' experimental work, since, for all they know, sickly plants could be purple, blue, or any other color.

When Aleisha says "I think it's that color because it doesn't get that much light, and, it—it has—and plants grow with light, so" she is probably answering the first question, but another problem arises. Aleisha's claim can have two quite different epistemological footings. It could be a common-sense lifeworld claim ("Everyone knows light makes plants grow") or it could be an empirical claim based on the report the girls have given and the experimental work they have done. Of course, it is one of the points of this whole curriculum to get children to understand that, in a scientific Discourse of the sort being used in the class, people must make claims with the second, and not the first, epistemological footing.

There is yet another ambiguity at play in this discussion that centers around what the task is taken to be. Are the children supposed to be redoing descriptively what the girls have already, in a sense, done, namely state the connection between plant health and light and the grounds for it (this is what Aleisha and Go appear to be doing)? Or are they supposed to take the results of the girls' work for granted (light makes plants grow healthier) and explain the connection between light and plant health (this is what Michael and Anna appear to be doing)? There is even a third option: Perhaps, they are supposed to try to challenge the control conditions of the girls' experiments (this is what Karen and Will appear to be doing).

The danger in this type of discussion (which, of course, also has many good features) is that "the rich can get richer and the poor poorer." Children who understand which interpretations are pragmatically most appropriate, what epistemological footing is called for, and can expand the language into fuller forms are simultaneously learning and practicing science and scientific ways with words. Children who cannot may simply be confused or even led down a garden path.

I certainly do not advocate getting rid of face-to-face conversation-like discussions. But I think they need to be supplemented with discussions where children are asked to take longer turns, expand their language, and make clear their reasoning and its connections to what others have said. In such "mono-dialogical discussions" students need, also, to be overtly scaffolded in how they use and think about scientific social languages, interpretations, and arguments. For example, things like epistemological footing or the contrasting situated meanings of everyday words and the same-sounding words in a scientific social language need to be overtly discussed.

I believe, further, that learners also require expanded texts, displaying the fuller forms of the social language they are to acquire, placed in the midst of practice and discussion, not just assigned as reading outside these. Finally, students need to have "reading lessons" on such expanded texts, where people more expert than they model how they read such texts and engage the students in overt discussion about the language and genre conventions of such texts and how these conventions arise out of history and relate to current practices.

REFERENCES

Bakhtin, M. M. (1986). *Speech genres and other late essays.* Austin: University of Texas Press.

Barsalou, L. W. (1999a). Language comprehension: Archival memory or preparation for situated action. *Discourse Processes, 28,* 61–80.

Barsalou, L. W. (1999b). Perceptual symbol systems. *Behavioral and Brain Sciences, 22,* 577–660.

Bazerman, C. (1988). *Shaping written knowledge.* Madison: University of Wisconsin Press.

Brown, A. L., Collins, A., & Duguid, P. (1989). Situated cognition and the culture of learning. *Educational Researcher, 18,* 32–42.

Engeström, Y., Miettinen, R., & Punamaki, R.-L. (Eds). (1999). *Perspectives on activity theory.* Cambridge, England: Cambridge University Press.

Gee, J. P. (1991). Memory and myth: A perspective on narrative. In A. McCabe & C. Peterson (Eds.), *Developing narrative structure* (pp. 1–25). Hillsdale, NJ: Lawrence Erlbaum Associates.

Gee, J. P. (1992). *The social mind: Language, ideology, and social practice.* New York: Bergin & Garvey.

Gee, J. P. (1994). First language acquisition as a guide for theories of learning and pedagogy. *Linguistics and Education, 6,* 331–354.

Gee, J. P. (1996). *Social linguistics and literacies: Ideology in discourses* (2nd ed.). London: Taylor & Francis.

Gee, J. P. (1999a). *An introduction to discourse analysis: Theory and method*. London: Routledge.

Gee, J. P. (1999b). Reading and the new literacy studies: Reframing the National Academy of Sciences report on reading. *Journal of Literacy Research, 31*, 355–374.

Glenberg, A. M. (1997). What is memory for. *Behavioral and Brain Sciences, 20*, 1–55.

Glenberg, A. M., & Robertson, D. A. (1999). Indexical understanding of instructions. *Discourse Processes, 28*, 1–26.

Halliday, M. A. K. (1994). *An introduction to functional grammar* (2nd ed.). London: Edward Arnold.

Halliday, M. A. K., & Martin, J. R. (1993). *Writing science: Literacy and discursive power*. Pittsburgh: University of Pittsburgh Press.

Hanks, W. F. (1996). *Language and communicative practices*. Boulder, CO: Westview Press.

Hutchins, E. (1995). *Cognition in the wild*. Cambridge, MA: MIT Press.

Latour, B. (1999). *Pandora's hope: Essays on the reality of science studies*. Cambridge, MA: Harvard University Press.

Lave, J. (1996). Teaching, as learning, in practice. *Mind, Culture, and Activity, 3*, 149–164.

Lave, J., & Wenger, E. (1991). *Situated learning: Legitimate peripheral participation*. New York: Cambridge University Press.

Martin, J. R. (1990). Literacy in science: Learning to handle text as technology. In F. Christe (Ed.), *Literacy for a changing world* (pp. 79–117). Melbourne: Australian Council for Educational Research.

National Reading Panel. (2000). *Report of the National Reading Panel: Teaching children to read*. Washington, DC: www.nationalreadingpanel.org

Rosebery, A. S., Puttick, G. M., & Bodwell, M. B. (1996). *"How much light does a plant need?": Questions, data, and theories in a second grade classroom*. Portsmouth, NH: Heinemann.

Snow, C. E., Burns, M. S., & Griffin, P. (Eds.). (1998). *Preventing reading difficulties in young children*. Washington, DC: National Academy Press.

Tomasello, M. (1999). *The cultural origins of human cognition*. Cambridge, MA: Harvard University Press.

Wenger, E. (1998). *Communities of practice: Learning, meaning, and identity*. Cambridge, England: Cambridge University Press.

CHAPTER ONE METALOGUE

Situating Identity and Science Discourse

James Paul Gee
Gregory J. Kelly
Wolff-Michael Roth
Randy Yerrick

Randy: Jim, I understand your claim that "A face-to-face conversational framework is problematic for the acquisition of scientific academic languages" in the sense that there are many obstacles in relying upon the discourse to be transparent enough to simply be appropriated by novices by observing and jumping in. I hope I am not narrowing your claim too severely by this overgeneralization. However, there were many science curriculum efforts in the past few decades (especially in the late 1960s and early 1970s) in which the operating premise seemed to be recounting the actual historical data and dilemmas to engage students in the historical debates *of others*. There have been many endeavors that sought to build historical cases for scientific arguments (e.g., Copernicus vs. Ptolemy). Among them were esteemed science education curricula such as Harvard Project Physics, Science: A Process Approach, and others. While there was much great historical and philosophical work that went into such efforts, there were no profound studies later released which demonstrated their effectiveness over others for learning science. For the most part I am very impressed with the sociological, philosophical, and historical treatment given to such projects, but I have not encountered any studies that suggest these curricular efforts make any profound impact on students' abilities to construct arguments or engage in scientific discourse. Why do you suppose the historical science reform efforts had less than a transformative effect on science teaching? My assumption was that the students not only were unable to engage in the specific discourse patterns desired by the authors but also were detached from the artifacts and their importance within the historical context in which they were generated. After all, it wasn't really *their* problem. That is why I enjoy reading the work of Cindy Ballenger (see chap. 6, this volume) and her colleagues, who situate science talk within children's lives.

Greg: For me, your issue is the interesting one: We really need to consider how affiliation and identity co-develop. Here's the idea. People tend to like what they do. Through practice people come to affiliate with certain social groups and develop the specialized languages of the group. For example, Shinichi Suzuki (1981) was both able to develop musical discourses of particular genre types in young learners and able to export a teaching method to diverse groups of learners all over the world. These learners are unlikely to recognize the gains of this social language through their initial participation (initial violin squeaks are intrinsically valuable for few). Rather, the beauty of this social language ("the gain") is developed along with competence. The initial gains may only be "satisfying a parent" or "mimicking an older sibling," or something of the sort. Still, many graduates of the Suzuki method come to love music. I suspect that you may not disagree that the initial perceived gains for learners are likely to be nondisciplinary and have more to do with their identity in relationship to the many others interacting with their learning than necessarily any assessment of the cognitive value of the discourse. This suggests that initial experiences will need to consider how students orient to the activities and the enjoyment that they receive from the experiences of inquiry into their natural world. While I'm sympathetic to your desire to develop the academic languages of science, care must be taken to consider how this can be done without alienating those speaking their lifeworld language. There is some reason to believe that students do find scientific discourse as manifest in schools alienating (e.g., Lemke, 1990).

Jim: First, I am not suggesting that the way to acquire early forms of academic social languages is via studying texts, historical or otherwise. I have suggested here and elsewhere that acquiring any academic social language, even school-based forms, requires: (a) immersion in practice wherein learners can learn how to situate meanings of words and phrases in practice (and, thus, not be stuck with just general or vague meanings), (b) multiple models of the social language in speech and writing well integrated with practice, and (c) mentors who use the social language in rich and redundant enough circumstances so that learners can match form to function and who coach learners on when and where to pay attention to how the particular social language works in practice.

Michael: And this is where I think schools fail, as I show in my chapter because with one teacher, students are never really immersed, they hardly ever learn situated meaning and phrases in practice. Even

CHAPTER 1 METALOGUE: SITUATING IDENTITY

when a teacher models a better phrasing for one individual, many other students in the class never benefit because, for one reason or another, they have not been attuned to the current conversation. But Jim, you had another point.

Jim: Yes, my second point was that my chapter readily admits that many learners find academic social languages alienating (that's the point of the hornworm example). Such learners will only lower their "affective filter" and begin to acquire such social languages if they see some gain in doing so. To see such a gain in learning a new academic social language, they must, I would argue: (a) believe they can now or will in the future be able to function with this social language to accomplish worthwhile goals of their own (even if this is just getting into college), (b) be able to make (and be helped to make) bridges between other identities and forms of language they bring to the classroom and the new social language, (c) trust that the discourses associated with the new social language will not denigrate or oppress them or people "like them," and (d) see themselves as becoming an accepted and valued member of a group of people who use and value the social language. If we "situate science talk" in children's lives as a way to bridge to and enhance the acquisition of one or more academic forms of language, that is a good thing. If we do so in such a way that these children cannot later read or discuss texts in content areas and can't get out of high school, while more privileged children can, that's a bad thing.

Michael: I agree with you and I would also stress, "*if* we 'situate science talk' in children's lives." For many in the field of science education, this means to *talk about* scenes from life outside school. You, Jim, and Cindy Ballenger (chap. 6) show that "situating science talk in children's lives" means accepting their lifeworld talk, which is a quite different and much more challenging task.

Jim: Let me also point out that young people learn new and technical social languages outside of school all the time. For example, in recent work I have studied children and teenagers using a specialist social language to play, manipulate, critique, and discuss video and computer games. We certainly need to know a great deal more about why it is that schools, in many cases, do so poorly at getting students to learn new social languages when students are so good at it outside school, when and where the conditions I discussed above are almost always met. New forms of language arise to help people accomplish new functions—as in the case of video games—and people who value and want to accomplish those functions don't reject such languages (and other tools).

Randy: Jim, you seem to say that these face-to-face situations are problematic for actually *acquiring* not *learning about* a secondary discourse like science. Isn't it the best we currently have for an instructional venue? I am referring of course to your essay "What is Literacy?"

(Gee, 1989) and the process of learning versus acquiring discourses that are not practiced in the home. Are there other, less problematic venues in which to appropriate scientific discourse of which you are aware?

Greg: I think Jim claims that "Lifeworld language is problematic for science," but he does not specify what aspect of science serves as the referent for this claim—although the citation to *Shaping Written Knowledge* (Bazerman, 1988) suggests a reliance of a very specific genre in science. After all, the book is subtitled "The Genre and Activity of the Experimental Articles in Science." Elizabeth, Jill, and Philip are in a sense-making phase; they are not trying to persuade a distant, skeptical audience of the veracity of their experiment. So, while I believe that you are correct that the lifeworld language of students is not generally valued by the representatives of science (teachers, tests, books) in legitimizing institutions, I argue that the formalized genre of the experimental article does not represent the totality of scientific activity, particularly ubiquitous sense-making activities. Although there are many examples of discourse processes in science settings, a most interesting example was provided by Garfinkel, Lynch, and Livingston (1981). This ethnomethodological study of the discovery of an optical pulsar contrasts the multiple discourses of the scientific processes, including the discovery announcement in a published scientific journal, notebook entries, and transcriptions of face-to-face conversational talk in the laboratory. The analysis shows how the lifeworld discourses of the laboratory work exhibit similarities to students' talk, in particular as contrasted with the subsequent published account. In this example, we learn how scientists are able to translate their lab banter to persuasive texts of published accounts. They are thus multilingual in the discourses of their work.

Jim: I am not taking any one branch or type of science as paradigmatic. There are a great many different social languages connected to many different types of science and other content areas. Some of these social languages share certain features (e.g., certain sorts of explicitness, use of technical terms, nominalizations, emphasis on relational verbs). Throughout their school years, children are faced with a number of different social languages, and they must, in fact, become good at learning new social languages. Of course, after school, people are also faced throughout their lives with new language demands. And, as I said earlier, outside of school, many young people are quite adept at learning (and transforming) new social languages.

Michael: The big issue for me was always why children no longer seem to learn when they come to school. I now know that there are some fundamental structural problems in schooling and in what schools value as *knowing*.

CHAPTER 1 METALOGUE: SITUATING IDENTITY

Jim: As I say in the chapter there is nothing "wrong" with what Jill and Philip are doing for some, perhaps many, purposes. But if children don't go beyond such forms of language, their talk will not help others to learn new things (by exposing, in their talk, their thinking in lucid ways), and they and these others will eventually not be able to handle the demands that later grades, high school, and college make on them in regard to reading, writing, and talking more academic or specialist forms of language. People who encourage poor children to stay with their vernaculars while allowing rich children to master more and more styles of language (and their associated functions) are not mitigating social injustice in our society—they are helping to enhance it (unless they don't just talk about, but actually change how schooling works in regard to things like textbooks and other texts written in academic forms of language).

Michael: This is exactly one of those structural problems I was referring to—as long as the social languages of the middle and upper classes are taken as a reference for succeeding, "situating science talk in the lives of children" will only aggravate the marginalization of children from poor neighborhoods. At the same time, the rich are inducted into the club "science."

Jim: Scientists use a variety of different social languages in their work. So, for example, they write grants differently than they write research papers; they write pieces for journals differently than they do for more popular science articles; they talk science differently than they write it and they talk differently in lectures than they do in discussions. Even in their "lab banter" they do not speak an everyday vernacular but, rather, still draw on representational resources crucial to their "form of life."

We should not confuse scientists' using informal styles of talk about science with people (including scientists in other settings) using an everyday vernacular (and the vernacular itself has more and less informal and formal styles). Elinor Ochs's work on physicists engaged in a discussion shows them using informal or "unplanned" speech, but their speech, in this setting, still has many features that are not in "everyday" people's (including the scientists') vernaculars (Ochs, Gonzales, & Jacoby, 1996). Children, too, in school, need to learn to recruit different styles of academic languages for different purposes.

Michael: I am also interested in Randy's question about acquisition and learning. Given that schools are places that usually promote learning (not acquisition), could you address this issue as well?

Jim: Well, children need to be able to produce, not just consume, academic forms of language and, thus, must not just learn about them, but acquire some degree of control over them, at least enough to write and speak them in school (again, unless we radically change schooling). Of course, it is a crucial question (but an empirical one) as to how much control is required.

Michael: In my chapter, I suggested that even after almost 8 years in a French immersion program, these English-speaking students still only had but a rudimentary control over the French.

Jim: We are all akin to nonnative speakers, readers, and writers for most or all of the areas we studied in school. In a polyglot world of multiple specialist discourses that seek to control and transform our world, that is, nonetheless, an important feature for informed citizens. To return to my earlier example of video games: I am myself very much a nonnative (not very fluent) speaker and reader of some of the specialist languages used by gamers. Nonetheless, I can accomplish my purposes, participate in the "community of practice" (to the extent I wish), and know enough about what is going on to engage in some degree of analysis and critique. Furthermore, if I need more fluency than I have now I have the resources to get it.

REFERENCES

Bazerman, C. (1988). *Shaping written knowledge: The genre and activity of the experimental article in science*. Madison: University of Wisconsin Press.

Garfinkel, H., Lynch, M., & Livingston, E. (1981). The work of a discovering science construed with materials from the optically discovered pulsar. *Philosophy of the Social Sciences, 11*, 131–158.

Gee, J. (1989). What is literacy? *Journal of Education, 171*, 18–25.

Lemke, J. L. (1990). *Talking science: Language, learning and values*. Norwood, NJ: Ablex.

Ochs, E., Gonzales, P., & Jacoby, S. (1996). "When I come down I'm in the domain state": Grammar and graphic representation in the interpretive activity of physicists. In E. Ochs, E. A. Schegloff, & S. A. Thompson (Eds.), *Interaction and grammar* (pp. 328–369). Cambridge, England: Cambridge University Press.

Suzuki, S. (1981). *Ability development from age zero* (M. L. Nagata, Trans.). Miami, FL: Warner Books.

CHAPTER TWO

Telling in Purposeful Activity and the Emergence of Scientific Language

Wolff-Michael Roth
University of Victoria

> *There is no such thing as language, not if a language is anything like what philosophers, at least, have supposed. There is therefore* no such thing to be learned or mastered. *We must give up the idea of a clearly defined shared structure which language users master and then apply to cases.*
>
> —Davidson (1986, p. 446, emphasis added)

In educational settings, the term *language* is mostly used in the sense of code, which can be translated and switched as needed, and which can be acquired by learning the proper words and syntax.[1] Thus, language instruction makes use of dictionaries and grammar books and usually ignores the social and cultural dimensions in which language is embedded and that are, in part, constituted by language (McKay & Wong, 1996). The introductory quote suggests that knowing language in the sense of pure code does not exist. Rather, knowing a language is tantamount to knowing one's way around a particular part of the world (Hanks, 1996). Equally important, language is tied up with who we are: Language is constitutive of our identities, which change even if we only switch from nonstandard to standard forms of a language (Derrida, 1998). Thus, changing my pri-

---

[1] In this chapter, I follow Gee's (1990) use of language and discourse. "Language" refers to an abstract system in the sense of Saussure's "langue." Similar to Saussure's "parole," "discourse" refers to what is communicable about some topic at a given time, place, or social, cultural, or institutional setting.

mary language from German to English and having a third language (French) move into second place was coextensive with profound changes in my identity and values (Roth & Harama, 2000). Similarly, my acquisition of social sciences discourses after having been trained and having worked as a physicist was accompanied by fundamental changes in the way I look at and understand the scientific and technological "advances" of our times. That is, my acquisition of another language, whether cultural or discipline-specific, not only involved the addition of another code but also changed who I am in relation to others and how I understand myself. Taking my cues from pragmatic and postmodern scholars, I propose a way of thinking about language in schools, and particularly about science as language, in a very different way than is usual in the field of science education.

When Randy Yerrick originally invited me to contribute to the symposium out of which this book evolved, he asked me to address the topic "Explicit guidance or negotiated purpose?: The utility of common knowledge and existing classroom discourse toward alternative outcomes." The central thesis of this chapter is that much more fundamental issues to learning must first be addressed before discussing desired outcomes, including the purposeful nature of activity and identity. I propose that educators think about language as being normally used in nonrepresentational ways, to get things done and to modify one's world. We normally tell in purposeful activity rather than represent the world in thought. Such a position "erase[s] the boundary between knowing a language and knowing our way around the world more generally" (Davidson, 1986, p. 446).

Posed in the way Randy put it to me, the question about guidance implies that language is something homogenous that can be learned in unambiguous ways. Explicit guidance requires something like a known terrain as well as definable outcomes. Our experience of language is different. Although the words in a language may be finite, there are virtually infinite possibilities of using and reusing formulations. The issue then is not to guide students to specific discourses, which has a lot to do with cultural reproduction and the associated cultural domination by those who have most of the cultural capital (Foucault, 1979). It has more to do with allowing students to learn to find their own ways around the world, that is, to find their own ways of participating in culture and to create their own forms of talking.

My reframing makes apparent a problem within much of science education today. Rather than being an invitation to participate (peripherally or centrally) in some discourses of our society, current science teaching indoctrinates to a particular way of looking at the world—including all the implicit messages that this way is the best and often only way of dealing with the problems that the world faces today (Roth, 2001). In its attempt to indoctrinate children to a particular worldview, science education is also exclusionary: Few students ever appropriate and continue to use scientific forms of talk after their initial experiences in elementary and middle school. I therefore suggest that the "problem" of science as language has to be decentered. In this chapter, I do this by focusing on phenomenological and pragmatic aspects of science conceived as language. I argue that the idea of science as language has not been theorized sufficiently; a full articulation of the idea brings to the foreground major issues (learning as by-product, identity) that encourage a radical rethinking of current instructional practices.

This chapter unfolds in four stages. First, I present a phenomenological and pragmatic perspective of language-in-use that differs from the ways that *language* and *discourse* are generally employed in science education. I articulate a view of language-in-use as nonrepresentational resources for getting around in and changing the lifeworlds that students and teachers inhabit. Second, I exemplify the perspective on language in the context of a best-case scenario in which upper-middle-class students who are engaged in collaboratively building concept maps develop proto-scientific languages. Third, I provide vignettes from two 7th-grade units on simple machines (one in French immersion, the other in English as first language) to illustrate how the conversations of immersion students are considerably handicapped despite 8 years of French as their school language. Fourth, I discuss the interaction of language, Self, and the purposes of everyday activities and provide an example of a discussion of a contentious issue in one community. I view such interactions as forms of engagement that could allow students to engage in authentic ways with the science-related concerns of their communities.

LANGUAGE: A PHENOMENOLOGICAL AND SOCIOLOGICAL PERSPECTIVE

For many decades, science educators focused on "concepts" and "conceptions" that were assumed to exist independently of students' mother tongues and cultures. Thus, student talk about certain phenomena has been classified in terms of (mis-, alternative, naïve, etc.) conceptions and concepts irrespective of culture and language. Take the case of conceptions about *earth* and *moon*, for example: The existing research includes studies in Greece, the United States, Israel, Taiwan, and Nepal (Pfundt & Duit, 1994). These countries are characterized by very different cultures, and the respective languages provide different kinds of semantic and syntactic resources for talking in context. In contrast to such a traditional approach in science education, studies now show how the talk about a topic is highly contingent: The interviewer's language and the artifacts present in the situation constitute resources and contexts that shape the responses of the children (Schoultz, Säljö, & Wyndhamn, 2001). Some scholars therefore suggest it is more appropriate to study concept *talk*, particularly the structural resources language itself provides and the deployment of these structural resources by individual speakers (Edwards, 1993). The individual student's talk during an interview situation is a concrete realization of a generalized talk about the phenomenon that always and already exists as a possibility at the sociocultural level. The talk is therefore never entirely the student's: It is always both individual and general, produced by him or her but always also for (and therefore belonging to) the other (Derrida, 1998).

To elaborate, take the following thought experiment. A child who has never considered the relationship between sun and earth is asked about their relative movement. Drawing on everyday ways of talking, he or she can nevertheless produce a response by situationally drawing on available linguistic resources. Thus, everyone—even physicists—marvel at and talk about a "sunrise," a sun that is "just coming up over the horizon," "sets in the west," or "is going down." It does not take much for our child to engage in similar talk and, if pushed, to attribute agency to the sun and describe the earth as something stationary. Similarly, it only takes the child's pointing around itself to constitute the inhabited world as a more or less flat plate, limited by the more or less circular horizon. It is reasonable to talk like this *because* language in situational use is all we have rather than stable conceptual structures and theories engraved in our minds. In fact, neuroscientists think that we do not need to represent the world because we know, from experience, that it is continually available to our perception whenever we need it (O'Regan & Noë, 2001). Drawing on available resources, the child is *telling* its world rather than reading out its conceptual structures with a content- and context-independent linguistic code.

In language, individuals not only draw on and thereby *reproduce* existing linguistic resources but also continuously *produce* semantic and syntactic variations. Simply think about the continuous production of new ways of talking in youth cultures. Some of these variations are never taken up by other members of the culture and may quickly disappear. Other variations are, both consciously and unconsciously, taken up by others who, for one reason or another, appreciate the resources these variations provide them in getting around in their worlds. Such variations and new ways of telling cannot be predicted a priori, but they *emerge* leading to the *nonteleological evolution* of language. Scientific languages undergo the same type of process. Thus, for example, the Copernican revolution was not based on inferences from and decisions about telescopic observations, the center of the universe, macroscopic behavior and microstructural motions (Rorty, 1989). Over a period of about 100 years, during which scientific discourse was more like an "inconclusive muddle" than an ordered set of propositions, European scientists found themselves using a language in which a network of hierarchically ordered propositions was taken for granted. New forms of language thus emerge as speakers gradually acquire the habit of using them while other forms gradually fade away as certain words become unfashionable. The special-purpose languages of science, as that of 20th-century European culture more generally, "are a contingency, as much a result of thousands of small mutations finding niche (and millions of others finding no niches), as are the orchids and the anthropoids" (Rorty, 1989, p. 16). Science educators, too, may find that participation in the creation and use of special-purpose languages that focus on relevant and contentious issues in the community is a much better way of thinking about language than learning a particular professional science discourse. Here, I use *Sabir* as an analogy for that in which I want students to become competent. Historically *Sabir*, the original lingua franca, was an auxiliary hybrid (Italian as its base with an admixture of words from Spanish, French, Greek, and Arabic) language employed in Mediterranean ports to allow sailors and merchants who spoke mutually unintelligible tongues to communicate with one another for commercial purposes. The notion of *Sabir* is consistent with the idea of creating and participating in "islands of rationality" characterized by multiple and situationally appropriate linguistic resources about (for my purposes) science-related issues concerning humanity today (Fourez, 1997).

At the individual level, new forms of language also *emerge* rather than being the result of conscious design. Detailed case studies have shown that new languages for talking about electromagnetic phenomena, quaternions, or quarks were the emergent result of the dialectic of human agency and existing structures (culture, including language), which literally constituted a surface of emergence (Pickering, 1995). Close analysis of science talk in different high-school physics classes also suggests that *emergence* (Roth, 1996b; Roth & Lawless, 2002) and *evolution* (Roth, 1996a) are much better metaphors for learning to talk science than *appropriation* or *acquisition*. But emergence also means that outcomes cannot be predicted or forced to occur with any precision—There is no such thing, as Davidson asserted in the introductory quote, as mastering a language and applying it to cases. This view is nonteleological and flies in the face of most educational thought and practice designed to bring all students to the *same* level of (linguistic) competence in a range of (curricular) domains.

The notion of *discourse* highlights that language is not just a code that allows a person to switch from one to another by simple translation. Agreement in language is agreement in a form of life (Wittgenstein, 1958), which comes to be shared explicitly in and through discourse (Heidegger,

1996). Much of everyday talk is not a rational activity but *telling*: We talk rather than putting words in fixed sentence structures much in the same way that we walk rather than placing our feet. There is a background understanding, a meaningful whole that comes with simply being in a world always and already shot through with meaning—the background *is* the meaning. Absorbed everyday, telling articulates, carves up, and names this background common to all those who have been socialized into a particular community; but such telling is always a function of the current orientation of a project. It is always *for the sake of something*. Words in general and, in the present context, science words more specifically do not get meaning from somewhere but rather *accrue to an already meaningful whole*. As we engage in activity, new articulations (joints) of the meaningful whole emerge from the dialectic of agency and structure, action and response to action. This telling, part of the common background, makes language as a linguistic phenomenon and communication possible. The notion of discourse implies culture and ways of seeing, which *presupposes* an existential understanding.

Traditionally, the term *scientific discourse* has been used to refer to the special-purpose language employed by scientists in their laboratories or, perhaps more accurately, in their formal papers, journal articles, and textbooks. This language is explicitly devoid of values, emotions, aesthetics, responsibilities, and so on (see Gee, chap. 1, this volume). For example, it allows scientists to develop and talk about genetically modified organisms without engaging in a discussion of long-term effects on individual, collective, and environmental health; it allows scientists to develop the atomic bomb and attribute its (mis-) use to politicians and evil dictators. Science teaching similarly focuses on issues presented out of context and without purpose and, in the process, indoctrinates those students who buy into science into a particular worldview (Roth, 2001). High-school students learn about the modification and reproduction of genes but do not simultaneously engage in the ethical debate about the risks of genetic manipulation; students are asked to make inferences about data without being familiar with the data's origins and collection methods.

Such a disposition of science and science teachers is evident in the following example from a classroom where the teachers provided pairs of eighth graders with a map subdivided into sections, each containing two numbers: one for light intensity, the other for plant density. Students were asked to indicate whether a pattern existed in the data and to provide evidence in support of their claims. One teacher overheard Andrej say that the question could not be answered because he did not have enough information.

Teacher: What did you say, "You don't know"?

2. THE EMERGENCE OF SCIENTIFIC LANGUAGE 51

Andrej:	Well, if you don't know where the ecozone is....
Teacher:	Where it is? From the information given....
Andrej:	Well, it might be near a lake, it might be near—it might be near a nuclear power plant.
Teacher:	But how would that help you if you knew where the ecozone is?
Andrej:	Well, it might affect the results.
Teacher:	But from the information given?

This student, like his peers, searched to obtain more contextualizing information; that is, he attempted to establish situated descriptions so that he could evaluate the suitability of his familiar language to explain the situation. The teacher attempted in these and several more turns not reproduced here to convince the student to work just with the information provided, that is, with the map and the numbers. Whereas Andrej attempted to articulate a concrete situation and thereby to integrate the representation to his lived-in world, the teacher tried to disallow such integration of the "problem" to the familiar world. This move seems to me typical of science (and therefore science education), which is engaged in developing a discourse void of and disconnected from everyday life. Scientists not only use a monoglossic repertoire but seek to impose it on public discourses as well (Roth, Riecken, et al., 2004). They claim to create knowledge but not to be responsible for how engineers and society employ what they create, and therefore have no reason to take responsibility for the tragedies arising from the use of nuclear warheads or chemical and biological warfare. Public debates of particular contentious issues, by contrast, are generally characterized by a heteroglossic language that addresses specific concerns and incorporates the different ways of talking to and about the issue (Lee & Roth, in press). These debates give rise to forms of *Sabir*, special-purpose languages drawing on many different relevant languages, that evolve to become increasingly appropriate for an issue at a given time. As a critical science educator, I would like students to be able to participate in the creation and use of such *Sabir*, such issues-oriented special-purpose languages.

To summarize, I argue for taking language not as code for expressing thoughts and concepts but as a nonrepresentational way of acting in and modifying the world. Language seen in this way not only constitutes a resource for getting things done but also constitutes a resource for doing identity work. In a situation in which speakers of different languages and disciplinary discourses come together, participants usually evolve a special-purpose language, a lingua franca or *Sabir*, that allows them to do their work without actually acquiring complete competencies in other languages or discourses. As a way of thinking about the purposes of science education, I suggest that it is more important to enable students and citizens to participate in the elaboration of contextually appropriate *Sabir* (whatever their root language and special purpose languages they master) than to learn scientific discourse as it is spoken in laboratories.

EVOLUTION OF LANGUAGE IN SCIENCE CLASSROOMS: A BEST-CASE SCENARIO

We become competent speakers of a language when we participate in using it for some purpose rather than when we learn it for its own sake. Students learn science talk when they participate in doing something that involves talking science. In such situations, students do not "construct" a language or discourse as if driven by a rational homunculus that tells them how to put together a language. Rather, students usually engage in some task; telling is part of coping and getting on with whatever students understand the task to be. That is, they draw on language but do not represent it, reflect on it, or use it as a vehicle to share private thoughts.

Throughout my career as a science teacher, I have asked students to articulate the design of experiments, subsequent observations, and explanations of outcomes. Collective concept mapping was one of the tasks that allowed me to come closest to the notion of "guidance" because the paper slips imprinted with concept words provided constraints and opportunities for talking about the curriculum topic at hand. The task requires students to organize the concept words into hierarchically related groups from *most inclusive* (top) to *most specialized* (bottom); concept words of equal inclusiveness or specialization are to be placed on the same level. Students make as many propositions as possible by linking pairs of concepts using a suitable verb or preposition of their choice. My ultimate purpose for this activity, however, was not the construction of these maps: These were but tools for allowing students to talk science, and they provided a specific purpose for this talk. Concept mapping was therefore a reflexive task because students developed language as they constructed propositions that become part of the language.

I recorded the following episode in my own classroom while teaching a senior-level physics course in a private, all-boys school. Ken, Miles, and Ralph were constructing a concept map on the duality of light. The complementarity principle states that the nature of light as observed depends on the particular experiment conducted: In some situations, light behaves like a particle, but in other situations it exhibits wave character. For several weeks, the students had read the relevant chapter, done experiments, and solved textbook problems. Nevertheless, it took them nearly an entire 60-minute period of concentrated effort to produce a map from the 25 words. The episode was recorded about halfway into the session

2. THE EMERGENCE OF SCIENTIFIC LANGUAGE

(turn 231 out of 549) and after "complementarity" had already been discussed repeatedly (16 turns at talk used the word).[2]

Ken:	But then, where does this one (*points to* "complementarity") come in, should we scrawl this across? The complementarity has to go somewhere.
Miles:	Yeah, that's why I liked it, because we're gonna use it, then we could put the wave down-
Teacher:	Then quantum and wave have to be side by side.
Miles:	Yeah.
Teacher:	Because of the complementarity principles-
Miles:	They have equal status.
Teacher:	That's right.
Ralph:	Does quantum mean its- Um- a particle- a wave [or a particle?
Miles:	[We had it right before.
Teacher:	That means that it is a particle.
Ralph:	And then we could say that energy is like particle.
Teacher:	No, it has particle character, it's bundled, like a piece of mass. See it (*points to textbook*) says, "particle-like entity or quantum," but you can also have phonons.
Ken:	Yeah, I know that's why quantum should be over-
Teacher:	Quantum is over-
Ken:	But that's-
Teacher:	I really don't know, because you are talking about light and X-rays.
Ken:	But because you're giving us that example with phonons. Then that's like Ralph was saying, "waves also have a lower, more detailed description within itself" and that's why it should be above quantum. Because quantum is only a subdivision of waves, [no?
Teacher:	[But then-
Miles:	Because if it's-
Ralph:	For [example-
Miles:	[Light's a wave, but we don't care for that particular concept map. What we are concerned about is the complementarity and

Diagrams shown alongside:

COMPLEMENTARITY
├── QUANTUM
└── WAVE

WAVE
├── QUANTIZED
│   └── QUANTUM
│   └── PHONON
└── NOT QUANTIZED

---

[2]The following transcription conventions are used:

[— square brackets in consecutive lines indicate where two speakers overlap;
points — italics indicate actions;
(heavy) — parentheses enclose nonspoken comments and actions;
its- — hyphen ending (part of) a word indicates sudden stop in utterance.

therefore we need the wave down here. Yeah, we could put it up here, but we don't need it. But we need it here, otherwise where are we gonna put it, we get down here and-

The transcript demonstrates the difficulties that these students experienced in producing a web of propositions from the concept words provided. In turn, they tested one proposition after another ("energy is like a particle," "light's a wave," "quantum means it's a particle," or "quantum is a subdivision of waves"). Some of these propositions survived, and students appropriated them into their way of talking and writing on the topic, as 1-week and 7-week delayed posttests showed (Roth & Roychoudhury, 1993). Other propositions were tried, put aside, tried again, and ultimately discarded as if the students had tested their palatability, with a final spitting out of those that were judged unsuitable.

The transcript exhibits the dual, reflexive nature of language: It was both the tool students used to get the task done—building a concept map—and the topic. The three students, occasionally assisted by me, constructed a system of propositions, that is, a discourse about light and its dual nature expressed in the principle of complementarity. But the students in this episode "constructed" a language only apparently. They did not construct the language in the way architects design houses from well-known modular pieces or carpenters construct houses from materials of given dimensions and with a small set of familiar tools. These students did come with a root language, the one that they speak every day at home and with their peers. In the process of students ordering the concept words and joining pairs of them with linking words into propositions, a language evolved but without student awareness of what the final product might look like. The language that they built *emerged* from their interactions and, as such, had unpredictable qualities. After the fact, like Whig historians, we might be tempted to say that they constructed their scientific language, that it had taken a prescribed or necessary path; but this would be misconstruing the central feature of students' language about the complementarity of light and how that language came about.

To say that students negotiated and socially constructed a discourse would be overstating what they actually engaged in and experienced. Ken, Miles, and Ralph each proposed certain propositions, sometimes pronouncing them repeatedly and then accepting or discarding them. The real learning happened *in the course of producing the map*, as the students participated in talking and thereby facilitated the evolution of a language. Creating a language to talk about photons and quanta was not the result of successfully putting together the pieces of a puzzle. The language was an emergent result of the interaction when words and sentences became associated with meaning as they found their place in a coherent language, itself an outgrowth of an already existing language.

Here, the language that evolved from the interactions within the group, with the teacher, and with other students in the class was the outcome of a historical and contingent process. The natural world did not tell the students which language to use, nor was there an existing community where telling and talking about complementarity was a regular, everyday feature. For the purposes of this class, the teacher was the social instance for making decisions about the acceptability. If the students eventually came to say that their previous language was inappropriate for talking about light and quantum in physics classrooms, they really meant that, having learned the new language, they were able to handle these issues and context more easily.

An important outcome of these lessons was the role that my interventions had on the concept maps and the language that students created in the process. The result of such interventions could never be predicted when we analyzed the tapes from the perspective of a first time through (Roth & Roychoudhury, 1992). In some cases, the students used the content of my talk to settle a conflict (quantum vs. photon). In other cases, students used what I had said in support of propositions with which I would have never agreed. Short of telling students which propositions were allowable and which were not, my interventions failed to constitute the kind of "explicit guidance" that Randy Yerrick had asked me to address. Furthermore, students frequently heard more than what I thought I had said. Students often used comments such as "he wants us to . . ." or "he is saying . . ." to elaborate what some statement meant beyond a literal hearing.

This section is partially entitled "a best-case scenario" because I believe that although using concept mapping in that particular context worked, it did so because of the upper- and upper-middle-class origins of the students, who brought substantial discursive resources to the school and were accustomed to the value system typical of schools. These students had a root language that allowed them to handle the concept-mapping task and to evolve a language about light and complementarity that was sufficiently close to pass as authentic scientific discourse. These students were a prime example of the "jocks" that succeed because school discourse and primary discourse, including all the cultural values that they embody, already coincide (Eckert, 1989). But even in this school, I reported on difficulties when students faced contradictory discourses, such as when some students' religious discourses and values conflicted with the disembodied and decontextualized discourses of school science (Roth & Alexander, 1997). That is, even students from bourgeois origins become disadvantaged if their primary discourse is some variant of middle-class English that embodies very different values (e.g., the debate about evolution and creationism as cast by the religious right in the United States). An interesting case to consider is that of middle-class students who have done science in a second language ever since they entered school but continue to speak their mother tongue or yet another language at home and in the streets. I will use the difficulties these students face to learning science *in* their second language as a vehicle for reflecting on difficulties students may face when learning science *as* a second language.

SCIENCE AND SECOND LANGUAGE

Since *Talking Science* (Lemke, 1990), science educators have begun to conceptualize science as a special-purpose language rather than merely a set of connected concept words. However, the consequences of thinking about science and scientific knowledge in terms of language have not been worked out. For example, in science classrooms, students spend a lesson or two, even a week, on a particular topic. The students' textbooks highlight a few content words, which teachers often test knowledge of by asking students to write definitions. (They do so even though there is evidence that students do not learn words from definitions and, in fact, use them to make inappropriate sentences and statements [Brown, Collins, & Duguid, 1989].) In this approach, scientific language is merely a code used for moving the concepts in mind to the outside world. My research in French immersion and regular English classrooms suggests that this practice of making students learn a few words at a time falls short of permitting students to acquire a language together with its culture. Because French immersion students have a much lower level of command in the language of instruction, that is, only rudimentary competencies in the root language, their ability to get on by using language in the telling mode (unreflective and nonrepresentational) at school is jeopardized. I have witnessed breakdowns in their usual ways of coping, getting on, and making sense as these ways were interrupted, changing their forms of participation and therefore their learning. I exemplify these issues in the following episode from one of my studies in an elementary school in a large urban center of Western Canada.

Here I provide excerpts from seventh-grade classrooms in which students learn science *in* a second language and science *as* a second language, allowing us to put in relief issues associated with rethinking science education in terms of language and discourse. First, an increasing number of children move to different countries, with different cultures than their own, where they have to learn school subjects in a language other than their mother tongues.[3] This has consequences that science ed-

---

[3] In Vancouver, Canada, more than 55% of families speak a language other than English in their home. Educational researchers in Toronto and Montreal have to send letters to the parents in 20 or more different languages (Laura Winer, personal communication, March 8, 2002).

ucators have to consider. Second, science itself is a second language, but one that has some family resemblance with the mother tongue.

Science in a Second Language

The first set of excerpts derive from a French immersion classroom, made up largely of very good to exceptional students who, in contrast to their peers in normal classes, used public transportation to be able to attend this special program. At the moment of the episodes, the students were in their 8th year of attending the program, which requires them to speak French throughout the school day, and all instruction but that in language arts is provided in French. Despite this immersion over a long period of time, the students had difficulties in articulating pertinent issues, even in making very simple statements that pertained to their everyday lives. Furthermore, their language was riddled with errors: Students made up words, sentences and tense constructions were inappropriate, gender of articles and corresponding nouns did not match, and so on. I do not make these characterizations to mock these students but rather to show that even after many years of intense involvement with a particular language, articulating even simple phenomena was difficult. It therefore should come as no surprise that these students, some more frequently than others, fell back on their native English as soon as the teachers moved out of hearing range.

Caitie, the leader of a group of five students, presented a machine that they had constructed according to a plan that she had earlier sketched in her notebook (see diagram in the transcript). Caitie had noted, "I think that that (phonetically close to *pourrait*, it could) be a [masc.] very simple idea [fem.]." They built a lever that could be used to lift an object placed on one side when sufficient weight (here marbles) was added on the other.

Caitie: *On avait essayé cinquante gram* (Engl.), *puis on essayé* (PP), *on a pensé*
One [we] had tried fifty grams then we tried, we have thought
que si on mettait des billes dans cette boîte-là, ça vait (non-word) *comme,*
that if we were to put the marbles in this box-there that like
ça vait (non-word) *fait comme-*
that does/done like-

Teacher: *Ça va soulever?*
That is going to lift?

Caitie: *Oui.*
Yes.

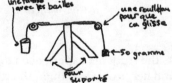

The design of Caitie's group for the machine presented.

Caitie used an indefinite personal pronoun to refer to her group that, while normally used to refer to individuals or groups rather vaguely, can be used colloquially to refer to specific persons. Caitie freely and indistinctly employed different tense forms and used the constructions "*comme ça*" (like this) and "*ça vait comme*" (it [goes?] like), which were used by the students whenever they could not express themselves to refer the listener to things in the situation or to actions they conducted (the "deictics" described by Gee, chap. 1, this volume). As the subsequent teacher question indicated, Caitie wanted to say that her machine would "lift" the 50-gram reference weight already present in her diagram.

A few moments later, the teacher asked what kind of system Caitie and her group had developed and called on Jonathan. Again, Jonathan, Joël, and Tyler in turn had difficulties talking about simple everyday objects and events.

Jonathan:	*Hum, comme qu'est-ce que tu as pour mesurer*
	Um, like what do you have? for measuring
	[*le-*
	[the
Teacher:	[*Ça s'appelle quoi ça le poids?*
	[That is called what that the weight?
Jonathan:	*Une balance.*
	A balance.
Teacher:	*Voilà une balance, donc un espèce de système de balance. Qu'est-*
	That is a balance, so a type of balancing system. What is
	ce qu'il y a dans la vie dehors, quand vous étiez jeunes, qui
	there in the world outside, when you were young, that
	fonctionne un peu pareil, que vous adoriez faire . . . Joël?
	works a little similar, what you liked to do . . . Joël?
Joël:	*Le teeter totter* [now-
	The teeter totter [now (in Engl.)
Teacher:	[*Comment on dit ça en français?*
	[How does one say that in French?
Tyler:	*Le trottoir.* (*There are five seconds of laughter in the classroom.*)
	The sidewalk.
Teacher:	*Bon essai Tyler! Une petite balançoire, une balançoire on*
	Nice try Tyler! A small balançoire [teeter totter], a balançoire one
	dit en français.
	says in French.

As the students experienced difficulties describing the machine that Caitie and her group had built, the teacher constantly requested naming devices and entities in French to avoid referring to everything using "*ça*," this or

2. THE EMERGENCE OF SCIENTIFIC LANGUAGE 59

that. She even interrupted students (here Jonathan and Joël) to introduce the appropriate word. It is also notable that much of the students' French resembled English in structure and many expressions were literal translations. For example, Jonathan used the expression *"qu'est-ce que tu as,"* which translates as "What do you have?," rather than the correct *"ce que tu as"* (that [which] you have). Frequently, students used English words in the French sentence. As a result, the language actually spoken by the students was a mixture, which sociolinguists call *pidgin*. In contrast to *Sabir*, a mixture of languages developed for specialized purposes and contexts (such as the one used in Mediterranean merchant ports), pidgin is a general-purpose language in which the grammatical structures of one language are filled with words of another. But at the moment, rather than providing them with an advantage, the pidgin actually threw them back because it was not designed to allow students to progress with the tasks at the level that they needed. To appreciate what students at this age level can do, we need to look at a similar class doing a similar unit. Fortunately, I was able to teach a split sixth- and seventh-grade class in the same school but with students who did not participate in French immersion, in part because many of them were officially labeled as "learning-disabled" students.

Science as Second Language

When students learn science in their mother tongue, they can draw on family resemblances between the new, scientific and the root, everyday language as a resource, which scaffolds their learning despite the irreconcilable differences between their semantic structures. Compare the previous episode with the following in a similar unit on simple machines, but in a classroom where students spoke their mother tongue. Although many students in this class had disabilities and needed special services to cope with the demands of regular schooling, the conversations that emerged were much more extensive. Although students still wrestled with evolving ways of talking about mechanical phenomena, they were, having their root language as a resource, not as handicapped as the French immersion students. Ultimately, these students developed highly competent ways of talking about simple machines, exceeding the levels that other students in their own and higher grades normally achieve (McGinn, Roth, Boutonné, & Woszczyna, 1995; Roth, McGinn, Woszczyna, & Boutonné, 1999).

Natasha:	This is more than a hundred, and it goes down so this (*points to the load*) comes up.
Teacher:	But how do you get your machine to work again? Like if you want to put up a second load?

Maryam:	We will, we take this load off (*the heavy one*) and then put up another one-
Chantelle:	How would you, bring it- How would you lift it up because, because it would be too big to just hang it down on it.
Teacher:	So there is a problem with picking up that side. Go ahead Carla.
Carla:	How do you . . . ?
Natasha:	After you bring something down heavier to make it go get another load.
Maryam:	Why don't we just put something heavier-
Natasha:	-and put it here after one this is done to bring it down to get another.
Maryam:	This is the hundred grams.
Natasha:	We can take this (*weight*) for instance and put it there (*opposite side of 100-g weight*) but we would have [another one.
Carla:	[How would you get it up?
Leanne:	If you lifted a hundred grams, and you needed to come back down, and you hang, and you have to jump up, how can you hang- actually hang something on it?
Don:	What would you be using for the lever?
Carla:	How do you like when you want to come down and pick up something? And bring up, they sort of take off that side, but if it's big, how do you take it off?
Natasha:	There are such things as big machines, you know!
Carla:	Where is it?
Teacher:	There are some questions about the design in that, I think they probably have an excellent idea here in using the lever, all of you might have seen the lever, but there are some problem that the class tries to address- you want to add to that, Leanne?
Leanne:	If there were to, on the 100-gram side, if they were to put a string hanging from that, then they could pull it down, if they had a long string.

Even though these students are at early stages of their development of a scientific language, and even though Maryam, Natasha, and Chantelle experienced difficulties in most of their schoolwork, they were able, collectively, to produce a conversation in which the basic principles of simple machines became salient. At issue were the basic principles that the task had asked for, using a lever and the design for a machine that Natasha and Maryam proposed and presented. The machine was like that designed by Caitie and, rather than using the lever principle, simply used pulleys to redirect the string (see diagram on p. 57). Carla, Leanne, and Chantelle were very critical of the design, questioning how the machine would be returned in the original position to be used again once a heavy weight had moved the load upward to the desired position. There was a barrage of questions, making it difficult for Natasha and Maryam to access the speaking floor.

The point here is that although the students in this class as a whole were, according to the school, far less scholastically able than the French immersion students, they competently participated in, developed, and sustained conversations over salient issues. Even though, from a scientific perspective, the talk was very much muddled, the conversation in its structure had some resemblance with other discussions that the students might have had about nonschool topics with which they were very familiar and in which they were highly competent. Their existing language provided a much greater surface of emergence for the new, scientific language than the existing French competencies in the other classroom. Thus, the French immersion students often became frustrated because they could not articulate their background understanding in the way that they might have done in English. According to their teacher, a native French speaker, the French immersion students did not develop the same conceptual understanding as their peers in English-based classrooms at the same school.

Comments on Science in or as Second Language

The main point of this article is that language is not a system of representation but a resource for acting in the world. Working in their everyday language, the students in the regular English (mother tongue) class were enabled not only to talk about scientific objects but, more importantly, they were enabled to engage and maintain a conversation, respond to or question others, and so on. They had all the resources required to constitute the events in the classroom and expand their competencies into new areas that had scientific objects and events as their topic. The French immersion students, by contrast, were like visitors in an unfamiliar land. Because they worked in an unfamiliar language, they were unable to make a conversation emerge, to engage seamlessly in question and answer sequences, that is, to construct many aspects that normally characterize a conversation.

These two examples provide a good analogy for thinking about what science instruction possibly can achieve. If the metaphor of science as language holds, the episode from the French immersion classroom forces us to rethink what we want and can achieve by means of science instruction in a regular classroom. That is, we delude ourselves if we think that students can achieve anything that resembles scientific discourse, at least, if we take as a reference what scientists or even science-popularizing journalists write. As a norm, the students will never speak as more than the occasional tourist who travels in foreign countries with a dictionary telling him or her how to ask for directions or the names of certain foods.

The French immersion students were not completely lost, however. Although literal translations are often inappropriate, the immersion students were at least in a position to use English noun words within a largely French sentence because material entities frequently have an exact equivalent. Furthermore, because the sentence structure was often more akin to English, others in this French immersion classroom did get a sense of what the speaker wanted to say. The students understood each other at least partially because they navigated the words while drawing on the same resources using, for example, French words in an English structure. Conversely, the sentences were frequently meaningless and incomprehensible from the French side: A French research assistant with a limited English repertoire often found it impossible to transcribe the lessons because to understand what was said one needed to know English. That is, if these French immersion students had tried to communicate with a French-only person, they would have had a much more difficult time making an interaction work.

Speakers of a language can tell when a sentence is correct or incorrect, even without formal knowledge of grammar. Because science was not their native language, students had few resources to help them decide whether an expression was scientifically appropriate or merely babble. That is, even though these students were in an immersion class, they were not immersed in a culture in the same way that they would have been if they had gone to France for a year. Learning to speak a language in a context where everyone else speaks it provides ample opportunities for hearing and receiving feedback that do not exist in the immersion classroom. It lies in the very structure of schooling that such learning opportunities do not exist. To a large extent for this reason I have argued for some time now that science instruction has to move into the community, so that students can become truly immersed in discourses about salient issues (e.g., McGinn & Roth, 1999). In the process of participating in purposeful activity, students not only learn a language but also, and more importantly, have opportunities to develop identities as contributing members to society.

LANGUAGE, SELF, AND PURPOSE

Insofar as traditional science instruction (including that proposed by recent reform documents) is concerned, I do not see how the science education community can make much headway toward the goal of *Science for All Americans* (American Association for the Advancement of Science [AAAS], 1989). First, to be familiar with a language, one has to have many opportunities to speak it with other native speakers. Second, because scientific language, as all other languages, comes with a whole set of cultural values and implicit assumptions, one might question whether it is in fact democratic or ethical to force *all* students to adopt the same set of values and ideology that come with the language.

I am also disheartened because other relevant aspects of language do not become salient in discussions within the science education community. Outside schools, people are always involved in some project, do things for some purpose; motives are integral aspects of the objects of activity (Holzkamp, 1985). Becoming more competent and learning are byproducts that can be observed from ongoing conversations. The language that articulates relevant aspects of the situation is always bound up with and used for the sake of some goal and in order to move the project ahead (Hanks, 1996). In schools, however, learning science concept words is often an end itself rather than a means to an end. As my example from the French immersion classroom shows, when language is an end in itself, even years of instruction do not lead beyond a rudimentary mastery. Even though the students were from the middle and upper classes, they experienced many frustrations because they lacked the level of control over their environment, the levels of telling and coping, that characterized their interactions when they were able to use their first language (English).

I hinted at the second and related issue in the introduction. Language is more than a code; it is part of a form of life and therefore is a central aspect of our identity, a means of establishing who we are in relation to others and how we understand ourselves. Science educators rarely consider the learner as someone whose identity development is profoundly bound up with language and culture (for exceptions see Brickhouse & Potter, 2001; Lemke, 1990; and Roth, Tobin, et al., 2004). In conversation and in *purposeful* activity directed to things other than learning itself, our identities are perpetually at stake, continuously emerging from interactions with others. This issue also arises when the first language is one of the various nonstandard forms of English, such as those spoken by African Americans or the inhabitants of Caribbean islands (Roth & Harama, 2000). To theorize science as language without also addressing the profound influence a new language has on identity is fundamentally unsound, a continuation of the Cartesian project of disembodied, value-free, and emotionless cognition. Leaving out identity and purpose from theorizing science as language means the elimination a priori of what others have recognized as the central issues of the relation of Self and Other, a relation always brought about and maintained through language (Ricœur, 1992).

I therefore propose to rethink what a science for all might look like—certainly different from the extreme form represented by laboratory science talk. In particular, all students may not need to be able to talk about the same content areas, such as atomic models, the role of mRNA in the reproduction of genetic material, or the Krebs cycle. After all, how many readers are able to converse with a car mechanic in detail about their engine problems or can understand the most rudimentary shoptalk of photocopy repairers? Furthermore, in this rhetorical question I have not even considered the distinction between talking *about* a practice and competently participating *in* it.

As long as there are (democratic) structures that support the participation of individuals with quite different experiences, backgrounds, and institutional locations, there is no need for every individual to be competent in a specialized (monoglossic) language. Rather, new hybrid heteroglossic languages can be created for specific purposes that are richer than any individual language on which they draw. Perhaps we should think along the lines of *Sabir*, a lingua franca that draws on many different languages and is developed for dealing with particular problems. The *Sabir* allows participants to engage in telling, making issues salient and revealing pertinent frames and solutions. Scientific language would be but one of the resources on which the creators of the *Sabir* would draw.

Preparing students to contribute in the creation of such special purpose languages cannot occur by making them learn a few definitions. To learn to contribute to an ongoing conversation and debate controversial issues, students need to participate in ongoing conversations and debate issues that *they*—not some curriculum developer or government official—have picked and framed. I now draw on my ethnographic studies of science in the community to project an image of the kind of situation in which I want students to competently participate.

SCIENCE IN AND FOR THE COMMUNITY

To think about and develop science curriculum, I take my lead from issues-oriented conversations (debates) rather than from the language used by scientists in formal presentations and their writings or, for these matters, the discourse of the reform documents (e.g., AAAS, 1989; National Research Council, 1996). When it comes to contentious issues that affect their lives, people begin to take positions and constitute their lifeworlds even before being informed more broadly. There are always fundamental concerns at work that motivate activity at a social level and individual participation. People learn new ways of talking, new languages, not because of language per se but because of the issues at stake. Learning, then, is discernible from the changing participation in praxis. In these interactions, new hybrid and heteroglossic discourses (i.e., *Sabir*) emerge that embody concerns from different root languages. I would like to see schools foster students' competencies to engage in conversations more broadly; to contribute to the emergence of *Sabir* that embody multiple concerns, values, and perspectives; and to access the speaking floor for making contributions. The following exchange from a public meeting in my community shows how even ordi-

2. THE EMERGENCE OF SCIENTIFIC LANGUAGE

nary citizens can successfully challenge scientist, although it also shows how scientists can successfully stop a citizen's participation.

This is the issue: The citizens of Salina Drive are not connected to the watermain and therefore draw water from individual wells, which are biologically or chemically contaminated whenever the groundwater levels are low in the summer and fall. The mayor and town council attempted to block a connection to the watermain and used the assessment by one scientific consultant (Lowell) as a basis for their decision. In the meeting, Lowell was questioned by the citizens of the community (in this excerpt, Naught). As a result of this questioning, a better understanding of science methodology and the nature of the problems emerged, even though many residents had little or no formal training in science. But in the pursuit of a fundamental concern, their form of interacting with scientists was changed.

Naught: Just a minute. You just said that it was a direct result of water and we've just had a record rainfall and it doesn't affect it? Well, there's something missing here.

Lowell: That means that only a certain amount of the rainfall can get into the aquifer being the heavy rains are running off. That's my interpretation.

Naught: Well, it could [also-

Lowell: [there's a limiting factor as to how much can get down into the-

Naught: -Well, well, it's, okay this is true but the thing is, is that what we've experienced is, rainfall in the order of 522% on average uh as far as monthly averages are concerned increase over the summer months. In other words what we've got through the winter period, through the 5 months previously preceding your test results, there was- If you took that and compared that to an average summer month, a month through that period, it is, there's, there were 522% more. Now, it would seem to me that we're probably not dealing on an average result with your tests, we're probably dealing on the hydrostatic head feeding that aquifer up in the higher, very much higher ends, so that the readings that you're getting are very much diluted.

Lowell: The um hydrograph that we have shows that uh the water levels are average in late April, early May and I put the average water level on the hydrograph here and the, [um

Naught: [Could it be an error? Could you be in error here?

Lowell: Well, I don't take the water level read[ings-

Naught: [No, no, no, but I [mean-

Lowell: [but I take the Ministry of Environ[ment-

Naught: [Well, you mean to say that on these particular aquifers

Lowell: out in Saline that they're taking the readings? And, [and- [the Ministry of Environment produced these readings.

Naught: And could there be an error? Could they be for example, relevant for [some-

Bishop: [I'm not, I'm really, I really don't want to get into, I hate to cut you off. . . .

Naught questioned the scientist. Even though he is not a scientist, Naught engaged in a conversation as part of which the scientist's methodology came to be questioned. Having experienced the water problems for many years, Naught had hired his own consultants for conducting water tests, had investigated possible explanations for the source of the problems, and had meetings with the neighbors on his street. Naught's objective was not to learn a scientific discourse for its own sake; rather, he wanted to change his conditions of living, which at the time forced him to haul drinking water from a gas station five kilometers from his home. The meeting format allowed this interaction with Lowell, in the process of which assumptions about the quality of the data and interpretation of longitudinal water data came to the fore.

Differences in the language and embodied assumptions were also discernible from the longer talks presented by scientists and local people. For example, Lowell drew on a scientific repertoire that articulated health problems in terms of specific biological parameters that had been exceeded. Health problems did not include other parameters somewhat euphemistically described as "aesthetic."

> The aesthetic objectives from the Guidelines for Canadian Drinking Water Quality were exceeded for some of the wells. Aesthetic objectives is- certain parameters in the water may cause the water to be corrosive, deposit forming, or unpalatable. These are given a separate category because they are not a health concern.

In contrast to the scientists' descriptions, Naught and other residents of Saline Drive addressed diverse problems, including the dangers arising from the concurrence of salt and chromium and the problems to plant and animal health. They described, as Gee (chap. 1, this volume) shows, the problem in terms of their lifeworld rather than in a decontextualized fashion, involving themselves as agents.

> We have the problem with staining, we have the problem that we can't drink, or use the water for any purpose, for gardening, because it destroys everything. When we first moved in my teenagers instinctively realized that the water was bad for their bodies and my teenage daughter when she came

home she said, "Oh, you know there's no way that I'm going to bathe in this. It leaves a scale on your skin."

The residents' discourse included evaluative statements ("was bad for their bodies," "destroys everything") of the kind that scientists disallow in their own language but that was consistent with a much broader discourse about human and environmental health (Roth, Riecken, et al., 2004).

As a final issue I raise the point that Naught and Lowell did not just debate some scientific issues (e.g., how to interpret the graphs that presented water levels over time) but were also involved in a battle over the right to speak. The town engineer (Bishop), who was also chairing the meeting, cut Naught off rather than allowing the debate to unfold until all the pertinent issues were on the table. Where do science students currently learn to remain in conversations and have their voices heard rather than be excluded from discussions of pertinent issues?

There are at least two lessons for science educators that I want to draw from these case materials. First, I do not argue that Naught knew a special science or that he had some special competence. I want to decenter the discussion of language and suggest that we need to rethink scientific language. We should stop calling *science* those special languages used by the inhabitants of laboratories and use it instead for the conversations that arise in the public sphere where people with many different languages and knowledge come to debate contentious issues. The heteroglossic hybrid languages that arise can embody and address many concerns that the monoglossic traditional scientific language alone cannot and is unwilling to satisfy.

Second, important for my rethinking of science education is neither Lowell's competence in doing the scientific analyses nor Naught's competence in raising questions about Lowell's method. Rather, I focus on their current competencies to engage one another and thereby develop a new form of language that is better adapted to solve the water problem. In this case, there was little effort made to evolve such a collective lingua franca, a *Sabir*, although I know of numerous situations in which such special-purpose languages emerged despite considerable initial differences (e.g., Ehn, 1992). This lingua franca is something that they need to evolve collectively, taking into account all the special languages that possibly could make a contribution. The excerpts presented here also raise important issues never discussed in schools: Discourses are always part of activity systems and therefore determined by the object and access to conversation. Naught was directly concerned by the quality and quantity of water that he could get from his well. He was not just interested in talking about the chemistry and biology of water quality or about the geology of groundwater in the area. Rather, he had a particular concern that was the central constitutive element of his motives.

CODA

In this chapter, I argue that language is normally used in a nonrepresentational way. Knowing a language means knowing to navigate and act in the world generally. Scientific language is one of the many special-purpose languages necessary for navigating and acting in special-purpose places, such as laboratories. As with all languages, the special-purpose language of science embodies values and concerns in addition to its special topics. But there are too many special-purpose languages for anyone to know them all. I therefore argue that it is more important to be able to participate in creating an issue-oriented special purpose language, a *Sabir*, on a just-in-time and as-needed basis. Such new languages allow us to navigate and act toward the issues, whether these concern the building of nuclear power plants, the unknown dangers of GMO in the food chain, the contribution of modern society to global warming, or simply getting access to drinking water in a community.

I did not always think about science and scientific language in this way. When I started my career as a teacher and subsequently as a researcher, I talked and wrote science education in the same teleological ways in which I heard/wrote and still hear/read my colleagues talk and write. I thought that we needed to provide students with special guidance to arrive at a specific way of talking about and looking at the world. There was only one way—the scientific way. Over the years, spending thousands of hours watching classroom videotapes, I have changed. Thinking about science as language, I am now constantly reminded of and guided by a discussion of language and culture, multiculturalism and colonialism by Jacques Derrida (1998). In my own curriculum design efforts, I encourage students to use multiple discourses and forms of representation (e.g., Roth & Lee, 2002) because I believe that forcing all students to appropriate the same (science) language (recall the slogan, Science for *All* Americans!) amounts to little more than a special form of colonialism and includes the sometimes explicit, sometimes tacit censorship by means of which other languages are discouraged and excluded from serious consideration. Hegemony means that students are both subjected to this form of colonialism and actively buy into and reproduce it in their daily praxis.

In my understanding of democracy, we need to allow future generations of adults to develop ways of participating in ongoing salient issues. Because of increasing specialization, no individual can have even rudimentary knowledge in more than a small number of domains. Participation to support the emergence of a collective *Sabir* is therefore more important to me than whether any one individual can talk about details like the differences between the atomic models that historically existed or the particular mechanisms underlying the transport and accumulation of

PCBs in the natural environment as a function of their molecular structure and physical characteristics. Supporting the emergence of situations in which democratic conversations (access, distributed power) about controversial issues can occur by developing and drawing on a *Sabir* especially suited to deal with the issues (problems) at hand is more important to me than any particular curricular content. Above all, I would like to see schools foster students' competencies for contributing to the emergence of heteroglossic *Sabir* that allow the presence of different motivations, values, and expressive forms that can undermine any hegemony at work.

I conclude with a reflexive comment about special-purpose languages that are not shared. If an educator submits an article with as many new words as science students face in a single chapter from their textbook, reviewers would probably reject it because of the "jargon." Why do science educators reject the work of their peers if there are a handful of new and unfamiliar terms but consider it normal to have their students encounter as many words on a daily basis? Is it unreasonable to assume that students reject science because they are tired of continuously dealing with jargon? Science educators interested in discourse and language might do well to consider the consequences deriving from the implications of such a move.

ACKNOWLEDGMENTS

The data for this chapter have been collected with the help of Grants 410-93-1127 and 410-99-0021 from the Social Sciences and Humanities Research Council of Canada. I am grateful to the assistance provided by Sylvie Boutonné and Stuart Lee in the taping and transcribing of the data.

REFERENCES

American Association for the Advancement of Science. (1989). *Science for all Americans: Project 2061*. Washington, DC: AAAS.

Brickhouse, N. W., & Potter, J. T. (2001). Young women's scientific identity formation in an urban context. *Journal of Research in Science Teaching, 38*, 965–980.

Brown, J. S., Collins, A., & Duguid, P. (1989). Situated cognition and the culture of learning. *Educational Researcher, 18*(1), 32–42.

Davidson, D. (1986). A nice derangement of epitaphs. In E. Lepore (Ed.), *Truth and interpretation* (pp. 433–446). Oxford: Blackwell.

Derrida, J. (1998). *Monolingualism of the other; or, the prosthesis of origin*. Stanford, CA: Stanford University Press.

Eckert, P. (1989). *Jocks and burnouts: Social categories and identity in the high school*. New York: Teachers College Press.

Edwards, D. (1993). But what do children really think? Discourse analysis and conceptual content in children's talk. *Cognition and Instruction, 11*, 207–225.

Ehn, P. (1992). Scandinavian design: On participation and skill. In P. S. Adler & T. A. Winograd (Eds.), *Usability: Turning technologies into tools* (pp. 96–132). New York: Oxford University Press.

Foucault, M. (1979). *Discipline and punish: The birth of the prison.* New York: Vintage Books.

Fourez, G. (1997). Scientific and technological literacy as a social practice. *Social Studies of Science, 27,* 903–936.

Gee, J. (1990). *Social linguistics and literacies: Ideologies in discourses.* London: Falmer Press.

Hanks, W. F. (1996). *Language and communicative practices.* Boulder, CO: Westview Press.

Heidegger, M. (1996). *Being and time* (J. Stambaugh, Trans.). Albany: State University of New York Press.

Holzkamp, K. (1985). Grundkonzepte der Kritischen Psychologie [Fundamental concepts of critical psychology]. In Diesterweg-Hochschule (Ed.), *Gestaltpädagogik—Fortschritt oder Sackgasse* [Gestalt pedagogy—Advance or cul de sac] (pp. 31–38). Berlin: GEW.

Lee, S., & Roth, W. -M. (in press). Monoglossia, heteroglossia, and the public understanding of science: Enriched meanings made in the community. *Society and Natural Resources.*

Lemke, J. L. (1990). *Talking science: Language, learning and values.* Norwood, NJ: Ablex.

McGinn, M. K., & Roth, W. -M. (1999). Preparing students for competent scientific practice: Implications of recent research in science and technology studies. *Educational Researcher, 28*(3), 14–24.

McGinn, M. K., Roth, W. -M., Boutonné, S., & Woszczyna, C. (1995). The transformation of individual and collective knowledge in elementary science classrooms that are organized as knowledge-building communities. *Research in Science Education, 25,* 163–189.

McKay, S., & Wong, S. -L. (1996). Multiple discourses, multiple identities: Investment and agency in second-language learning among Chinese adolescent immigrant students. *Harvard Educational Review, 66,* 577–608.

National Research Council. (1996). *National science education standards.* Washington, DC: National Academy Press.

O'Regan, J. K., & Noë, A. (2001). A sensorimotor account of vision and visual consciousness. *Behavioral and Brain Sciences, 24,* 883–917.

Pfundt, H., & Duit, H. (1994). *Students' alternative frameworks and science education.* Kiel: IPN.

Pickering, A. (1995). *The mangle of practice: Time, agency, & science.* Chicago: University of Chicago.

Ricœur, P. (1992). *Oneself as another.* Chicago: University of Chicago Press.

Rorty, R. (1989). *Contingency, irony, and solidarity.* Cambridge, UK: Cambridge University Press.

Roth, W. -M. (1996a). The co-evolution of situated language and physics knowing. *Journal of Science Education and Technology, 5,* 171–191.

Roth, W. -M. (1996b). Thinking with hands, eyes, and signs: Multimodal science talk in a grade 6/7 unit on simple machines. *Interactive Learning Environments, 4,* 170–187.

Roth, W. -M. (2001). 'Authentic science': Enculturation into the conceptual blind spots of a discipline. *British Educational Research Journal, 27*(1), 5–27.

Roth, W. -M., & Alexander, T. (1997). The interaction of students' scientific and religious discourses: Two case studies. *International Journal of Science Education, 19,* 125–146.

Roth, W. -M., & Harama, H. (2000). (Standard) English as second language: Tribulations of self. *Journal of Curriculum Studies, 32,* 757–775.

Roth, W. -M., & Lawless, D. (2002). Signs, deixis, and the emergence of scientific explanations. *Semiotica, 138,* 95–130.

Roth, W. -M., & Lee, S. (2002). Breaking the spell: Science education for a free society. In W. -M. Roth & J. Désautels (Eds.), *Science education for/as socio-political action* (pp. 65–91). New York: Peter Lang.

Roth, W. -M., McGinn, M. K., Woszczyna, C., & Boutonné, S. (1999). Differential participation during science conversations: The interaction of focal artifacts, social configuration, and physical arrangements. *The Journal of the Learning Sciences, 8*, 293–347.

Roth, W. -M., Riecken, J., Pozzer, L. L., McMillan, R., Storr, B., Tait, D., Bradshaw, G., & Pauluth Penner, T. (2004). Those who get hurt aren't always being heard: Scientist–resident interactions over community water. *Science, Technology, & Human Values, 29*, 153–183.

Roth, W. -M., & Roychoudhury, A. (1992). The social construction of scientific concepts or the concept map as conscription device and tool for social thinking in high school science. *Science Education, 76*, 531–557.

Roth, W. -M., & Roychoudhury, A. (1993). The concept map as a tool for the collaborative construction of knowledge: A microanalysis of high school physics students. *Journal of Research in Science Teaching, 30*, 503–534.

Roth, W. -M., Tobin, K., Elmesky, R., Carambo, C., MacKnight, Y. -M., & Beers, J. (2004). Re/making identities in the praxis of urban schooling: A cultural historical perspective. *Mind, Culture, and Activity, 11*, 48–69.

Schoultz, J., Säljö, R., & Wyndhamn, J. (2001). Heavenly talk: Discourse, artifacts, and children's understanding of elementary astronomy. *Human Development, 44*, 103–118.

Wittgenstein, L. (1958). *Philosophical investigations* (3rd ed.). New York: Macmillan.

CHAPTER TWO METALOGUE

Understanding, of Science and Teaching

James Paul Gee
Wolff-Michael Roth
Randy Yerrick

Jim: Michael's chapter centers on a paradox that is, as far as I am concerned, central to the "New Literacy Studies" when it applies to content learning (e.g., learning science in a classroom). The paradox is this: to "really," "conceptually" understand area X you need to understand the representational resources (including the specific social language or register) area X uses, since these resources constitute both the substance of X conceptually and are part and parcel of the social practices through which X is instantiated in the world. Acquiring these representational resources is akin to language acquisition and any form of language acquisition requires immense practice, much immersion in relevant activities, and dialogical interaction with native speakers, in this case, people who have mastered the representational resources of area X.

Michael: You raise an interesting issue that has become part of my research in science. What does it take, for example, to understand and be able to interpret a graph? My research shows that even experienced scientists have difficulties interpreting graphs from an introductory university textbook in their own field, ecology, unless they are professors and teaching in the area. It turns out that a standard interpretation or reading of a graph requires familiarity with (a) situations or phenomena that the graph might represent, (b) data collection and instrumentations that lead to suitable data, and (c) rules of transformation to get from the data to the graphical representation. One of the problems for scholarly discourse is the ambiguous nature of the term *understanding*, which sometimes refers to intelligibility, some-

73

times to the ability to explain concepts and theories, and sometimes to the ability to relate concept words to concrete situations. In my phenomenologically grounded approach, I think about understanding as emerging from our being in the world that gives rise to a massive background of implicit relations most of which are not and cannot be articulated. I think this is the underlying reason for the requirement of "immense practice" that you articulate.

Randy: Michael, I think you also raised a very important issue surrounding how we as science educators remain stuck in this problem. We are not very good at defining how to best raise and nurture young science students because we continually defer to the scientists' collective interpretation of what should be talked about in school and how children can most expertly talk about it in a standard, universally applicable way. I totally agree with you that we need to stop calling science discourse that which is represented by the inhabitants of laboratories and the special "monoglossic" discourse they use. What we need to aim for should be nestled within the lives, experiences, and language of children. I recently observed a high-school science lesson in which a very knowledgeable and experienced teacher was trying to teach students about the genetic engineering workshop she had attended in the summer. I was struck not only by how the students misused terminology and constructs for application to real-world phenomena but also how disparate these experiences would be for students in lower track classes just across the hall. It reminded me of the history of post-Sputnik space curriculum development (DeBoer, 1991; Duschl, 1990) when it was presumed that scientists understood better than teachers what was most important to teach children (like the genetic engineers who offered training workshops for this teacher). Furthermore, we continue to have scientists exert undeserved political privilege in deciding what kind of talk is most acceptable for meeting certain testable standards to which all students can be held accountable. Are we not allowing insiders to the scientific community tell us as outsiders what children are able to do and what they should be doing and, thus, aiding in the process of devaluing our own research and judgments?

Jim: Helping all students to acquire representational as well as dialogical command over scientific discourse is simply not going to happen. This is especially true for specific branches of science (e.g., genetic engineering) for most students in most classrooms—and certainly not for poorer students. In most cases, there is simply not enough time or enough teachers who know X well enough at the level of representational resources, and, for many poor children, they will not have had a head start on these resources in their homes and communities.

Randy: These are places and instances in which science becomes the separator of privileged from the unprivileged. Because I know you and believe I understand where you are coming from, I know that is not what you are about. But could you address the issue raised about equity in science? Not from the popular perspective from gender or other categorical treatments we often see but rather from the perspective of identity and voice. How does one balance the need to maintain integrity and authenticity of children's contributions and genres and the need to specifically teach about "rules of power" of scientific others (Delpit, 1988)?

Michael: I agree with both of you that the poor—but also women, First Nations people, working class, and other groups—are disadvantaged, but they are so in the way science is conceived of even in the recent reform efforts that want science to be for *all*. We would get similar disadvantages, this time for different groups, if every student had to take 100-meter sprinting as a school subject and everyone was measured on the same scale. I have recently argued two points. First, it is inappropriate to use laboratory science for the image of school science if it is to be for *all*. Second, I argued that rather than looking for scientific literacy, we should understand it as something that is achieved collectively, where people bring differing and different levels of expertise to the same conversations together with the willingness to listen. In the chapter, I frame this in terms of sabir, a composite of different languages evolved for the purpose of dealing with particular problems and issues.

Jim: You bring up another dilemma here, as well: Even if we could get all children engaged in an authentic process of language acquisition in regard to the representational resources of (some branch of) science, wouldn't this amount to "colonizing" them in terms of the values and norms of the scientists who "own and control" (to some extent) these representational resources? You offer one answer to this dilemma—concentrating on the mix of social languages in the public sphere as scientific and technical issues are debated and advocating that students and the public acquire a "sabir"—a type of pidgin (for most linguists)—for dealing with these debates. While I think that there is much merit to this view, there is, in my view, still more that needs to be said about classrooms if we are to speak to salient equity issues (e.g., children must, unless we substantively change schools, come to control forms of academic registers, at least the sorts of social languages that are used in their textbooks and on their tests, if they are to get out of high school, go to college, and compete on equal terms in society). In any case, it is the paradox that you point to so well that anyone interested in the intersection of language and content needs to speak to.

Randy: It obviously takes a significant amount of knowledge of science, history, language acquisition, and teaching to make such strides with one's students. How are teachers to take such steps with an admittedly limited knowledge of such areas? Should they first concentrate upon scientific knowledge as many have argued for decades or is there another defensible place for them to begin to change their practices with children? Should they begin by *finding* the science in the lives and language of children? Should they begin by recognizing their own discourse patterns in their classrooms and the limitations and constraints as a way to make themselves more willing to move beyond their dissatisfying situation that doesn't yield their ultimate goals anyway? What would you suggest for a teacher who really wants to change? Where should they begin?

Michael: You are raising lots of questions to which I don't have answers. I have always been interested in finding out whether some way of being in the classroom is feasible, that is whether I can successfully pull it off with a given set of children rather than in the question whether teachers in general *ought* to teach like this or how to get more teachers to teach in that way. (This is, perhaps, because I abhor schools, which I consider to be tools producing inequality, differential labor forces, and therefore social divisions.) I have never thought about myself in terms of acquiring any specific form of knowledge then coming to teaching with a particular set of attitudes, which would allow me to make the most appropriate curricular decisions at any one moment. Attending to and caring for the situational needs of individuals rather than a rigid set of rules, being open to and on the outlook for differences in perspectives, implementing democratic values and social justice, having students participate in and have control over the events and processes of being in the classroom.

I have to admit that I don't know where teachers might want to begin, given their different biographical trajectories and the identities that they have developed. Some time ago, I taught an advanced elementary science methods course the last and capstone course in a fifth-year program. I used *Designing Communities* (Roth, 1998) and the video of my teaching a seventh-grade science class that contributed to the writing of the book. The students in my class were impressed but suggested that they recognized it was possible for me to teach like this, but they wanted methods that really worked for everyone not just for me. Perhaps it comes as no surprise that I recently theorized teaching from a first-person perspective, to understand teachers' *Being and Becoming in the Classroom* (Roth, 2002) and the tremendous role working *On the Elbows of Another* (Roth & Tobin, 2002) plays in allowing other teachers to become a little more like me—if they so desire.

Randy: I understand that questions about teacher education do not have easy answers. I pose them so that we might hear first-hand from teachers like you, who contribute to the conversation as you have earlier about stimulating important changes within your locus of control. I find our science education community, however, in a kind of conundrum when it comes to discussing goals for preparing expert teachers. I have learned since venturing into this book with the two of you that our science education community (I would not necessarily call Science Education a *discipline* in the strictest use of Schwab's parameters) tends to talk at annual meetings as if we share a common meaning when we use phrases like "science talk" or "scientific understanding" or "appropriate use of the concepts." Clearly we do not, but I do not know how to forward this kind of conversation beyond the relatively few who are already interested in having it. Perhaps this volume's collections of voices will help the larger audience discuss these issues more deeply.

REFERENCES

DeBoer, G. (1991). *History of ideas in science education: Implications for practice.* New York: Teachers College Press.

Delpit, L. (1988). The silenced dialogue: Power and pedagogy in educating other people's children. *Harvard Educational Review, 58,* 280–298.

Duschl, R. (1990). *Restructuring science education: The importance of theories and their development.* New York: Teachers College Press.

Roth, W. -M. (1998). *Designing communities.* Dordrecht, Netherlands: Kluwer Academic Publishing.

Roth, W. -M. (2002). *Being and becoming in the classroom.* Westport, CT: Ablex.

Roth, W. -M., & Tobin, K. G. (2002). *At the elbow of another: Learning to teach by coteaching.* New York: Peter Lang.

CHAPTER THREE

Discourse, Description, and Science Education

Gregory J. Kelly
University of California, Santa Barbara

In this chapter I put forth arguments supporting the value of description for science education. In particular, I present reasons to consider the ways that science is constructed interactionally among participants in various communities through discourse processes. Research in science education has shown a shift in focus toward the intersubjective processes of knowledge construction, representation, appropriation, and evaluation. This shift situates educational processes in the intersubjective spaces centered on language and social processes (Kelly & Chen, 1999; Roth & McGinn, 1998) and suggests a number of questions about how science is interactionally accomplished in educational settings. Ways that knowledge is formulated, communicated, critiqued, appropriated, and evaluated define *what counts as* science for the particular social group under consideration (Kelly & Green, 1998). To consider what counts as science for educational purposes, I turn to the field of science studies. Science studies are the emerging multidisciplinary fields (e.g., sociology, anthropology, rhetoric, history, philosophy of science) that make scientific knowledge, practices, and communities the subject of investigation (for overview of the field, see Bowen, chap. 4, this volume; Jasanoff, Markle, Petersen, & Pinch, 1995). I argue that empirical, descriptive studies—largely from the sociological and anthropological points of view—of social interaction and cultural practices in science settings offer important intellectual frameworks for understanding school science. To understand the relationship of science and education, I consider problems with interpretation of description as articulated in the form of epistemological critiques of such empirical studies of scientific practice. As documented by this collection and others elsewhere, educational reform concerns ways of addressing a set of normative questions about students, their educational experience, and knowledge. These questions may be informed by descriptive studies of everyday classroom life. Thus, this chapter provides a rationale for ethnographic and sociolinguistic descriptions of everyday life in science classrooms.

I formulate three arguments in defense of descriptive studies and, in particular, descriptive studies from an ethnographic and sociolinguistic point of view. I argue that such description contributes to understanding of meaning of educational constructs as actualized in social practices, provides examples of intellectual problems in educational settings, and through a focus on the everyday practices of people offers a basis for discussion and critique of such practices. Finally, I caution against reading descriptive accounts normatively.

Before I begin to argue this line of reasoning, I describe what I mean by *descriptive* and *normative*. Normative arguments are based on a moral point of view and focus on the ideals or norms that guide social practice (Fuller, 1992; Merton, 1973; Strike, 1989). Normative accounts are typically constructed from first principles on philosophical grounds. An example of a norm of this sort (*ideal* norm) might be that science values evidence in adjudicating among competing ideas. However, the norms of any social group are signaled through social action and are thus subject to investigation by social science research. Therefore, under certain circumstances the norms of a social group can be described. An example of a norm of this sort (*conventional practice* norm) might be the social practices entailed in anything from the relationship of students to their scientist mentors during their graduate school apprenticeship, to the structure of peer review, to the assessment of experimenter competence, to institutional policies concerning conflicts of interest (e.g., Collins, 1985; Mukerji, 1989; Traweek, 1988). Descriptive accounts focus attention on the actual practices of a social group, regardless of whether these accounts conform to stated norms or rules for behavior. And indeed, science studies often show how norms in practice are more complex than purported ideals (Mitroff, 1974; Mulkay, 1975).

Descriptions of science and science education as social practices are plentiful (e.g., Kelly, Brown, & Crawford, 2000; Knorr-Cetina, 1999; Latour, 1987; Lynch & Macbeth, 1998). Such descriptive accounts are not immune from the value-considerations typical of normative accounts. As noted by Rorty (1982), description often contains some value judgment: "*Whatever* terms are used to describe human beings *become* 'evaluative' terms" (p. 195). Therefore, the normative-descriptive aspects of accounts of human social groups are not in all cases mutually exclusive, and the distinction made throughout this chapter will be made with this in mind. As a centrally moral enterprise, education will always have normative issues to consider (Strike, 1982): What are the aims of education? What/whose knowledge is of most worth? What educational experiences are most worthy of deliberate actions? Nevertheless, descriptive studies of science and science education offer potentially useful perspectives worth considering in the development of educational programs.[1] Therefore, before returning to the role of sociolinguistic description, I shall review the debates in science studies regarding description and relativism.

SCIENCE STUDIES, DESCRIPTION, AND EDUCATION

A central concern for science education since the 1950s has been understanding and characterizing the nature of science in order to draw implications for instructing students (DeBoer, 1991; Lederman, Wade, & Bell, 1998; Schwab, 1978). Current reform efforts in the United States (American Association for the Advancement of Science [AAAS], 1989; National Research Council [NRC], 1996) place student understanding about scientific practice, formulation of evidence, inquiry processes, and communication of knowledge alongside conceptual understanding and science process skills as central goals for science education. Thus, the nature of science can be interpreted to be both an area of knowledge to be learned by students as well as a referent for achieving other goals regarding communication, assessment of knowledge claims, and understanding the role of science in society. Furthermore, the latest generation of science education reforms focuses on achieving competence in the form of scientific literacy for all students, rather than imparting knowledge to an elite few (Bianchini & Kelly, 2003).

Science education has had a long history with philosophy and history of science (Duschl, 1994; Matthews, 1991) and has more recently concerned itself with a broader range of disciplinary perspectives researching science including sociology, anthropology, and rhetoric, among others (Kelly, Carlsen, & Cunningham, 1993; Roth & McGinn, 1997; Roth, McGinn, & Bowen, 1996). Science studies presumably speak to how science is understood and defined, and thus may have applications for education. However, post-Mertonian sociological and anthropological studies of scientific practice have generally steered clear of normative questions about how

---

[1]As Rorty (1982) noted, policymakers often wish for answers to questions like, "if we create more/less student centered learning, science achievement will increase/decrease" (p. 196). A description of classroom practices can thus be frustrating to the audience of policymakers and speaks to a different set of questions and responses—questions and responses I address through this chapter.

science ought to be practiced, how quests for truth provide for rigor in inquiry, or how community ideals lead to certified knowledge.[2] Rather, these ethnographically oriented studies have focused on the actual practices of scientists in everyday situations (e.g., Garfinkel, Lynch, & Livingston, 1981). Therefore, the ways descriptions of scientific practice can have implications for education remain obscure. This is aggravated by the orientation of some of these studies that purposefully set out to be symmetric in terms of explanation of true and false beliefs (Bloor, 1976), and explicitly relativist with regard to the role of the natural world in the construction of scientific claims (Collins, 1981). These (largely sociological) studies are agnostic regarding scientific truth in order to study in an unbiased manner the social processes that lead scientists to determine the plausibility of knowledge claims. This naturalistic attitude does not, however, deny scientific truth; rather the investigators remain agnostic purposefully for methodological reasons.

The methodological orientation of these studies of scientific practice has generated pointed criticisms from scholars of various fields[3] as to how such studies consider the cognitive content of scientific theories, to the epistemic status of scientific theory, as well as to the implications of such studies for education (e.g., Gross & Levitt, 1994; Koertge, 1998a). There is a general concern about the resultant dangers of relativism and postmodernism:[4]

1. Science studies of particular sorts potentially (seek to) undermine the validity and utility of science with spurious, often methodologically dubious critiques that, if valid for some small subsection of science, fail to make the case for science generally;
2. Science studies of particular sorts seek to undermine the epistemological validity of science by focusing on irrelevant social aspects of science and neglecting the cognitive content of scientific knowledge; and
3. Science studies operating from the point of view of methodological symmetry (failure to commit to a true–false distinction of relevant science prior to investigation) suggest that science exhibits no ra-

---

[2]An exception to this general trend would be feminist scholarship of science (e.g., Alcoff & Potter, 1993; Longino, 1990; applied to education, see Barton, 1998).

[3]This debate over science has been dubbed "science wars" in the mass media (Begley, 1997).

[4]Lyotard (1984) defined *postmodern* as "incredulity toward metanarratives" (p. xxiv). *Postmodernism* is defined variously (perhaps indicative of the movement) by both adherents and detractors; however, there is often a general notion that there exists a heterogeneous field of multiple discourses, that the constructed world is fragmented and diverse, and that self-legitimizing discourses are to be viewed with suspicion.

tional means to adjudicate among competing ideas. Taken as a group, these criticisms pose problems if implications of science studies are to be drawn for education.[5]

Harding and Hare (2000) presented a set of arguments against the relativism they perceived in certain constructivist perspectives in science education. They argued instead for open-mindedness, scientists' own definitions of *truth*, the evolving nature of theories over time, and the important role of the community of scientists. Harding and Hare were particularly concerned that relativistic views would be promulgated in schools and that these views contrasted how scientists interpret their work. For example, they contrast *truth* defined by scientists—"at a practical or specific level dealing with individual theories, surrounded by the particular evidence that supports them" (p. 226)—with ways *truth* gets defined out of context by science educators—as eternal, applied to all theories, and hence unsupportable given changes in scientific knowledge. Similarly, Matthews (1997) raised concerns derived from science education's overreliance on "relativism, constructivism, and postmodernism" (p. 309). Like Harding and Hare, Matthews was worried that particular views of the nature of science might be translated into pedagogies that indoctrinate students into only these views. Matthews argued for a liberal, impartial view of education (i.e., respecting students' autonomy and agency to decide for themselves regarding the nature of science) and argued forcefully for truth and reason.

Stepping beyond the dichotomy of positivism versus postmodernism, Loving (1997) focused more centrally on how educators draw conclusions based on their reported evidence. In her review of the state of science education, Loving cautioned against polarization of perspectives and suggested the need for a "knowledgeable and reasonably balanced view of the nature of science" (p. 424). She identified a set of problems generated by overly relativistic views of science, often characterized as postmodern, that tend to focus on the role of personal, political, and cultural influences of scientists on the outcome of science. Loving's argument sought to identify ways that such social processes have been exaggerated in science studies and to provide a view that is more realist and cognitivist in orientation. She noted that one central difficulty with studies undermining the epistemological authority of science is the epistemological shortcomings of these very critiques, often along methodological lines. And like Harding and Hare, Loving advocated a particular normative view, that is, a "balanced view of the nature of science" in schools.

---

[5]In its most strident articulation, Koertge (1998b) put education on the list of "civilian casualties of the postmodern perspectives on science" (p. 255) but offered little evidence of any supposed harm in the form of empirical studies of classroom practice.

The normative considerations of Harding and Hare, Matthews, and Loving, articulated as arguments for the need for open-mindedness, impartiality, and balance in portraying science,[6] offer a point of entry for descriptive studies. For example, Harding and Hare's argument for consideration of *truth* as defined by scientists could be informed by the systematic examination of the practices that come to define *truth* across sciences and over time.[7] Rhetorical and historical analyses of the changing nature of scientific communities, the relationships of scientists to their respective potential audiences, and the demands of the experimental research article genre go far to identify the social dimensions of scientists' definitions (Atkinson, 1999; Bazerman, 1988). Thus, the empirical investigation of definitions of *truth, reason*, and so on may offer insights into the social, rhetorical, and cultural practices that come to define such terms in the first place (Lynch, 1993). By focusing on the ways people accomplish the work of their everyday lives, whether in schools, laboratories, conferences, or through texts, science studies potentially offer much for education (Kelly, Chen, & Crawford, 1998).

Educators can eschew the dangers of relativism by reading science studies from an educational point of view: That is, as educators, we can be agnostic about the absolute truth of science studies and focus on what they deliver in terms of findings about science. For example, Latour's anthropology of science (1987; Latour & Woolgar, 1986) has offered ways of reconsidering inscription in scientific practice by focusing on how data inscriptions are translated in various texts to lesser or greater degrees of abstraction. Roth and McGinn (1998) effectively applied this perspective to science education. The results produced by Latour's relativism need not be taken as an entire package of ontological or epistemological goods. Our normative responsibilities in education require judicial readings of this and other research of this kind. Nevertheless, if researchers seek to demystify scientific practice and make available to students ways of engaging in disciplinary inquiry, then we need to go beyond identifying the value of truth and reason and investigate how truth and reason are accomplished. For students to become competent in science it is not enough to declare the goal of discovering truth; rather the knowledge and practices embedded in the intermediate steps of establishing a claim as true need to be made explicit. The processes of establishing claims, as well as other epistemological is-

---

[6]I do not advocate a position contrary to open-mindedness, balance, and impartiality in interpreting science in educational settings. These are core values worth supporting and Loving, Matthews, and Harding and Hare offer various justifiable reasons for their respective points of view.

[7]One criticism of Harding and Hare's idealized view of scientists is offered by Brickhouse (2001), who noted how what counts as doing science can be constrained by such valorization.

sues (justification, evaluation, legitimation, etc.) foreground the role of language and the relevant practices of a discourse community.

DESCRIPTION IN SOCIAL SCIENCE, EXTENDING WITTGENSTEIN'S ADVICE

Learning science involves centrally epistemic activities (Duschl, 1990; Strike, 1982). Before turning to specific case studies, I digress slightly to consider ways that description speaks to understanding these epistemic activities. Wittgenstein's (1953/1958) *Philosophical Investigations* changed the nature of philosophy by providing alternatives to the metaphysical quest for the true essence of meaning. Rather than argue on traditional philosophical grounds, Wittgenstein examined the conditions for uses of language in everyday contexts. If educational researchers choose to follow Wittgenstein, the goals of social science will not be to identify the one true account of science, schooling, or other social institutions. For example, science studies will not seek to identify the true essence of the nature of science, to find the essence of the scientific method, or to finally resolve the realist–instrumentalist debate. In education, we will expect science studies to deliver neither the authentic view of science nor a small set of generalized best practices for science teaching. Rather, our empirical projects will have less ambitious goals, as we recognize the contingencies of our own and others' languages (Rorty, 1989). Paragraph 66 of *Philosophical Investigations* presented the issue as follows:

> Consider for example the proceedings that we call "games". I mean board-games, card-games, ball-games, Olympic games, and so on. What is common to them all?—Don't say: "There *must* be something common, or they would not be called 'games' "—but *look and see* whether there is anything common to all.—For if you look at them you will not see something that is common to *all*, but similarities, relationships, and a whole series of them at that. To repeat: don't think, but look! (p. 31)

Wittgenstein's advice "don't think, but look!" offers a role for social science: Researchers can look and question how "inquiry approaches," "back-to-basics," or "hands-on science" (substituting for "games" in paragraph above) are being invoked, constructed, construed, modified, and practiced as meanings of a group of people engaged in some activity.[8] From this point of view, empirical research of the ethnographic variety contributes to ongoing conversations about how to interpret and under-

---

[8]Arguing along similar lines, Martin (1997) substituted "science" for "games" in Wittgenstein's paragraph 66 in a keynote address to the Society for the Social Study of Science.

stand what counts as disciplinary knowledge and practices, in various contexts, from various points of view. In this way the multiple language games characteristic of the postmodern condition (Lyotard, 1984) are investigated through inquiry.

Rorty (1982) followed Wittgenstein by suggesting that social science introduces new vocabularies for "redescription, reinterpretation, manipulation" (p. 153). Thus, a respecification of the uses of language and their connections to concerted activities may be achieved through the vocabularies of social science research. Descriptive accounts of scientific practice need to consider who is conceptualizing the phenomena, for what purpose, and how.

Wittgenstein's advice is captured well by Michael Lynch (1992), who referred to epistemological dimensions of scientific activity:

> Ethnomethodology's descriptions of the mundane and situated activities of observing, explaining, or proving enable a kind of rediscovery and respecification of how these central terms become relevant within particular context of activity. Descriptions of the situated production of observations, explanations, proofs, and so forth provide a more differentiated and subtle picture of epistemic activities than can be given by the generic definitions and familiar debates in epistemology. (p. 258)

Lynch's argument for "a more differentiated and subtle picture of epistemic activities" provides a reading of how science studies can speak to education. Description can illuminate the knowledge embedded in social practices central to educational processes: Learning involves knowing how and what to observe, weighing relative merits of proposed ideas, considering evidence for particular points of view, constructing and acquiring relevant knowledge, and so on. Thus, we need to investigate not only what people come to know, but also how people come to know in various settings. This suggests a key implication for education: Science studies can serve as analogous cases for similar investigations in educational settings. Simply put, a central lesson of science studies is not conclusive evidence about the true nature of science, but rather the methodological orientation to investigate science education from an empirical, descriptive point of view as it is created through social activity (Kelly et al., 1998; Lynch & Macbeth, 1998).

ETHNOGRAPHIC AND SOCIOLINGUISTIC DESCRIPTIONS OF SCIENCE IN EDUCATION

To this point, I have argued that emerging visions of science among educators have placed intersubjective processes of representation, communication, evaluation, and legitimation at the center of their debates. Furthermore, as suggested by Lemke (1990) and others (Carlsen, 1991; Kelly & Crawford, 1997), teaching inscribes a view of science through the discourse processes of science classrooms. Teaching science in any form communicates images of the disciplines to students. As a broad vision of science is part of the normative goals of science education reform, I have argued that despite criticisms, science studies potentially have much to offer to education, substantively and methodologically. I have also argued that epistemic practices can be specified through the descriptive study of situated activities. One way to focus attention on the description of classroom discourse is to take an ethnographic and sociolinguistic perspective.

The field of sociolinguistics informs the descriptive accounts that I offer. Gumperz (1982) defined *sociolinguistics* as a "field of inquiry which investigates the language usage of particular human groups" (p. 9). Similarly, Hymes (1974) argued for descriptive, multidisciplinary approaches to investigate the organization of language as part of the overall communicative conduct of a community. A sociolinguistic perspective examines not only how language is used situated in particular contexts, but also what gets accomplished through choices of language use by speakers and hearers. For example, a sociolinguist can examine the shift of student discourse from peer talk to a scientific register for its lexical and syntactic changes as well as for how slotting into a new register defines a social identity for the speaker given a particular audience, chain of actions, and norms for interacting.

I adapt aspects of sociolinguistic analysis for the description of everyday life in science classrooms for two central reasons. First, sociolinguistics clarifies the prominence of language in defining the cultural norms of a social group. Science classrooms develop norms for interaction over time that come to define science through particular linguistic practices (Kelly, Crawford, & Green, 2001; Lemke, 1990; Sutton, 1996). Furthermore, understanding the ways science is positioned in educational events requires understanding how science gets spoken and written in different contexts (Kelly, Chen, & Prothero, 2000). Second, sociolinguistics works under the assumption that understanding the meaning of particular social events requires ethnographic description. Thus, social events are viewed as the product of a history of social practices and, in turn, as constructing social practices through interaction. Its central focus on language use and ethnographic orientation thus make sociolinguistics well suited for interpreting science classroom events.

I now provide three examples of the value of sociolinguistic description for science education from my own research conducted over 2 academic years, first in a third-grade class, and in the subsequent academic year in a fourth/fifth-grade combined class with the same teacher in a public elementary school in California (Crawford, Kelly, & Brown, 2000; Kelly,

Brown, & Crawford, 2000; Kelly & Brown, 2003). I use these examples to explicate the three unique roles of description in qualifying science embedded in classroom discourse.

First Role of Description: Specifying Meanings Through the Study of Social Practices

Wittgenstein's advice "don't think, but look!" directs social science researchers to examine meaning through everyday practical action. Descriptive accounts of everyday (physical, discursive) action provide concrete examples of the situated meanings associated with the sometimes vague constructs in science education like *nature of science, inquiry, hands-on science*, and *constructivism* (DeBoer, 1991). Such an approach offers valuable reference for discourse about education with its multiple, fluid, and ever-changing meanings of key ideas. To illustrate how situated practices define meanings, I present the case of *anomaly* as a public, recognized, witnessable, and socially constructed event in inter-psychological space, rather than as cognitive artifact.

The events in this case centered around an investigation of the effects of sunlight on algae. The teacher and her third-grade students, under the guidance of a participating marine scientist and the educational research team, devised a simple experiment using two samples of algae. These samples were observed by the classroom members and then subjected to two treatments. One sample was covered with aluminum foil and placed in a closet; the other was placed near a window, thus receiving a number of hours of direct sunlight each day.

The class made multiple observations of both samples over the course of weeks. The first set of results were as follows:

sample	*observation 1*	*treatment*	*observation 2*
algae sample 1	green	sunlight	clear
algae sample 2	green	darkness	green

A third-grade student, Kathryn, described this result to the class. She first provided an account of the experimental process of putting one sample in the light and another in the dark (lines 1.1–1.10).[9] She then mentioned a potentially confounding contingency (that they had put more algae in the "sunlight treatment" sample, lines 1.12–1.14; 1.26) before reporting the fact about the differences in observation (line 1.35).

---

[9]The classroom discourse was transcribed with one message unit per line (Green & Wallat, 1981). Message units are the minimal meaning unit of communicative delivery, identified through contextualization cues: pitch, stress, intonation, pause structures, physical orientation, proxemic distance, and eye gaze (Gumperz, 1992).

3. DISCOURSE, DESCRIPTION, AND SCIENCE EDUCATION   89

Kathryn:	1.1	oh
	1.2	_____ we put
	1.3	allergy [*algae*]
	1.4	in
	1.5	these cups [*goes to the windowsill and points to a cup*]
	1.6	and we let one
	1.7	we left one
	1.8	in the su-
	1.9	like
	1.10	out here in the sun
	1.11	and
	1.12	we accidently put a little more
	1.13	in
	1.14	the one where it's supposed to be dark
	1.15	and then
	1.16	uh
	1.17	on Monday
	1.18	I think
	1.19	it was Monday=
Lori:	1.20	=two weeks later
[Teacher]	1.21	=cause= we went away for the holiday
Kathryn:	1.22	=yeah=
	1.23	two weeks later
	1.24	umm
	1.25	we got the one in the dark
	1.26	cause we put a little more in there
	1.27	and we got the one in the dark
	1.28	and
	1.29	compared it
	1.30	and the one in the dark had more
	1.31	than
	1.32	this one
Lori:	1.33	and how much more
	1.34	did it have?
Kathryn:	1.35	it had a lot more

Kathryn's description of the results set off a series of events as the students and teacher sought to better understand the observational evidence. (They were aware by this time of the scientific explanation of photosynthesis and plants' need for sunlight.) The continued dialogue brought on by Kathryn's recognition of the unexpected results and experimental contingency offered the ethnographic research team a way to reconsider the participatory nature of the classroom discourse patterns. The sociolinguistic analysis included not only what was said but also how the class members defined norms for interaction that made open discussions a routine event. (A detailed analysis is provided elsewhere, Kelly, Brown, & Crawford, 2000). Kathryn's statement and the subsequent conversation were salient not only because of the ways that this event became recognized as anomalous but also because such a sequence was the consequence of the discursive work established by the members over time.

Description of the events provided a way to respecify anomalous data as interactionally recognized, acknowledged, and accomplished through classroom discourse. To date, much work concerning anomalies in science education has focused on either their role in conceptual change as analogous to theory change in science (Strike & Posner, 1992) or the attribution of various psychological responses of students to individual characteristics (Chinn & Brewer, 1993, 1998). The alternative focus on social interaction defines anomalous data differently, as a topic to be developed through discourse processes involving student participation in science. Thus, such description offers empirical examples for discussion of the epistemic shift in studies of scientific cognition and practice from individual minds to communities and groups (Duschl & Hamilton, 1998). The social processes leading to the recognition of a discrepant event became a recurring feature of the organizing dialogues and activities. This suggests that students need to learn when an event (or event described as data) is to count as consistent with, or in contradiction to, expectations for a particular social group. Coming to learn how to participate in a social event that defines unexpected results as significant is thus an educational experience for newcomers to science. In professional communities, learning to observe from a particular point of view typically requires a period of socialization through talk and gesture as learners come to perceive and focus on aspects deemed worthy of attention within a specific discipline (Goodwin, 1994, 1995). In this case, *anomaly* was situationally specified through the discourse processes of the classroom participants.

Second Role of Description: Providing Puzzles for Comment and Critique

Descriptive accounts can contribute to science education by identifying the epistemic practices entailed in scientific activity. Although description is not immune from problems of reference and representation,[10] it can

---

[10]Problems of reference and representations in ethnography are longstanding and becoming increasingly more complex. They are occasionally referred to as the "representational crisis" (see Britzman, 1995; Denzin, 1997). Smith (1996) while acknowledging the validity of certain aspects of the poststructural/postmodern critiques of reference and representation, provided a rationale for continued sociological inquiry despite such problems.

nonetheless make visible the practices involved in constructing and learning scientific knowledge. I shall provide an example from the previously mentioned ethnographic study. (A detailed analysis is provided elsewhere, Crawford et al., 2000).

In this case, the beginning of a formal mathematics lesson became an impromptu conversation about science (i.e., a science lesson) through some creative teaching. Over the course of 2 academic years, an elementary teacher included in her classroom a marine science observation tank with live animals and plants. Just prior to the onset of the mathematics lesson (in the 2nd year of the study), a number of her fourth- and fifth-grade students were using the time allotted for a "3-minute break" to observe the activities at the marine tank. Two students, Mark and Joe, noticed what was to them unusual behavior of the animals: A whelk snail's foot was attached to the tank's glass and a sea anemone seemed to be attached to the back of the whelk snail. The students further reported their observation of a "slimy" substance emanating from the animals. Diverging from her planned lesson, the classroom teacher, Lori, took the students' observations as an opportunity for the whole class:

Lori:	2.1	I don't understand what's going on [*To whole class*]
	2.2	with the
	2.3	anemone
	2.4	and the whelk [*Students' hands go up with "oh oh"*]
	2.5	do any of you
	2.6	have a theory?
	2.7	who was
	2.8	observing back there today?

Significantly, the teacher began the episode by identifying herself as not understanding (2.1–2.4) and invited student comment (2.5–2.8). A sociolinguistic analysis of the classroom discourse considered ways that the teacher positioned herself in the community. In this case, and elsewhere, she identified herself as "not knowing" and opened the floor to student comment and interpretation. By following the students' interests and inquiries, the teacher allowed conversation to wander into subject areas of uncertainty. Evidence from the ethnography suggested that the students recognized the teacher's statement regarding her lack of understand as legitimate: They understood her as posing questions to frame their investigation, rather than to converge on a pat answer.

In this case, the teacher's questions led to a whole-class scientific investigation in which many members of the class constructed various arguments about the ongoing episode. Analysis of the classroom discourse (Crawford et al., 2000) identified a range of student concerns about the animals, the phenomena, and the multiple interpretations of the events.

As was typical for these students, they sought advice from sources outside their classroom. They suggested calling the marine scientist who had helped them set up the tank and had provided the animals in the first place. The nature of the conversations then turned to how they would pose the questions to the scientist. In the example provided below, Steve identified the importance of some relevant details:

Lori:	2.9	kay [*To Steve*]
	2.10	what would you ask her?
Steve:	2.11	I don't know
Lori:	2.12	wait a minute [*To whole class*]
	2.13	hands down
	2.14	Steve's talking
Steve:	2.15	I would
	2.16	ask her
	2.17	you know
	2.18	where the um
	2.19	the mouth of the
	2.20	uh
	2.21	whelk is
	2.22	cause some people think
	2.23	it's in front of the foot
	2.24	and some people think it's
	2.25	on the foot
	2.26	on the (side) of the foot

Steve's question (lines 2.15–2.26) can be thought of as properly scientific: The location of the mouth could be used as evidence as to whether or not the whelk may have been eating the anemone—an issue of concern for the students. This was but one of the issues the students saw as relevant to the situation. They also questioned whether snails and anemones are "enemies" and if human intervention would cause damage. Overall, the students' series of questions indicated that they were concerned primarily for the well-being of the animals and secondarily with learning about the animals from the situation.

The whelk snail and anemone episode became an inquiry event for the classroom participants. For these students such activities constituted doing science; that is, the participants considered these activities "science." The ethnographic and sociolinguistic investigation raised a number of issues for education beyond the particulars of the classroom events. The argument put forth by us as ethnographers of the classroom concerned ways the teacher used an orientation to students, time in the classroom, and participation of outsiders (animals, ethnographers, marine scientist) to provide ways of engaging students in scientific activity (Crawford et al.,

2000). Thus, science was accomplished by a teacher who took up the role of fellow investigator, allotted time for students to explore materials outside formal science instruction, and sought to engage the whole class in an inquiry process. Inquiry thus was accomplished without the stereotypical "hands-on" activities, student small-group work, posing a hypothesis, or following the proverbial scientific method. Science happened during a "break" from instruction, an often-overlooked aspect of teaching.

This example describing the events of the classroom on a particular day identified the epistemic practice of engaging in inquiry into animal behavior. The students observed potentially fruitful events, focused on relevant aspects of the animals' behavior, posed a series of questions concerning needed information to make decisions about the sorts of actions that would allow for developing knowledge, considered the ethical dimensions of curiosity versus the animals' well-being, and sought background knowledge relevant to their investigation. The ethnographic description of these social actions identified how the classroom teacher was able to use the students' interest to gently move them into more sophisticated scientific understanding (i.e., through making the students' observations public, requesting student ideas about possible actions, and following the students' suggestion of bringing in the relevant subject-matter expert). Through description, the teacher's engagement with her students in such inquiry as an opportunity for student learning is preserved for further discussion and debate.

Third Role of Description: Bringing Focus to the Everyday Lives of People in Real Situations

A third reason for description derives from the work of feminist sociologist Dorothy Smith (1996). Smith's approach draws from Mead and Bakhtin and suggests a focus on the everyday social practices of actual people. Smith views social science inquiry as concerned with:

> the ongoing coordinating or concerting of actual people's activities. Consciousness, subjectivity, the subject, are hence always embedded, active, and constituted in, the concerting of people's activities with each other; concepts, theories, ideas, and other terms identifying operations of thought are themselves activities or practices and enter into the coordination of action. (p. 172)

While acknowledging the importance of reflexivity and the seriousness of problems of representation brought by postmodern thinking, Smith (1996) provided a way to "tell the truth" by recognizing that people exist in social organization and that everyday situations are reasonable places to begin social science inquiry. One dimension of Smith's social theory of knowledge is most relevant to uses of description: "Theory must formulate referring, representing, inquiry, and discovery as the locally organized social practice of actual people" (p. 173). Smith's focus on the social practice of actual people in everyday situations provides for levels of understandings of others. To take Smith's argument further, description of the concerting of actual people's activities aligns researchers' ways of sense making with the ways humans make sense generally. Ethnographic studies can make such sense making more methodical and reflexive.

The illustrative example I provide is from the first year of the 2-year ethnography. In this case a bilingual third-grade student is describing the functioning of his group's solar energy device (Kelly & Brown, 2003). The example shows how student sense making, *in situ*, reveals student knowledge. Such knowledge was not available to be recognized and acknowledged without some work of the participants (researchers included). This is thus an example of how elementary school science is constructed through locally organized social practice, as suggested by Smith (1996).

The transcript is taken from an episode when a third-grade student, Javier, had taken the role of spokesperson for his group of five students. The students had worked for a number of weeks creating functional solar devices. Like the other five groups in the third-grade class, Javier and his partners were making a presentation of their solar energy projects to their classmates and teacher. The presentation began with four group members in front of the class. Elizabeth stood behind the project holding it for the class to see. The project consisted of a parabolic reflector centered in an aluminum covered cardboard box, flanked by aluminum foil reflectors. Tom and Joel stood to the left and right of the project. Javier stood just to the side, between the project and the class members. Here Javier explains how the solar-powered fan was energized by a photovoltaic cell:

Javier:

Line #	Transcribed talk	Researcher comments
3.1	okay	
3.2	now	
3.3	what we're gonna do	
3.4	kay	
3.5	we're gonna have a bigger mirror than this	
		Javier holds small compact mirror.
3.6	and what we're going to do	
3.7	is this here	Javier moves mirror to demonstrate reflection of light on the parabolic reflector. "It" refers to light.
3.8	you can see	
3.9	a little bit	
3.10	it starts to reflect	

Line #	Transcribed talk	Researcher comments
3.11	on the satellite	Javier points to "satellite" (parabolic reflector) demonstrating points of reflected light.
3.12	bout right there	
3.13	and um	
3.14	so what we're gonna do	
3.15	is the sun hits here	Javier motions to show the direction of sunlight.
3.16	and then	
3.17	and then	
3.18	the sun hits here	Javier points to top reflector, then repeats in conjunction with pointing to bottom reflector.
3.19	and then it'll go down to here	
3.20	then it'll reflect on that	Javier is referring to the parabolic reflector. "this" refers photovoltaic, which Javier places in center of the parabolic reflector.
3.21	this will be on there	
3.22	and this will reflect on that	
3.23	and then we get the solar power	
3.24	it'll um come through this wire	Javier holds wire (i.e., lead output from photovoltaic).
3.25	right here	

In this first part of his description, Javier began by noting that the group would have a bigger mirror than that in his hand and demonstrated how the mirror would reflect light onto their parabolic mirror (called "the satellite" by Javier) situated in the center of a cardboard box (lines 3.5–3.11). He next traced the path of the sunlight, motioning with his hands from a point outside of the box (line 3.15), to the top and bottom reflectors (lines 3.18–3.19), to the parabolic reflector (3.20), and finally to the photovoltaic cell (lines 3.21–3.22). This series of indexical phrases was "made sense of" through the pointing and gesturing of Javier.

Next, he related how the solar energy would go from the photovoltaic to the circuit (lines 3.23–3.25), explaining that the wire needed to be plugged into the terminals on the fan (lines 3.26–3.27). He noted that in the first attempt the fan did not turn; he described this lack of current as due to "a completely dead battery" (lines 3.31–3.32), holding a battery for the class to see:

Javier:

Line #	Transcribed talk	Researcher comments
3.26	and this is gonna be plugged in	"this" refers to wires from photovoltaic
3.27	this	
3.28	if we have enough	
3.29	this thing will go on	Javier holds fan, "this" refers to fan. "this" refers to the fan.
3.30	and then	

3.31	we're gonna do our first try	
3.32	is a completely dead battery	
		Javier holds battery in hand, shows class.
3.33	and	
3.34	what we're going to do	
3.35	is put that right there	
3.36	and then we're gonna put	
3.37	one side right there	Javier places photovoltaic back in center of parabolic reflector.
3.38	and the other side right here	Javier demonstrates connections from photovoltaic wires to two sides of battery to make complete circuit.
3.39	and we're gonna see	
3.40	if that'll	
3.41	if that'll um	
3.42	charge it up	"it" refers to battery.
3.43	with this wire	Referring to wires from photovoltaic
3.44	right here and right here	Javier repeats connection for complete circuit.
3.45	and if it'll charge up	"this" referring to fan.
3.46	then it'll make this go	

Javier proceeded to again use a set of indexical statements (e.g., "this," "there," and "right here") referring to the different components of the electrical circuit, the fan (lines 3.29, 3.46), the photovoltaic cell (line 3.35), the two terminals of the battery (lines 3.37–3.38, 3.44), the battery (lines 3.42, 3.45), and the leads from the photovoltaic (line 3.43).

Reading just the middle column of "transcribed talk" leaves ambiguity and the reader's sense making at a loss. Outside the context of the situated meaning of the particular actors (researchers included), with shared history and shared knowledge, and without the relevant visual cues, the scientific merits of Javier's explanation are difficult to assess. However, by filling in the relevant background knowledge and shared assumptions through sociolinguistic analysis (Gumperz, Cook-Gumperz, & Szymanski, 1999), the student's talk demonstrates an understanding of important scientific ideas, including the path of sunlight, the reflection of light, the angles of incidence and reflection for reflected light, the transfer of energy, the need for energy sources for production of work, the production of current, and the necessity of complete circuits for current to pass. Thus, understanding this conversation from the indigenous point of view (Emerson, Fretz, & Shaw, 1995) shows how within the particular situated speech situation, the student was able to successfully complete the academic task of presenting the design and function of his group's solar energy project. Nevertheless, this understanding had yet to be formalized into the abstract, distant language of scientific explanation.

Currently there is much discussion of accountability, standards, and student understanding of scientific content in educational reform (AAAS, 1989; NRC, 1996; for discussion, see Apple, 1998; Carlsen, Cunningham, & Lowmaster, 1995; Rodriguez, 1997; Strike, 1998). California is among the states that have defined grade-level science content standards. For California students this means a set of propositions that they are to know (for review, see Bianchini & Kelly, 2003). Consider a relevant example from California's *Science Content Standards Grades K–12* (California Department of Education, 2000):

> Light has a source and travels in a direction. As a basis for understanding this concept:
>
> > *Students know*: light is reflected from mirrors and other surfaces.
> > (Grade 3, Physical Sciences, #2b, p. 8)

In this case, the standard is literally that: "*Students know*: light is reflected from mirrors and other surfaces." In the context of Javier's classroom, he can be reasonably interpreted as articulating the knowledge relevant to this standard. Nevertheless, he articulated this knowledge *in situ* and in the context of the functioning of his group's project. Although he never actually makes a definitive statement of the sort specified in the standard ("light is reflected from mirrors and other surfaces"), he twice suggests the idea: In line 3.10, "it starts to reflect" and in lines 3.18–3.22, "the sun hits here/and then it'll go down to here/then it'll reflect on there/this will be on there/and this will reflect on that." He does not say, "Light reflects." His talk must be unpacked and examined to be seen as counting as evidence for the standard. Such an interpretation needs to consider whether he knows the statement of fact manifest in his talk and gestures about the path of the sunlight. It remains to be seen whether pedagogy aimed at providing students with knowledge of such statements will include student-initiated projects related to socioenvironmental issues. Nevertheless, the description of the concerted activities that together frame and posit meaning to Javier's experience in science offer observers ways of understanding what students can understand through such activities. Furthermore, subsequent interviews with Javier and examination of his school products identified a strong affiliation on his part with science (see Crawford, 1999).

The sociolinguistic description of the concerted efforts of the classroom members identified how Javier was scientifically competent for the relevant social group. The example showed how, through a focus on the discourse processes, students could be seen as engaging with science. The multiple ways in which the teacher framed classroom activities invited students to "talk science" and thus begin to experience ways of understanding and articulating their knowledge. The description of the events of actual people provides examples of the consequences that the inviting discourse had for student participation and affiliation in this classroom. These examples are relevant to discussions of science education reform because as a subject matter science has been shown to be alienating to students due to use of specialized linguistic features of scientific discourse (Halliday & Martin, 1993; Lemke, 1990). Therefore, to create more accessible forms of science instruction across settings the discursive work of the teacher to involve students needs to be made visible. Such pedagogy faces challenges to its survival given increasing calls for standards-based teaching and assessment.

CONCLUSION

What content should be communicated as science has been and continues to be of serious discussion and debate in educational circles (AAAS, 1989; DeBoer, 1991; NRC, 1996). The argument I have developed suggests that science education has much to learn from the study of everyday actions that define for participants (e.g., scientists, students, teachers) key constructs (e.g., *science, rigor, accuracy, evidence, truth*) within various communities (e.g., research teams, scientific communities, science classrooms, student small-groups). This argument includes recognizing the value of science studies in two ways: Science studies give us more detailed information about scientific practices, thus expanding a possible repertoire of what counts as science education (Cunningham & Helms, 1998; Roth et al., 1996), and studies of scientific practice provide a methodological rationale for similar studies of science in school settings (Kelly et al., 1998; Kelly & Crawford, 1997). This argument further requires that we read such studies from an educational point of view and are careful not to translate description accounts of science into normative considerations for education situations (the naturalistic fallacy). For example, many practices evidenced in science, such as agonistic debates and *ad hominem* attacks may not be appropriate for learners of science. Our use of intellectual referents in education should not be aimed at a metaphysically secure account on which to base our science education programs. Our focus should rather be on how people construct science through social interactions in various contexts and in various ways.

A key contribution of science studies is not merely that we study everyday life in schools—this too has a history in educational ethnography (Green & Wallat, 1981; Mehan, 1979). Rather, science studies suggest an openness to explore school science as it is constructed without prejudging (at least initially) whether such activities counts as science beyond the local community (Lynch & Macbeth, 1998). Thus, this point of view suggests the study of school-science-in-the-making (following Latour, 1987) by examining the concerted activities of members of a relevant community (Kelly et al., 1998). The research point of view of descriptive studies of this sort differs significantly from immediate questions about how science in schools should be taught.

Although there are many ways to describe everyday life in science classrooms, I have offered an ethnographic and sociolinguistic framework as a plausible approach. The examples described in this chapter demonstrate some of the ways this approach can contribute to normative considerations for educational reform. First, a review of science in various settings evinces the important role a discourse community has for defining relevant cultural practices. The cultural practices that *count as* science for a group are defined in and through social interaction, including, importantly, uses of language in particular ways for particular purposes. This suggests that in analyzing the educational opportunities for students, educators need to consider the linguistic resources made available and how students are positioned to engage with such resources. Second, knowledge claims are proposed, debated, evaluated, accepted, or rejected within a relevant discourse community. The social processes of such knowledge legitimization, as well as the culture norms defining legitimate epistemic practices, can be demystified through empirical investigation. Demystifying science and leaving open multiple ways of knowing have implications for broadening what counts as science for students (particularly as interesting alternatives are becoming evident; see Aikenhead, 1997; Watson-Verran & Turnbull, 1995). Finally, from a methodological point of view, ethnography and sociolinguistics provide ways of understanding the language forms of students, teachers, scientists, and community members, as they collectively construct school science. Sociolinguistics offers ways of understanding the differences between students' ways of talking and the discourse of more formalized epistemological communities.

The focus on the ways people construct science through concerted, coordinated activity suggests a role for descriptive studies. My argument for description included three normative reasons to continue research of this sort: Description specifies uses of language not available without empirical investigation, description offers new puzzles to consider, and description by focusing on the everyday lives of people potentially develops ways of understanding and empathizing, and thus improving human conditions

(Rorty, 1982; Smith, 1996). Furthermore, I suggest we do not read description normatively: Depiction of everyday conditions does not suggest ways things ought to be or ways education programs should be formulated. Rather than providing descriptions of the way science *really* is, anthropological and sociological studies of science contribute to on-going conversations about science and provide a methodological counsel: Focus on the everyday practices of members of scientific communities.

ACKNOWLEDGMENTS

The research reported in this chapter was supported in part by a grant from the National Academy of Education's Spencer Postdoctoral Fellowship program. The data presented, the statements made, and the views expressed are solely the responsibility of the author.

An earlier version of this chapter entitled, "Descriptive studies of science and education: Relativism, representation, and reflexivity" was presented at the annual meeting of the American Education Research Association Seattle, WA, April 10–April 14, 2001.

I would like to thank Teresa Crawford and Candice Brown for help with the ethnographic data presented in this chapter. In addition, I would like to thank Julie Bianchini, Charles Bazerman, Judith Green, and James Heap for their helpful comments on an earlier version of this manuscript.

REFERENCES

Aikenhead, G. S. (1997). Toward a First Nations cross-cultural science and technology curriculum. *Science Education, 81*, 217–238.
Alcoff, L., & Potter, E. (Eds.). (1993). *Feminist epistemologies*. New York: Routledge.
American Association for the Advancement of Science. (1989). *Science for all Americans*. New York: Oxford University Press.
Apple, M. W. (1998). The politics of official knowledge: Does national curriculum make sense? *Teachers College Record, 95*, 222–241.
Atkinson, D. (1999). *Scientific discourse in sociohistorical context: The philosophical transactions of the Royal Society of London 1675–1975*. Mahwah, NJ: Lawrence Erlbaum Associates.
Barton, A. C. (1998). *Feminist science education*. New York: Teachers College Press.
Bazerman, C. (1988). *Shaping written knowledge: The genre and activity of the experimental article in science*. Madison: University of Wisconsin Press.
Begley, S. (1997, April 21). The science wars. *Newsweek, 129*, 54–57.
Bianchini, J. A., & Kelly, G. J. (2003). Challenges of standards-based reform: The example of California's science content standards and textbook adoption process. *Science Education, 87*, 378–389.
Bloor, D. (1976). *Knowledge and social imagery*. London: Routledge & Kegan Paul.

Brickhouse, N. W. (2001). Embodying science: A feminist perspective on learning. *Journal of Research in Science Teaching, 38,* 282–295.
Britzman, D. P. (1995). "The question of belief": Writing poststructural ethnography. *Qualitative Studies in Education, 8,* 229–238.
California Department of Education. (2000). *Science content standards for California public schools: Kindergarten through grade twelve.* Sacramento, CA: Author.
Carlsen, W. S. (1991). Subject-matter knowledge and science teaching: A pragmatic approach. In J. E. Brophy (Ed.), *Advances in research on teaching* (Vol. 2, pp. 115–143). Greenwich, CT: JAI Press.
Carlsen, W. S., Cunningham, C. M., & Lowmaster, N. (1995). But who will teach it? [Review of the book *Benchmarks for scientific literacy*]. *Journal of Curriculum Studies, 27,* 448–451.
Chinn, C. A., & Brewer, W. F. (1993). The role of anomalous data in knowledge acquisition: A theoretical framework and implications for science education. *Review of Educational Research, 63,* 1–49.
Chinn, C. A., & Brewer, W. F. (1998). An empirical test of a taxonomy of responses to anomalous data in science. *Journal of Research in Science Teaching, 35,* 623–654.
Collins, H. M. (1981). Stages in the empirical program of relativism. *Social Studies of Science, 11,* 3–10.
Collins, H. M. (1985). *Changing order: Replication and induction in scientific practice.* London: Sage.
Crawford, T. (1999). Scientists in the making: An ethnographic investigation of scientific processes as literate practice in an elementary classroom. *Dissertations Abstracts International, 61*(01), 89A. (University Microfilms No. AAT 9960301)
Crawford, T., Kelly, G. J., & Brown, C. (2000). Ways of knowing beyond facts and laws of science: An ethnographic investigation of student engagement in scientific practices. *Journal of Research in Science Teaching, 37,* 237–258.
Cunningham, C. M., & Helms, J. V. (1998). Sociology of science as a means to a more authentic, inclusive science education. *Journal of Research in Science Teaching, 35,* 483–499.
DeBoer, G. E. (1991). *A history of ideas in science education.* New York: Teachers College Press.
Denzin, N. K. (1997). *Interpretive ethnography: Ethnographic practices for the 21st century.* Thousand Oaks, CA: Sage.
Duschl, R. A. (1990). *Restructuring science education: The importance of theories and their development.* New York: Teachers College Press.
Duschl, R. A. (1994). Research on the history and philosophy of science. In D. Gabel (Ed.), *Handbook of research on science teaching and learning* (pp. 443–465). New York: Macmillan.
Duschl, R. A., & Hamilton, R. J. (1998). Conceptual change in science and in the learning of science. In B. J. Fraser & K. G. Tobin (Eds.), *International handbook of science education* (pp. 1047–1065). Dordrecht, The Netherlands: Kluwer Academic Press.
Emerson, R. M., Fretz, R. I., & Shaw, L. L. (1995). *Writing ethnographic fieldnotes.* Chicago: University of Chicago Press.
Fuller, S. (1992). Social epistemology and the research agenda of science studies. In A. Pickering (Ed.), *Science as practice and culture* (pp. 390–428). Chicago: University of Chicago.
Garfinkel, H., Lynch, M., & Livingston, E. (1981). The work of discovering science construed with materials from the optically discovered pulsar. *Philosophy of the Social Sciences, 11,* 131–158.
Goodwin, C. (1994). Professional vision. *American Anthropologist, 96,* 606–663.
Goodwin, C. (1995). Seeing in depth. *Social Studies of Science, 25,* 237–274.

Green, J., & Wallat, C. (1981). Mapping instructional conversations: A sociolinguistic ethnography. In J. Green & C. Wallat (Eds.), *Ethnography and language in educational settings* (pp. 161–205). Norwood, NJ: Ablex.

Gross, P. R., & Levitt, N. (1994). *Higher superstition: The academic left and its quarrels with science.* Baltimore: Johns Hopkins University Press.

Gumperz, J. J. (1982). *Discourse strategies.* Cambridge, England: Cambridge University Press.

Gumperz, J. J. (1992). Contextualization and understanding. In A. Duranti & C. Goodwin (Eds.), *Rethinking context* (pp. 229–252). Cambridge, England: Cambridge University Press.

Gumperz, J. J., Cook-Gumperz, J., & Szymanski, M. H. (1999). *Collaborative practices in bilingual cooperative learning classrooms.* Santa Cruz, CA: Center for Research on Education, Diversity, and Excellence.

Halliday, M. A. K., & Martin, J. R. (1993). *Writing science: Literacy and discursive power.* Pittsburgh: University of Pittsburgh Press.

Harding, P., & Hare, W. (2000). Portraying science accurately in classrooms: Emphasizing open-mindedness rather than relativism. *Journal of Research in Science Teaching, 37,* 225–236.

Hymes, D. (1974). *Foundations in sociolinguistics: An ethnographic approach.* Philadelphia: University of Pennsylvania Press.

Jasanoff, S., Markle, G. E., Petersen, J. C., & Pinch, T. (Eds.). (1995). *Handbook of science and technology studies.* Thousand Oaks, CA: Sage.

Kelly, G. J., & Brown, C. M. (2003). Communicative demands of learning science through technological design: Third grade students' construction of solar energy devices. *Linguistics & Education, 13*(4), 483–532.

Kelly, G. J., Brown, C., & Crawford, T. (2000). Experiments, contingencies, and curriculum: Providing opportunities for learning through improvisation in science teaching. *Science Education, 84,* 624–657.

Kelly, G. J., Carlsen, W. S., & Cunningham, C. M. (1993). Science education in sociocultural context: Perspectives from the sociology of science. *Science Education, 77,* 207–220.

Kelly, G. J., & Chen, C. (1999). The sound of music: Constructing science as sociocultural practices through oral and written discourse. *Journal of Research in Science Teaching, 36,* 883–915.

Kelly, G. J., Chen, C., & Crawford, T. (1998). Methodological considerations for studying science-in-the-making in educational settings. *Research in Science Education, 28,* 23–49.

Kelly, G. J., Chen, C., & Prothero, W. (2000). The epistemological framing of a discipline: Writing science in university oceanography. *Journal of Research in Science Teaching, 37,* 691–718.

Kelly, G. J., & Crawford, T. (1997). An ethnographic investigation of the discourse processes of school science. *Science Education, 81,* 533–559.

Kelly, G. J., Crawford, T., & Green, J. (2001). Common task and uncommon knowledge: Dissenting voices in the discursive construction of physics across small laboratory groups. *Linguistics & Education, 12,* 135–174.

Kelly, G. J., & Green, J. (1998). The social nature of knowing: Toward a sociocultural perspective on conceptual change and knowledge construction. In B. Guzzetti & C. Hynd (Eds.), *Perspectives on conceptual change: Multiple ways to understand knowing and learning in a complex world* (pp. 145–181). Mahwah, NJ: Lawrence Erlbaum Associates.

Knorr-Cetina, K. (1999). *Epistemic cultures: How the sciences make knowledge.* Cambridge, MA: Harvard University Press.

Koertge, N. (Ed.). (1998a). *A house built on sand: Exposing postmodern myths about science.* New York: Oxford University Press.

Koertge, N. (1998b). Postmodernisms and the problem of scientific literacy. In N. Koertge (Ed.), *A house built on sand: Exposing postmodern myths about science* (pp. 257–271). New York: Oxford University Press.

Latour, B. (1987). *Science in action: How to follow scientists and engineers through society*. Cambridge, MA: Harvard University Press.

Latour, B., & Woolgar, S. (1986). *Laboratory life: The construction of scientific facts*. Princeton, NJ: Princeton University Press.

Lederman, N. G., Wade, P. D., & Bell, R. L. (1998). Assessing the nature of science: What is the nature of our assessments? *Science & Education, 7*, 595–615.

Lemke, J. L. (1990). *Talking science: Language, learning and values*. Norwood, NJ: Ablex.

Longino, H. E. (1990). *Science as social knowledge: Values and objectivity in science inquiry*. Princeton, NJ: Princeton University Press.

Loving, C. (1997). From the summit of truth to its slippery slopes: Science education's journey through positivist-postmodern territory. *American Educational Research Journal, 34*, 421–452.

Lynch, M. (1992). Extending Wittgenstein: The pivotal move from epistemology to the sociology of science. In A. Pickering (Ed.), *Science as practice and culture* (pp. 215–265). Chicago: University of Chicago Press.

Lynch, M. (1993). *Scientific practice as ordinary action: Ethnomethodology and the social studies of science*. Cambridge, England: Cambridge University Press.

Lynch, M., & Macbeth, D. (1998). Demonstrating physics lessons. In J. Greeno & S. Goldman (Eds.), *Thinking practices in mathematics and science learning* (pp. 269–297). Mahwah, NJ: Lawrence Erlbaum Associates.

Lyotard, J. F. (1984). *The postmodern condition: A report on knowledge*. Minneapolis: University of Minnesota Press.

Martin, E. (1997). Anthropology and the cultural study of science. *Science, Technology, & Human Values, 23*, 24–44.

Matthews, M. (Ed.). (1991). *History, philosophy, and science teaching: Selected readings*. Toronto, Canada: Ontario Institute for Studies in Education Press.

Matthews, M. R. (1997). James T. Robinson's account of philosophy of science and science teaching: Some lessons for today from the 1960's. *Science Education, 81*, 295–315.

Mehan, H. (1979). *Learning lessons: Social organization in the classroom*. Cambridge, MA: Harvard University Press.

Merton, R. K. (1973). *The sociology of science: Theoretical and empirical investigations*. Chicago: Chicago University Press.

Mitroff, I. (1974). Norms and counternorms in a select group of Apollo moon scientists. *American Sociological Review, 39*, 579–595.

Mukerji, C. (1989). *A fragile power: Scientists and the state*. Princeton, NJ: Princeton University Press.

Mulkay, M. (1975). Norms and ideology in science. *Social Science Information, 15*, 637–656. (Reprinted *Sociology of science*, pp. 62–78, by M. Mulkay, Ed., 1991, Bloomington: Indiana University Press)

National Research Council. (1996). *National science education standards*. Washington, DC: National Academy Press.

Rodriguez, A. J. (1997). The dangerous discourse of invisibility: A critique of the National Research Council's national science education standards. *Journal of Research in Science Teaching, 34*, 19–37.

Rorty, R. (1982). Method, social hope, and social science. In R. Rorty (Ed.), *Consequences of pragmatism* (pp. 191–210). Minneapolis: University of Minnesota Press.

Rorty, R. (1989). *Contingency, irony, and solidarity*. Cambridge, England: Cambridge University Press.

Roth, W. -M., & McGinn, M. K. (1997). Science in schools and everywhere else: What science educators should know about science and technology studies. *Studies in Science Education, 29,* 1–44.

Roth, W. -M., & McGinn, M. K. (1998). Inscriptions: Toward a theory of representing as social practice. *Review of Educational Research, 68,* 35–59.

Roth, W. -M., McGinn, M. K., & Bowen, G. M. (1996). Applications of science and technology studies: Effecting change in science education. *Science, Technology & Human Values, 21,* 454–484.

Schwab, J. J. (1978). The nature of scientific knowledge as related to liberal education. In I. Westbury & N. J. Wilkof (Eds.), *Science, curriculum, and liberal education: Selected essays* (pp. 68–104). Chicago: University of Chicago Press.

Smith, D. (1996). Telling the truth after postmodernism. *Symbolic Interaction, 19,* 171–202.

Strike, K. (1982). *Liberty and learning.* Oxford, England: Martin Robertson.

Strike, K. A. (1989). *Liberal justice and the Marxist critique of education.* New York: Routledge.

Strike, K. A. (1998). Centralized goal formation, citizenship, and educational pluralism: Accountability in liberal democratic societies. *Educational Policy, 12,* 203–215.

Strike, K. A., & Posner, G. J. (1992). A revisionist theory of conceptual change. In R. Duschl & R. Hamilton (Eds.), *Philosophy of science, cognitive psychology, and educational theory and practice* (pp. 147–176). Albany: State University of New York Press.

Sutton, C. (1996). Beliefs about science and beliefs about language. *International Journal of Science, 18,* 1–18.

Traweek, S. (1988). *Beamtimes and lifetimes: The world of high energy physicists.* Cambridge, MA: Harvard University Press.

Watson-Verran, H., & Turnbull, D. (1995). Science and other indigenous knowledge systems. In S. Jasanoff, G. E. Markle, J. C. Peterson, & T. Pinch (Eds.), *Handbook of science and technology studies* (pp. 115–139). Thousand Oaks, CA: Sage.

Wittgenstein, L. (1958). *Philosophical investigations* (G. E. M. Anscombe, Trans., 3rd ed.). New York: Macmillan. (Original work published 1953)

CHAPTER THREE METALOGUE

Contrasting Sociolinguistic and Normative Approaches to Redesigning School

Gregory J. Kelly
Wolff-Michael Roth
Randy Yerrick

Randy: I think it is valuable to discuss scientific work from its social context and that such an interpretation is not divorced from the cognitive processes and epistemic values endemic to such work. Using children's interactions (especially between overt academic direction like the whelk) bring lofty academic goals into clearer focus. We probably can all identify at least one example from literature that is guilty of "throwing the baby out with the bathwater." I also appreciate your pointing out the futility of arguing anthropologically about the way science "really is" and emphasizing that the process must be interpreted from within the instructional or research context. You provide a methodological counsel: "Focus on the everyday practices of members of scientific communities. One lesson for education is to understand the phenomena of science education through the study of everyday actions of participants."

Having said all that, my question is: Given that most science teachers have not been scientists themselves, nor have they studied science from any historical or disciplinary knowledge perspective, "What kind of knowledge do expert science teachers draw upon to make classroom scientific activity more authentic and like that of scientists?" There are probably many answers to this question. How is your answer specifically different from those promoting argumentation or those who may promote some kind of *inquiry teaching*?

Greg: Your questions recognize that I am not advocating a univocal definition of *authentic science*, especially one established by making realist assumptions about social practice determined through empirical study. But you nevertheless pose the issue of what knowledge base is required of teachers to be able to create scientific activities that are more like those of scientists. From my viewpoint, we want to consider multiple points of view concerning scientific practice and should refrain from trying to make school more like science.

Michael: In this, you certainly make a move that some of the science studies folks—I am thinking of folks like Trevor Pinch, Michael Lynch, or Doug Macbeth—have been raising over the past several years when science educators showed up at their conferences. In the early 1990s, some science educators including myself had advocated a shift from taking scientists' definitions of science, nature of science, and epistemology of science to taking sociologists' and anthropologists' descriptions as normative frames for redesigning school science. You now argue for a more balanced, multiperspective approach.

Greg: Certainly! Scientists should have a say in establishing educational goals, but so should other interested parties including teachers, parents, students, and community members, among others. As not to be too evasive regarding Randy's question about teachers' knowledge, I shall say I think we need to develop knowledge of and about science among all those involved in science education, along with the litany of other important issues relating to schools, students, culture, and social justice. I believe some first-hand experience would be a good way to develop science teachers' views about science. This may include research experience as part of their professional development. Thus, knowledge of and about science, along with skills for handling the complexities of investigation in classrooms, and dispositions toward students as curious inquirers are likely to be part of the abilities needed to engage with the subject matter in interesting and robust ways. I do not know that my views are different than those promoting argumentation or inquiry teaching other than that I do not believe that any one approach to instruction should be advanced without considering the many contextual factors of particular circumstances.

Michael: I actually went one step further. After spending considerable amounts of time on teaching and researching "authentic science," meaning scientists' science, I spent similar efforts to suggest that we take not sociologists' and anthropologists' descriptions of science as it is practiced in the laboratory but science as it is contested in the community (McGinn & Roth, 1999). Now, in yet another turn, I believe that as long as we begin thinking science education from any description of any science, we end up with the same problems of the learning environment being irrelevant to students. I now advocate rethinking science in schools from questions of critical citizenship, which in part questions the very structure of societal reproduction by current forms of schooling (e.g., Roth, 2002).

Randy: You may be right there, but I want to take the discussion in a different direction. I understand that scientific norms cannot be taught as a direct means for accomplishing science standards, nor can scientific approaches be wholly discounted as inappropriate because they may evoke unsavory norms among students. I think it also good that it is left to the reader to interpret such a claim. Greg, you suggest that for science educators to "create more accessible forms of science instruction across settings the discursive work of the teacher to involve students needs to be made visible. Such pedagogy faces challenges to its survival given increasing calls for standards-based teaching and assessment." I wonder if you can articulate what it is about standards-based teaching that makes concentrating on changing the discourse of science classrooms such a challenge? Can higher education carve out for educational "leaders" a cogent and acceptable option for interpreting children's understanding in a more context-grounded way that would have a greater impact? If there is such an option, what might it look like and how can we make as much progress with these leaders as research journals have in the last decade?

Greg: Standards-based instruction poses any number of challenges, depending on its implementation. In California, the science standards are organized around a set of assertions that students are to know. These standards have been criticized in a number of places; in Bianchini and Kelly (2003), we bring these criticisms into focus. The California science content standards focus on scientific facts and investigation skills without regard to issues such as the nature of science, science–technology–society perspectives, and relationship of academic content to pedagogy. Thus, a large number of specified standards impose restraints on teachers interested in following creative paths to engaging students in the multiple discourses of science. Of course, standards-based approaches can be implemented in other ways in other settings. I do not believe that higher education can present a set of research results that will be persuasive given the highly political nature of educational decision making. However, through the multiple channels of communication (undergraduate education, teacher education, research, preparation of educational leaders, participation in public debates) higher education can serve to undermine any overreliance on blunt measures of student learning. The alternative argument would have to entail reliance on a professional teaching community dedicated to, and responsible for, contextually formulated views of children's understanding. Higher education can serve this end by providing examples of such analysis and including classroom teachers in the conceptualization of research studies of educational practice.

Michael: I find it quite interesting how U.S. educators can live with standards; worse, I cannot understand that educators actually contribute to the creation of standards. To me, all of this sounds and smells like the dark ages of behaviorism—how do you design environments that maximally determine what human beings know and how they behave. The best you achieve is the reproduction of an unjust society, with the poor getting poorer and the rich (financial, cultural capital) getting richer, traditionally middle- and upper-class students. This whole standards-based approach also does not take into account the work of critics, including Jean Lave (1997) and Klaus Holzkamp (1993), who showed that it is an illusion to think that learning can be planned administratively. Your argument to focus on the praxis of science in different settings as a referent for rethinking science education seems to require very different goals than those set by the standards rhetoric.

Greg: I do not wish to live with standards; I wish to live with students. I'm dedicated to improving science education in public schools, particularly for under-served students. As these students must live with externally imposed standards, I work with them and their teachers to create the best possible situation, given the restraints. Unfortunately, the state of California has organized and enforced state-mandated standards for science education. Therefore, in the chapter in this book I sought to identify some of the innovative ways of treating science subject matter and noted that such approaches risk elimination because of the narrowing of the curriculum by the state standards. So yes, I agree with Michael that my focus on the praxis of science implies very different goals than those set by the standards rhetoric. Getting back to Randy's question, as part of the higher education research community, I hope to create education that moves away from administratively planned, mandated curricula (often enforced with coercion) that serves to recreate an unjust society. In the interim, I cannot turn my back on the least powerful because I disapprove of the decisions of the most powerful.

REFERENCES

Bianchini, J. A., & Kelly, G. J. (2003). Challenges of standards-based reform: The example of California's science content standards and textbook adoption process. *Science Education, 87,* 378–389.

Holzkamp, K. (1993). *Lernen: Subjektwissenschaftliche Grundlegung.* Frankfurt/M.: Campus Verlag.

Lave, J. (1997). On learning. *Forum Kritische Psychologie, 38,* 120–135.

McGinn, M. K., & Roth, W. -M. (1999). Towards a new science education: Implications of recent research in science and technology studies. *Educational Researcher, 28*(3), 14–24.

Roth, W. -M. (2002). Taking science education beyond schooling. *Canadian Journal of Science, Mathematics, and Technology Education, 2,* 37–48.

CHAPTER FOUR

Essential Similarities and Differences Between Classroom and Scientific Communities

G. Michael Bowen
Lakehead University

Steven and Alicia walked into their seventh-grade classroom and sat down at the back of the room for their science period. Their homeroom teacher told the class to "quiet down" several times and then gave some instructions for the next day. Steven and Alicia quietly chatted to each other during this, sharing candies in their double desk. An overhead describing the water cycle was put up and students were told to open their binders and copy the note. Taking out their binders, Steven flipped to the first blank page; Alicia opened the divider marked "Science" and turned to her last note. They began copying the overhead as the paper partially covering it was slid down but essentially ignored the teacher's ongoing comments about the note because of their discussion about "tomorrow's visitor." Some students completed copying the note and became, in the teacher's words, "disruptive" while the slower students finished copying. Before some were done he removed the overhead, saying they could "borrow it later" to finish. He moved to a desk at the center-front of the room to model a lab activity. Taking a small empty yogurt container the teacher demonstrated putting a bit of water in the bottom and then inserted the bottom part of a Styrofoam cup and weighted it down with a few pennies. He then loosely placed thin plastic wrap over the top of the yogurt container, held it in place with an elastic band, and placed a penny on the center of the plastic wrap to form a depression. While describing his actions, he also discussed the relationship between this apparatus and the note on the water cycle that students had copied from the overhead. He instructed the students to work in groups of two, with one person getting up to get equipment so that they too could build the water-cycle model. Alicia and Steven each turned to another person with whom to work. Alicia's partner was chosen to go around the classroom and distribute the plastic wrap. Steven and his partner chatted away about another school activity but were not able to describe to the researcher what they were doing and why, or how the activity was related to the water cycle. Alicia, even though working alone, provided a clear description of what was being done, why it was being done, and how the activity related to the water-cycle note. Several days later, Steven suggested that the activity "didn't work" because it wasn't warm enough where the containers were kept, although he still wasn't sure how that was related to the water cycle. He did know that the little Styrofoam cup with the pennies was supposed to have water in it, but could not explain why. (Based on observations in a seventh-grade classroom, Spring 1998)

In the foregoing vignette constructed on the basis of my ethnographic work, the essential nature of science classrooms is apparent. In this setting, students were to use a "simple" activity as an experiential model for a relationship in their textbook that was a dramatic simplification of a complex process that is actually quite difficult to observe in its entirety. There are few differences between this activity and those experienced by science undergraduates. There are considerable differences between this type of activity and those that characterize the construction of science knowledge in science communities. In these settings, attempts are made to further elaborate the complexities of what is taken as "known" so as to generate more precise, complex, predictive, and explanatory models. The contrast between these activities, science-in-the-making (Latour, 1987) compared to science-in-the-classroom, renders understandable why classroom practices so poorly develop interest in science and student understanding of the complex relationships in the natural world.

Until recently, conversations surrounding what constituted scientific knowledge addressed either the substantive nature of scientific knowledge or the syntactic nature of knowledge construction (i.e., the "scientific method"). However, in this view of science, claims and conduct are based on a view of science practice derived from scientists' own historical accounts of their work in research. A very different view of science practices emerges from the studies by ethnographers who have conducted their research on day-to-day science practices in the research settings of scientists themselves (e.g., Latour & Woolgar, 1986). This work revealed that the operation of science and the development of scientific theories in science communities differ greatly from the frequently promoted notion that a standard scientific method is employed by scientists to make sense of the world. Instead, this ethnographic work demonstrated that scientists' research processes and theory development are messy, human, and, in many cases, remarkably subjective endeavors—the processes and resulting theories reflect the complexity and humanness of the scientists. Despite this, K–16 science classrooms often still present science as a rational and abstract endeavor that reflects the formal writing of disciplines rather than the actual processes by which scientists come to know (Traweek, 1988). The stark contrast between these two perspectives on science leads to the suspicion that teachers often unwittingly misrepresent the essence of scientific knowledge and activity to their students.

Approaches to teaching school science have remained relatively conservative throughout several reforms. How is it that school science took on its current appearance? What are the important similarities and differences between scientific and classroom communities that are revealed when discussing changing science classroom discourse and practices to that more consistent with scientific communities? Few resources exist to help teachers in accomplishing this task. A major unresolved question is whether classroom discourses have enough family resemblance with scientific discourses to allow such changes to occur.

Suggestions for science curriculum reform often assume that the culture(s) of scientific communities can be mapped, more or less directly, onto classroom cultures. However, there are substantial differences in the culture of the two endeavors, and there is some question as to whether it is actually desirable to replace the traditional classroom culture with a truly "authentic" science one. For instance, changing school discourses to scientific discourse runs the risk of further stratifying students. Yet, movement from classroom science as reflected in the vignette—the individuals who conducted the water-cycle activity are clearly distinguishable from those in a typical "science" research setting—toward approaches that develop a better understanding of concepts and practices are needed, as I will present a case for below. However, to offer advice on different classroom approaches that could achieve proposed reforms we need to better understand how competencies in science practices in professional communities themselves came to be.

To help understand the trajectories of competency in various science practices, I draw on a series of studies conducted over the past decade (often in collaboration with Wolff-Michael Roth) to examine the development of science discourse and practices of participants in a variety of "science" settings. This research included studies of seventh- and eighth-grade students conducting field research projects, ethnographies, and interviews with undergraduate science students in their classrooms and field experiences, and ethnographic studies of professional and academic biologists as they conducted their own independent fieldwork.

In this chapter I examine the contradictions students face in the things they do to learn "about science," and the things that scientists seem to do in their own practices. I describe aspects of undergraduate science education experiences, including lectures, laboratory investigations, and the resources such as textbooks and journals upon which they are able to draw to learn about concepts. I draw parallels with teaching K–12 science to speculate about the origins of these school practices (as the described undergraduate science experiences are those experienced by teacher candidates themselves). I present evidence on when and how individuals develop deep understanding about their area of science by examining their engagement in "authentic" research practices. I conclude with describing how similar "deep" understandings were developed in middle-school students. From this I sketch some insights that may pave the way toward constructing classroom environments that exhibit more structural similarity with scientific communities.

UNDERSTANDING PRACTICES

Both classrooms and scientific research sites constitute communities, which represent dynamic social milieus within which there are typical, socially organized practices (Lave & Wenger, 1991). They therefore represent "communities of practice" (Wenger, 1998). I developed the following theoretical framework (adapted from Denning & Dargan, 1996) as suitable for investigating science settings, which allows me to compare classroom and professional science in the making (Bowen & Roth, 1999a). By understanding the similarities and differences between the two, we can better understand how to gradually alter the ways in which science classroom teaching is implemented. The analytic framework for social practice contains the following components:

- *Standard practices*, enacted by members, by means of which the characteristic activities of the domain get done. For example, in science these include designing experiments and transforming and scaling graphs so that they support the rhetoric of the main text. Use of graphs and tables are central to the practice of science (both its conduct and the factual claims; Latour & Woolgar, 1986; Lynch, 1985).
- Ready-to-hand *material resources*, such as tools and equipment, that members use as part of their standard practices where they use the tool transparently, focusing on the task rather than the tool.
- *Linguistic resources* that members use to make distinctions important to competent and efficient work. For instance, important to pop-

4. CONTRASTING CLASSROOM VS. SCIENTIFIC COMMUNITIES

ulation ecology is the difference between *population size* and *population density*.

- *Breakdowns*, interruptions of standard practices and slow-down of an activity's progress that evolve from the breaking and absence of tools or changing of familiar contexts such that tools or language do not allow disciplinary practices to go forward.
- Sets of *ongoing concerns* of members include common missions, interests, and fears. For example, field ecologists are concerned with managing species for commercial exploitation or conservation.

My ethnographic research conducted in communities of scientists, undergraduate students, and eighth-grade students focused on these aspects of the communities. To discuss issues of science literacy I focus both on *ready-made-science* (e.g., concepts, theories) and *science-in-the-making* (Latour, 1987). However, readers should not take my separate discussion of these to mean that students could develop an understanding of one without developing an understanding of the other. Recent studies suggest that neither the structure of the natural world nor the nature of tools can be understood separated from the human practices and interactions within which they are relevant (Cobb, 1993; Roth, 1996). In the following sections I first examine the "typical" experience of undergraduate science students (including those who will become science teachers) and the various ways that scientific practices and claims are experienced in a decontextualized and reductionist manner. I then describe three case studies, detailing settings in which individuals engaged in "authentic" science practices and experienced the complexity of science practices and the difficulties in deriving knowledge claims.

THE UNDERGRADUATE (SCIENCE) EXPERIENCE

The common experience in undergraduate science programs consists of 4 years of lectures punctuated by associated "cookbook" laboratory activities. In these science courses inscriptions, such as graphs and tables, play a major role—in a typical 50-minute period of one 2nd-year ecology course, more than 25 inscriptions were presented (Bowen & Roth, 1998), which is consistent with sociological studies that highlight the centrality of inscriptions to the practice of science (Latour & Woolgar, 1986). In this section I first highlight university students' graph-related competencies and then extend my investigation to other aspects of the undergraduate experience.

Competencies in Science Practices

In a series of studies, the interpretation practices of undergraduate science students were examined to understand how they made sense of inscriptions (such as data tables and graphs), which are a core constituent of scientific literacy (Roth, Bowen, & McGinn, 1999). Despite their high exposure to inscriptions, science graduates interpret graphs much differently than do individuals who use graphs as part of their everyday work. For instance, when presented with a data task in which graphing would suitably summarize a data source, science graduates used this approach only infrequently (Bowen & Roth, 1999b; Roth, McGinn, & Bowen, 1998). In interviews, in-program science students discussed graphs without making those linguistic distinctions necessary to effectively interpret the graphs (Bowen, Roth, & McGinn, 1999). Overall, the university science students had learned to provide correct, rote answers to standard graphs but did not effectively integrate their usage into broader understandings of ecological issues. Presentations of graphs in their lectures and evaluation schemes in their courses led to a view that graphs have but a single viable interpretation. Yet, interviews with experienced graph users suggest there are numerous viable interpretations of a single graph depending on the experiential and conceptual referents interpreters bring to the tasks. Additionally, compared to experienced researchers, students used few resources external to the graph itself to construct their interpretation. Experienced researchers frequently made reference to experiences and concepts that were tangential to the graph topic as they worked at making sense of the graph. Students, on the other hand, usually only referred to components immediately available within the graph itself to construct their interpretation. There is little evidence that lecture-oriented courses with cookbook activities contribute to the development of graph-related competencies. This is reinforced by the observation from these studies that there were few differences in graph interpretation competencies between those teacher candidates with strong science backgrounds (i.e., BSc) and those who had much less of a science background (e.g., 3-year BA).

By engaging preservice secondary students in short-term outdoor projects of their own, it was anticipated that the difficulties they had with graph interpretation would be ameliorated. However, their project work suggested that the inscription interpretation problems demonstrated by undergraduate science students were connected to other issues involving the construction of knowledge claims in science. In the students' reports on their project work numerous nonstandard practices were apparent (in total, in 12 reports 90 problems were recorded). These included research questions unanswerable by the study design, constructs inappropriately operationalized, data reported in tables that were quite difficult to understand because of their structure, inappropriate transformations of data (such as in graphs), and claims that frequently did not match the research questions or the reported data. Surprisingly, although students were to conduct correlational studies, only 9 of the 24 research questions were actually stated as correlations. A further 13 questions were stated as causal relationships (although in almost all cases it was not possible to address causality in the time allotted to the study but only possible to infer it from some that were correlational in nature). Overall, the reports lacked the structures common to standard science reports. Clearly, these difficulties in conducting their own projects suggest that undergraduate science programs poorly prepare students to engage in independent science research activities. Why might this have occurred? It is not unreasonable to expect that science program graduates should be able to design, conduct, and report on small inquiry studies of their own design. To answer this question, it is perhaps relevant to examine the different resources used by science undergraduates to learn about their discipline—textbooks, lectures, and structured investigations.

The Textbook Experience

An examination of ecology journal articles and textbooks suggested that different inscriptions than those commonly found in the discourses of professional science dominated the resources used by students. Illustrations and pictures were the most frequent form of inscription in textbooks (with some variation between university and high-school textbooks) whereas journal articles were dominated by equations and graphs (Roth & Bowen, 2001b; Roth et al., 1999). The graphs found in textbooks are similar to those used in lectures and are frequently transformations of those that had previously appeared in journals. The alterations for inclusion in textbooks actually complicate the interpretability even for experienced researchers and students (Roth & Bowen, 2001b). In addition, textbooks provide less information to help readers interpret the graphs and tables than do journal articles, which provide information to guide the interpretation of inscriptions in the caption and in the main text. As a result, the journal article reader alternates attention back and forth between the main text, the caption, and the inscription—which makes it difficult for the interpretation to deviate from that desired by the author (Bastide, 1990). However, in textbooks the inscriptions may not be referred to in the main text, and they often have substantially shorter captions. Thus, the textbook reader gets little guidance as to how to interpret textbook inscriptions and are thus less likely to reach the interpretation originally intended by the producer of the inscription. This may explain the poor competency of science program graduates with inscriptions—one student noted that he essentially just ignored the various inscriptions when reading his textbook.

The Lecture Experience

To better understand the undergraduate experiences in lectures, I extensively studied a typical 2nd-year ecology class. In general, the experiences that help lecturers make sense for themselves of the inscriptions they present are generally unavailable to their students, which might partly explain why students have difficulty making sense of the inscriptions. In addition, there was a discontinuity between the gestures used to refer to the inscriptions and the comments being made by the lecturer. This low coordination between the verbal descriptions and the gestures indicating to what features of the inscriptions to attend further compounded student difficulties in understanding lecture presentations of inscriptions (Bowen & Roth, 1998). Apart from these problems, lectures presented a scant image of everyday science practices related to the use and interpretation of graphs. The normally existing mutually constitutive relationship (i.e., where each constructs the other) between phenomena and their graphical relations is often undiscussed in lectures, nor is there opportunity for students to make sense of the inscriptions by moving back and forth between the inscriptions and the textual descriptions of it (such as they can when reading a journal article) (Roth & Bowen, 1999a). Finally, the mundane examples chosen to explain inscriptions in lecture often obscure rather than clarify the relationships in the natural world depicted in the inscriptions.

Cookbook Investigation Activities

Apart from lectures, another common undergraduate science experience is structured (or "cookbook") laboratory investigations. The following description, drawn from my field notes of a typical outdoor activity in undergraduate science, helps make sense of the difficulties that science program graduates had in conducting simple outdoor research activities in their science education methods courses. Further notes suggest that students experience these activities as disconnected from the lecture component of their course and the research methods as inviolable, predetermined, and standardized. In the following activity students were bussed to an ocean shore to conduct a shoreline organism survey at low tide. A transect (essentially an extended linear sampling grid) had been laid out by the instructors from the top of the high-tide zone to the water's edge. The following note describes the experience:

4. CONTRASTING CLASSROOM VS. SCIENTIFIC COMMUNITIES

The first groups are sent out to set up a grid along the transect. The lab instructor provides instructions and descriptions of how beach transects are done while the rest of us are still marooned on the edge of the parking lot. We are told to lay out one-meter quadrats every few meters down the transect line; we are to measure the distance from the center of each to the transect line but we do not have a description why the quadrats should be at different distances from the line other than that the sites chosen were to be random. She tells us that we should record the substrate, time, and location (latitude and longitude) on the sheets she had handed to us and that we could "never have too much information." She tells my group to head down to the mid-tidal area, to "pick a random area," and to measure the distance to the line.

We proceed and stop at an area that was sort of clear (no others were working close by). The TA brings us a plastic frame. We are told to use mining tape to mark the corners of the quadrat and then pass the frame on. By eyeballing the distances, we turn the quadrat into 9 equal areas. I stand beside Nancy and was responsible for section C3. Nancy, holding the clipboard, discussed with others what should be looked at. (Field note, November 1998)

Each group of students only participated in one part of what was described as constituting a complete research activity. Each group collected its samples from one spot along the transect line and only later, after a sharing of data among groups, wrote their reports in which they examined the complete data set for patterns. The students experienced ecological field methods as a set of standardized procedures predetermined by an external authority that made all relevant decisions beforehand. It was difficult for students to identify organisms because they had not previously done any identification activities. Therefore, they not only had to learn new field methods but also had to identify organisms by relying on field identification sheets they had not seen before; this identification is in itself a difficult task (Law & Lynch, 1990). One result was that quite a number of the students were not particularly motivated to accurately identify the organisms in their quadrat and often settled on their first identification. They made numerous identification mistakes, perhaps unsurprising given the paucity of resources on which they could rely. Students' most immediate goal was to fill in the data sheet to the satisfaction of the teaching assistant so they could later write a laboratory report based on the data. Errors, even when caught, were often uncorrected—their orientation to the task was notably student-like, unlike researchers with a vested interest in defensible data collection and recording. Given their lack of control over the exercise, as well as being unsure about why they were engaging in it, this is hardly surprising (and remarkably similar to the opening vignette). There appear to be few structural differences between Grade 7 and undergraduate science experiences.

Understanding the Undergraduate Experience

Using the analytic framework presented earlier, how could we apply a sociocultural lens to understanding the undergraduate experience? Unlike the generative epistemological aspects of disciplinary science praxis, the undergraduate science experience is concerned with arriving at pre-established, normative answers—they are expected to (demonstrably) replicate outcomes of field research that have been previously captured in standardized exercises. But my research shows that this both poorly prepares prospective teachers to teach the inquiry-oriented methodology promoted in reform documents (American Association for the Advancement of Science, 1993; National Research Council, 1994) and poorly prepares science undergraduates to become scientists.

Undergraduates learn about the ready-made facts of science and engage in standardized and ritualized activities that are quite unlike those enacted in science disciplines. The formalized texts (textbooks, journal articles, lectures) used to communicate "science facts" to them often contain little information about the details of the practices that led to the findings (Bowen & Roth, 2002b). In addition, lectures and textbooks often present scientific problems (especially as represented in inscriptions) in a substantially decontextualized fashion that oversimplifies the natural world and the relations between it and the inscriptions. It is therefore unsurprising that in this environment students learn to primarily focus on the outcomes of the assessments in their course—which often means relating again, often on exams, the narratives of simple relationships found in the lectures and textbooks. In contrast with scientists discussing real-world relationships and claims among themselves, a standard aspect of discourse in undergraduate science programs is to speak very matter-of-factly, deterministically, focusing on what is known, while spending little time on what is unknown. While conducting structured investigation activities, broader questions of interest to scientists generally remain unnoted as students work at replicating the short-term practices (and related questions) that can occur within the time frame of a single laboratory period—for whose activities the parameters are often clearly defined, the methods predetermined, the variables already constructed.

In both the laboratory activities and lectures (and associated assessments) students are required to make the linguistic distinctions common in a discipline but often without having any experiences with the scenarios or activities to which the distinctions are related. What is most evident from the descriptions of the various settings of undergraduate science education is that competency at standardized practices, understanding of disciplinary concerns, and the opportunity and ability to make necessary linguistic distinctions do not seem to accrue from these programs. What follows is an argument that what is needed is not just engagement in investigation activities, but engagement in prolonged investigation activities with interpenetrated ambiguities and unknowns within a common community of interest.

TRANSITIONS TO DEEP SCIENTIFIC UNDERSTANDINGS IN PROFESSIONAL COMMUNITIES

Having identified the shortcomings in how science education is often enacted, insights into how to address them can be gained by studying settings in which those shortcomings are addressed. This forces a focus onto the necessity for learning practices in a community of those who are already successfully versed in them (or those related to them), which arises from the fact that the procedures and practices are underdetermined by description (Suchman, 1987). In this situation, context includes tacit and embodied knowledge that cannot be conveyed in principle and that must be gained through practical experience. An example for the process of appropriating tacit forms of knowledge through participation in the everyday practices of research science is provided in the following case studies. The three narratives articulate experiences of three new researchers engaging in their own research tasks as they struggle with trying to align their practices with standard practices in their science communities. Together, these case studies present an argument that competency in science emerges from prolonged engagement in the setting as part of a community.

Jose—An Undergraduate Honors Thesis Experience

Jose was a 4th-year undergraduate in a science program at a university in central Canada working on his independent inquiry honors thesis project. For his project he revisited a masters-degree project conducted 2 decades earlier in which the student had constructed detailed floral distribution maps through a great blue heron colony. At the time of the original study the center transect passed near the middle of a great blue heron colony where there was the most biological damage from the birds' excrements. Jose reconstituted the center transect and remapped the flora so he could estimate rates of the recovery of the flora to help understand if potential colony areas needed to be protected so as to ensure the sustainability of great blue heron populations. He needed to complete several tasks in thick undergrowth and woods. First, he had to relocate the original center transect boundaries and delineate them on the ground. Second, he had to subdivide that transect into 10- × 6-meter quadrats so he could then identify and enumerate the (changes in) tree species along the transect length.

The original researcher had placed metal stakes in the ground at the corners of the belt transect, but Jose was unable to find all of them; one was missing. Using a compass, Jose first marked the northern baseline and then the southern baseline. Two problems became obvious to him. First, in several locations distances from the edge of the quadrat to characteristic trees (those rare in number, large enough that they were much older than two decades, and located on the original map in the masters thesis) differed from markings on the original map. Second, he noted that there needed to be a bend in the transect for the stakes at the north and south ends to be the required distance apart. He identified the cause of the latter problem only after several days of discussion with me (his field helper) and after having walked back and forth across the transect area many times.

> The different transect lines require different lengths to get to the same point due to [changes in] elevation. . . . I never considered that. . . . The elevation differences would account for the differences in the two sides. . . .

He recognized that changes in elevation of the land meant that using straight line distances measured on the ground would result in areas that would be rectangular on the ground but not on a surface projection of the ground (such as a map). Jose solved this problem in the manner he felt would have the least impact on the accuracy of his data collection—by using uneven quadrilaterals in the middle section of the belt transect where there were fewer trees.

> If I am going to shift it into phase, this is the only place I can do it, right here. Because it is clear, like there are hardly any trees here, and there are hardly any trees two meters back this way. So I am not losing a great deal of the sampling truth. . . . There are no trees that would be included or excluded at the point where the shift is occurring . . . it won't nullify what I get from this experiment.

Jose resolved this difficulty in replicating the transect lines to his satisfaction such that he was confident that in using them he could address his original research interest. However the issue of missing trees was more problematic. Having established the transect lines, Jose proceeded to check their accuracy by confirming the location of characteristic trees. Despite these being accurate in some places, in others there were inconsistencies that he could not resolve.

And 1.2 meters from *that* American beech there should be another one. . . . I am actually convinced there are places where she is just wrong. . . . Or I could be in the wrong spot, which is the most frightening possibility, because there is not an American beech there. But this has happened before, where I could find certain trees that were definitively there, but now this, which should have been there, but isn't. . . .

Jose considered two possibilities, that the original map creator had errors on her map that he was unable to reconcile in his data collection or that he was considerably wrong in the location of his transect. The problem with the latter explanation was that he was reasonably certain where one of the transect lines was and even checking over a considerable distance was unable to find two American beech trees within 1.2 meters of each other. He even checked for decayed stumps. Ultimately, Jose resolved these contradictions by considering that rather than resampling the exact area, he was taking a representative sample through the same area as the original study. By considering his research as one form of a replication as opposed to another, he resolved the conflict around whether he had laid the transect incorrectly or whether the original author had errors in her reporting.

Mike—Masters Student Experience

How can we understand the transition from the reductionist perspective of undergraduate students to that of the working ecologist? In this episode I relate my own struggles with attempting to clean up real-world data to construct an interpretable inscription—a process that contributes to this transition. As discussed earlier, science graduates understand graphs (and science concepts) in a decontextualized fashion, disconnected from real-world examples. Unlike those with research experience, science students have discussions that indicate they expect data to conform to clear patterns—such as those seen in the data models found in lectures and textbooks. In the following excerpts I relate my experience of learning to graphically represent real data during my masters-level research. For an initial resource I had a toxicology textbook that depicted the possible relationships between mixes of pollutants as *additive, synergistic* (i.e., more-than-additive), and *antagonistic* (i.e., less-than-additive). I was attempting to determine the effect of sub-lethal mixtures of copper and zinc on fish behavior, and initially I expected that the relationship would clearly fit one of these categories. This expectation of a clear categorical conformity is consistent with my many interviews with undergraduate science students about data interpretation. The following notes detail a change in understanding from the naïve interpretation of one without research experience toward the nuanced understanding of one who perceives relations between the real world and inscriptions.

> I was comparing the effect of exposure to varying levels of two different metals at the same time on a measure of fish behavior. I first played at great length with the graphing program using my original data set. First using measured levels of the metals, then the expected levels (the added measured dose), back to measured, back to expected, the hope of using new measured values, finally settling on using expected values. Yet, despite all of the "playing" with whether to use measured or added values I consistently obtained a completely un-interpretable x-y-z graph.... The resultant surface was all over the place—not at all like the "clean" ones I had seen in the text—and essentially indecipherable from the perspective of determining which of the three conditions were in effect. I did this "playing" for quite a while, trying different contour intervals, data transformations, etc. Nothing looked useable so I resorted to constructing a data table based on the mathematical model I'd developed from the data.... And this generated a smooth 3-dimensional graph surface and 2-dimensional isocline representation which looked like that in the textbook and was therefore more interpretable. I no longer had to question the quality of my data, the equipment I had designed, or my data collection approach; I had demonstrated a relationship.

This episode relates four expectations, which I believe to derive from my undergraduate experience and which parallel those found in interviews with undergraduates interpreting graphs in the ethnographic studies.

1. Data collected in a study will result in "clean" inscriptions without any form of translation or transformation.
2. There is no difference between graphs that depict measured data and those that depict data models.
3. There *should* be a relationship between the variables that *could* be depicted graphically.
4. To be unable to generate typical inscriptions or depict typical relationships is a failure in the new researcher's methodology rather than an inappropriate method of inscribing the data.

I developed an understanding of the relationship and how to depict it through a process of going back and forth between different inscriptions. The graphical model was one that could be interpreted according to standard rules and led me to the realization that what had been graphed in the reference textbook was a model and not the data on which it was based. This developing understanding reflects the initial transition that graduate students undergo: They move from a reductionist view on relationships between variables to a more qualified, holistic understanding that model representations are eidetic images. The back-and-forth nature of the activities through which this understanding developed is typical of the types of interpretations of inscriptions by experienced researchers—and atypical of those of undergraduate students (Roth & Bowen, 2001b).

I gained further insights when interpreting the graph by moving back and forth between the mathematical model and the inscription. In doing so I subsequently realized that the effect of squaring on low numbers resulted in an inscription that represented additivity at low concentrations and less-than-additivity at higher concentrations. From this I realized that real-world scenarios do not necessarily conform to the clean categories represented in texts.

Expecting clear, discrete categories of relationships, such as those depicted in the text, when examining real data is a norm for undergraduate science students interpreting inscriptions. It was only through engaging in interpreting complex data, with interactions that lay between generalized categories, that an understanding of data as representing transitions of relationships, as opposed to just categories of relationships, developed. My initial struggles occurred because I analyzed the data from perspectives typical of undergraduate enculturation. Through attempting to represent complex data, movement along the trajectory from newcomer to professional researcher occurred. Only rarely are opportunities provided to K-16 students to engage in these sorts of struggles with data, although clearly these struggles help develop a better understanding between science talk and the actual conduct of science itself. The generation and interpretation of inscriptions is rarely as straightforward as it is presented in either lectures or in cookbook activities.

Sam—The Doctoral Student

Sam was a doctoral student engaging in her second season of dissertation research at a mountainous remote field setting in western North America where she was collecting data on the ecology and natural history of a lizard species. My ethnographic work focused on the construction of claims in field ecology (for details see Bowen & Roth, 2002a, 2002b; Roth & Bowen, 1999b, 2001a). In the following I describe how Sam developed a technique to quantify lizard color.

Sam saw the construction of a measure of skin color as a necessary component of understanding the lizard natural history. Noticing differences, in her first season she started to ponder how the color patterns of lizards could be quantified. Her first attempts were unsatisfactory. The use of color photographs resulted in colors that were inconsistent from one day to the next. She then attempted to develop an *in situ* absolute scale of comparison using paint chips from a local hardware store. However, the transitions between colors were too abrupt to allow visible differences in the lizard color to be captured; she also felt that the tool itself might be hard to defend in an academic context. However, between field seasons she discussed the idea of an absolute scale with others, until someone suggested that Munsell soil color charts could convert lizard colors into numbers.

By using this standardized tool Sam could literally measure the color of each lizard. The standardized aspect is important because that meant that she could then both test her system of measure for validity and reliability, and comparatively share her results with other researchers. Sam engaged in developing a methodology that, so she hoped, would allow her to obtain consistent measures of lizard color:

> Color analysis takes place in a white paper-lined box placed sideways with a clip-on lamp to maintain consistent light with the standard color sheets in the box. . . . Sam brings over the first lizard and asks Mercedes to record the numbers she calls out. Sam . . . laboriously examines the lizard against the color standards. Moving the lizard back and forth from hole to hole in the cards and switching cards back and forth three or four times, she finally announced three numbers (representing hue, value, and chroma) based on the closest match to the lizard. She then records the temperature of the room. Sam is concerned about the effect of dust/dirt on lizard color and spends a bit of time wiping it, then realizes it's about to shed, and decides that she's going to keep track of shedding schedules. . . . Sam is concerned about the color categorizing being "repeatable" and "consistent" and "without any observer bias on color measure." . . . Some of doing the color was difficult. The observer had to be consistent about where to look both generally (center back behind the shoulders) and specifically (the outer edges of some scales were more lightly colored). (Field note, July 1997)

This episode stands in contrast to the typical undergraduate experience. Sam developed interest in color as a variable during her engagement with research in the field and then collected relevant data without even being sure that it would be useful. Her understanding of color significance ultimately derived from her search for significant correlations with other measured variables, many of which also emerged from her longtime onsite engagement. She constructed arguments about the efficacy of the Munsell chart only after having found statistically significant correlations. In this manner, what was the appropriate tool emerged with the consideration of what was a defensible tool for a given correlation. In use, the tool and data each construct the other. The stronger the relationship, the higher is the validity that can be ascribed to the tool. Through these elaborated procedures, including comparing colors measured by different observers and blind presentation of animals, over two seasons Sam developed a methodology to measure individual lizards that she felt comfortable defending to others in her community while at the same time developing an understanding of the shortcomings of her methodologies (Bowen & Roth, 2002a, 2002b).

Implications of Studies of Developing Researchers

The three different experiences reflecting (the development of) deep understanding of science practices and concepts possess many common threads:

- All three individuals were operating in a setting with some external supports (both social and professional), but they often had little immediate community to rely on for support as they struggled with designing their study and analyzing their data.
- They had each designed (much of) the study themselves.
- They were attempting to address a research problem about which the answer was generally unknown. What their research was judged on was not some externally determined correctness of findings, but how convincingly and compellingly they constructed a series of arguments and data to lead to those findings.
- Their work was characterized by struggling over a prolonged engagement with methodological, analysis, and design issues that needed to be resolved to result in research results defensible to the broader community in which they participated.
- They accepted something as an answer when it led to the most convincing and compelling argument for defending their results.

Their struggles arose because all three found themselves immersed in the messiness of nature. Variables had to be defined, methodologies developed, and tools adapted to address the research questions; relationships that were weak needed to be presented convincingly, variations between natural settings and inscriptions used to present results had to be resolved, and field situations with a potential myriad of influences needed to be reduced to a series of variables that were addressable in the study. These processes characterize the process and knowledge growth of science.

Addressing and resolving these types of issues are common occurrences in science research settings but are in considerable contrast to the activities found in most school settings (K–16), in which students are attempting to produce an answer that somebody else already knows. This search for a known answer also reflects the common undergraduate experience of those of who will become K–12 teachers and is subsequently mirrored in most enacted K–12 science curricula. It is not surprising that science teachers tend to focus toward lectures, replicating the practices that they had previously experienced, and clearly the enacted K–12 curricula arise from an understanding of science that develops during the undergraduate experience. This then raises the question, how might we conceive of a K–12 science curriculum that derives from the experiences of those entering professional science communities in order to lead to more detailed understandings of scientific practices and interpretations in K–12 students? To answer this I turn to a description of student activities in an eighth-grade science class that engaged in prolonged independent field research.

ENGAGING MIDDLE SCHOOL STUDENTS IN COMPLEX AND UNFAMILIAR SETTINGS

Competent practices emerge as one participates with others in relevant activities; practical knowledge requires participation (e.g., Bourdieu & Wacquant, 1992). By understanding better how biologists know what they know, we develop better understanding of how we could teach students biology. The previous sections suggest that science knowledge cannot be transferred but must be built up through lived experience, particularly through participation with others in meaningful activities. The view of biology as practice characteristic of a community implies that we teachers need classes where students actually do biology over a long term. In traditional teaching of science and its practices teachers often engage students in series of short-term unconnected tasks with little opportunity for them to develop any deep understanding of how compelling scientific arguments develop through an accumulation of collected, connected evidence.

The following five conditions were taken as necessary in designing inquiry-style environments in which students can develop competency in the tool-based generative practices of science (see Roth, 1995; Roth & Bowen, 1995):

1. Students work and learn in contexts in which (some) problems are ill-defined.
2. Students experience uncertainties and ambiguities of scientific knowing and learning.
3. Learning is driven by the current knowledge of the students.

4. Students experience themselves as members of a community of learners in which knowledge, practices, and discourses are shared among members and develop out of their interactions.
5. Members in these communities can draw on the expertise of more knowledgeable others and on any other resource that could enhance their learning.

An outdoor investigation activity was designed for three Grade 8 classes using this frame. Student research activities were structured to parallel those of professional communities of scientists, offering the opportunity for the social interactions and generative frameworks of those communities to develop. Student pairs conducted self-designed studies in their 35-square-meter research area to examine the relationships between biotic and abiotic features. They recorded their observations and research questions in a field notebook. Each week they formulated one or more research questions, collected data at their site, shared their findings with other students, and wrote a report. Occasionally, there were methods classes where various research methodologies were taught. Students could choose from a wide range of equipment for their work including trowels, soil corers, soil thermometers, moisture meters, hygrometers, pH meters, soil nutrient test kits, and so on. They addressed questions such as "Is there a relationship between pH level, moisture level, and the height of the horsetails?" and "Is there a relationship between the amount of acidity and how [well] plants grow in those conditions?"

The intent of engaging students in this exercise was to develop both their understanding of the process of science (science-in-the-making from both a methodological and an epistemological perspective) and also their understanding of ecological issues. In the following episode we follow two students into their "ecozone" for the first time as they start looking at it and noting features around which they could design study questions.

Damian: We could record the temperature, you know, it's 14 degrees.
Ellen: Are you sure?
Damian: Yea.
Ellen: What else do you want to record? Height of stuff?
Damian: Yea. . . . I was wondering if there are any insects, have you found any?
Ellen: Well we should get basic [things], like the height and the width and the circumference of trees.
Damian: Ok. I wanna get some plants that are just born or just started to grow.
Ellen: I haven't seen any flowers down there yet, but I guess it's not the right time.

[Several minutes later]
Ellen: What is that, like a vine or something?
Damian: I don't know.
Ellen: Ok.
Damian: I'm gonna tie this [so it's marked as being measured].
Ellen: Ok, ahm, what do we label that as?
Damian: I'm not sure, we're just leaving as it is, we can number them first and when we go research them we're gonna find out what their names are.
Ellen: Ah, its kinda, like bush vine for now.

In their initial foray into their field research setting the students were surveying their site and making decisions about what baseline measurements they could take, which they could examine later for changes. They identified a common abiotic factor that is recorded in field settings (the temperature) and as they attend to different features of their ecozone proposed other items to measure. During the measuring of an unknown plant they recorded data about it but did not, at this moment, have a sense of what the long-term purpose was in recording that information.

This is similar to the work done by Sam as she identified and worked on defining "lizard color." After identifying lizard color as a salient variable, she defined a methodology and collected the data without being sure whether it was relevant to her interests. In a similar fashion, these two students were collecting information on the plant *because they could*, with the expectation that relevancy may develop later.

Having established that baseline data, a week later they were back in the field addressing some specific focus questions. They had noted differences in their ecozone both in the amount of grass and in the slope of the land. They constructed a methodology to address this:

Ellen: So um if we have three different sample jars should we take samples from three different locations, like see if the slope has anything to do with it, if it is more acidic on the top than on the bottom.
Bowen: Which field question are you addressing today, Ellen?
Ellen: Um, (looking into Field Notebook) the acidity levels on the grass. Ok, is there a relationship between the pH level in the soil and the amount of grass found? So Damian is getting the pH, we got samples and now we're just gonna measure moisture.
(Ellen works with the moisture meter.)
Ellen: Whow, wow, very dry, two point five on the dry scale.
Damian: We're gonna measure how deep you measure it so we can have it controlled everywhere.
Ellen: Yeah.

Damian: Is this enough?
Ellen: What?
Damian: That much soil?
Ellen: Oh, I don't think we need that much, but yea, that's probably enough. Should I push it all the way down?
Damian: But don't push it if it doesn't go any more.
Ellen: No.
Damian: Oh, so the deeper you push it the moisture it's moist down there.
Roth: Did it change?
Damian: Yea, it increased.
Ellen: Yea, let's see where it is half way. Ok, halfway in.
Damian: It's still dry.
Ellen: Two point five. And all the way, is. . . .
(A few moments later.)
Ellen: We got a pattern, because we measured one on the top, the middle, and the bottom again, or the moisture, the top was like three, it was dry, and the middle was the average range around, or a little dryer, and the bottom was very wet. And then we tried it on the field too. And it wasn't as wet as on the bottom of the hill, not as wet.

In the context of conducting their research activity the students chose to focus on more than a single possible variable for making sense of the correlations they were investigating. They collected observations on the amount of grass and then on different distances down the slope took measures of moisture and pH as well as collected soil samples for observation back at the school science laboratory. This sequence mirrors the efforts of Sam and Jose, both of whom collected data for a substantial number of (possibly) significant factors they observed in their field site, some of which they only analyzed in the university laboratory at a later date. In this sequence Damian and Ellen were balancing the attempt to reduce the messiness of the real world (the multivariate condition) into a number of variables that they could still competently deal with but that would represent the complexity of the relationship they were studying (i.e., what abiotic features are related to the difference in the grass). At the same time, they recognized the importance of standardizing their methodology as moisture levels vary with probe depth. This, again, mirrors Sam's efforts at standardization.

These two students were by no means unique in the class—their work was similar to that of most of their peers. It was common for them to ask *do-able* questions (Fujimura, 1992), examine multivariate relationships, adapt and adopt (new) tools to their research, and generate inscriptions in their reports (usually with fewer problems than those of the science program graduates who engaged in a similar task in their science education methods course). This extended class activity was unusual in that it had students use textual (both spoken and written) and inscriptional (graphs, tables) means to persuade their peers and teacher about the validity of their findings. There was no known right answer to their questions, but rather they participated in an environment in which they were expected to construct convincing and compelling claims about natural relationships. The effects of this were notable.

This work with eighth-grade students has been analyzed with respect to their use of inscriptions, which determined that they had considerable competency in their interpretation of data and graphs (Roth, 1996; Roth et al., 1998). So, not only does engaging students in independent inquiry-oriented activities help them develop their skills at making linguistic distinctions, material scientific practices, and scientific ongoing concerns and standardized practices, such student engagement also develops competency at the practices embedded within those areas in a manner similar to the experiences of older students such as Sam and Jose. Numerous other studies have also noted that students who have extensive experiences interacting with phenomena and representational means increase their competency at graph interpretation (e.g., Mokros & Tinker, 1987).

Taken together, this research suggests that students need more than instruction on the mechanical aspects of constructing independent research and graphs, but need extensive experiences allowing them to translate between the real world and the inscriptional tools they are using to report their studies. Clearly, it is possible to engage students who are younger than those commonly allowed to attempt independent inquiry activities within which they can gain considerable competency at science practices. Although the eighth-grade students were in a somewhat special circumstance (an upper-middle-class private school with class size of 22 and a teacher with considerable science experiences), they nonetheless demonstrate that these activities both are possible and can develop science skills in younger students.

DISCUSSION

In the previous sections, I drew out some similarities between the activities of eighth-grade students and more advanced college students. Nevertheless, I argue that the eighth-grade students were not participating in a community of scientists because of structural differences between their classroom and scientific communities. Most notable among these is that their community of practice occurs within only a single setting (the school they were in) and that the discourse about (acceptable) practices was bounded within that school and did not interact with members in the broader science communities. However, the setting provided many opportunities analogous with science communities such as the asking of questions after immersion in the research setting, prolonged engagement, defense of methods and interpretations, adapting and adopting tools, and so on.

In contrasting these different research studies, implications for both the preparation of teachers and the structuring of science classrooms can be drawn. These implications reflect the overall position that interpretive perspectives are mediated by individual experience such that the interpretation of a scientific claim only ends up reflecting that of the original claimant if various experiential similarities exist between the claimant and the reader. Ultimately, this alters the view of scientific literacy from one of being able to identify or describe various scientific theories or essential content to the view that it encompasses being able to participate in the interpretation of scientific claims, understanding them in ways similar to those who constructed those claims. Superficially these may appear similar, but the reflexive interpretations of scientific claims and representations by those with fieldwork and research experience and how these contrast with those who do not have that experience emphasize that the distinction is an important one. Clearly, decontextualized symbolic mastery (i.e., ready-made science) of scientific practices and tools is not particularly useful in the absence of practical (i.e., science-in-the-making) mastery. Both are clearly necessary to develop an understanding of scientific concepts, claims, and practices and to allow reflection about the practices of science.

AVOIDING SCIENTISTIC ATTITUDES

However even having competency in both symbolic and practical mastery is, I would argue, insufficient for K–12 science education. Symbolic knowledge of the practices and claims, even combined with experiential knowledge of the practices, could contribute to developing *scientistic attitudes* in which students come to unquestioningly accept the claims and methods of science without taking a critical disposition toward them—a rigidity and pathology of thought that takes the practices of science and its findings as being inclusive of all knowledge. Even if we were to take the graduate experience as an authentic enculturation to scientific practices in K–12 settings, we would generate the same problems as are found there—the development of a *naïve doxa* (Bourdieu, 1992, p. 248) in which reflection on practices and implications of them little occur. Clearly this is not what is desired. An enculturation into science practices—a developed competency in tool use, language use, and interpretive stances—needs to be balanced with an enculturation into a critical reflection toward those same scientific tools. Teachers therefore need to engage students in communities of practice that develop both an understanding of the tools and concepts as well as a meta-awareness of the shortcomings of those tools to explain the surrounding world. Studies in ecology settings, such as described with the eighth-grade students in which natural world relationships were so contextual and inconsistent, arguably provide a foundation for demonstrating both how science tools can be used to investigate the world but also how difficult it actually is to use those tools to describe that same world. It is apparent that this awareness needs to be developed explicitly; otherwise one is left with students, such as the 4th-year student Jose, who attribute the difficulty in conducting research and drawing clear conclusions to their own methodological shortcomings, not to the inherent difficulty in doing so. Thus, science classrooms need to both engage students in activities of the sort described as well as engage them in meta-conversation about the implications of the practices.

REFERENCES

American Association for the Advancement of Science. (1993). *Benchmarks for science literacy*. New York: Oxford University Press.

Bastide, F. (1990). The iconography of scientific texts: Principles of analysis. In M. Lynch & S. Woolgar (Eds.), *Representation in scientific practice* (pp. 187–229). Cambridge, MA: MIT Press.

Bourdieu, P. (1992). The practice of reflexive sociology (The Paris workshop). In P. Bourdieu & L. J. D. Wacquant (Eds.), *An invitation to reflexive sociology* (pp. 216–260). Chicago: University of Chicago Press.

Bourdieu, P., & Wacquant, L. J. D. (1992). *An invitation to reflexive sociology*. Chicago: University of Chicago Press.

Bowen, G. M., & Roth, W. -M. (1998). Lecturing graphing: What features of lectures contribute to student difficulties in learning to interpret graphs? *Research in Science Education, 28*(1), 77–90.

Bowen, G. M., & Roth, W. -M. (1999a). Biology as everyday social practice: Toward a new epistemology. In *Proceedings of the 1997 History and Philosophy of Science and Science Teaching Conference*. Calgary, Canada: Alta.

Bowen, G. M., & Roth, W. -M. (1999b, March). *"Do-able" questions, covariation, and graphical representation: Do we adequately prepare preservice science teachers to teach inquiry?* Paper presented at the annual conference of the National Association for Research in Science Teaching, Boston.

Bowen, G. M., & Roth, W. -M. (2002a). Of lizards, outdoors and indoors: Constructing lifeworlds in ecological fieldwork. In D. Hodson (Ed.), *OISE Papers in STSE Education* (Vol. 3, pp. 167–191).

Bowen, G. M., & Roth, W. -M. (2002b). The "socialization" and enculturation of ecologists: Formal and informal influences. *Electronic Journal of Science Education, 6*(3). Retrieved July 15, 2002, from http://unr.edu/homepage/crowther/ejse/bowenroth.html

Bowen, G. M., & Roth, W. -M. (2002c). Why students may not learn to interpret scientific inscriptions. *Research in Science Education, 32*, 303–327.

Bowen, G. M., Roth, W. -M., & McGinn, M. K. (1999). Interpretations of graphs by university biology students and practicing scientists: Towards a social practice view of scientific representation practices. *Journal of Research in Science Teaching, 36*, 1020–1043.

Cobb, P. (1993, April). *Cultural tools and mathematical learning: A case study*. Paper presented at the annual meeting of the American Educational Research Association, Atlanta, GA.

Denning, P., & Dargan, P. (1996). Action-centered design. In T. Winograd (Ed.), *Bringing design to software* (pp. 105–120). New York: ACM Press.

Fujimura, J. (1992). Crafting science: Standardized packages, boundary objects, and "translation." In A. Pickering (Ed.), *Science as practice and culture* (pp. 168–211). Chicago: University of Chicago Press.

Latour, B. (1987). *Science in action: How to follow scientists and engineers through society*. Milton Keynes, UK: Open University Press.

Latour, B., & Woolgar, S. (1986). *Laboratory life: The social construction of scientific facts*. Princeton, NJ: Princeton University Press.

Lave, J., & Wenger, E. (1991). *Situated learning: Legitimate peripheral participation*. Cambridge, England: Cambridge University Press.

Law, J., & Lynch, M. (1990). Lists, field guides, and the descriptive organization of seeing: Birdwatching as an exemplary observational activity. In M. Lynch & S. Woolgar (Eds.), *Representation in scientific practice* (pp. 267–299). Cambridge, MA: MIT Press.

Lynch, M. (1985). *Art and artifact in laboratory science: A study of shop work and shop talk in a laboratory*. London: Routledge & Kegan Paul.

Mokros, J. R., & Tinker, R. F. (1987). The impact of microcomputer-based science labs on children's abilities to interpret graphs. *Journal of Research in Science Teaching, 24*, 369–383.

National Research Council. (1994). *National science education standards*. Washington, DC: National Academy Press.

Roth, W. -M. (1995). *Authentic school science: Knowing and learning in open-inquiry laboratories*. Dordrecht, Netherlands: Kluwer Academic.

Roth, W. -M. (1996). Where is the context in contextual word problems?: Mathematical practices and products in Grade 8 students' answers to story problems. *Cognition and Instruction, 14*, 487–527.

Roth, W. -M., & Bowen, G. M. (1995). Knowing and interacting: A study of culture, practices, and resources in a Grade 8 open-inquiry science classroom guided by a cognitive apprenticeship metaphor. *Cognition and Instruction, 13*, 73–128.

Roth, W. -M., & Bowen, G. M. (1999a). Complexities of graphical representations during lectures: A phenomenological approach. *Learning and Instruction, 9*, 235–255.

Roth, W. -M., & Bowen, G. M. (1999b). Digitizing lizards or the topology of vision in ecological fieldwork. *Social Studies of Science, 29*, 719–764.

Roth, W. -M., & Bowen, G. M. (2001a). Of disciplined minds and disciplined bodies. *Qualitative Sociology, 24*, 459–481.

Roth, W. -M., & Bowen, G. M. (2001b). Professionals read graphs: A semiotic analysis. *Journal for Research in Mathematics Education, 32*, 159–194.

Roth, W. -M., Bowen, G. M., & McGinn, M. K. (1999). Differences in graph-related practices between high school biology textbooks and scientific ecology journals. *Journal of Research in Science Teaching, 36*, 977–1019.

Roth, W. -M., McGinn, M. K., & Bowen, G. M. (1998). How prepared are preservice teachers to teach scientific inquiry? Levels of performance in scientific representation practices. *Journal of Science Teacher Education, 9*(1), 25–48.

Suchman, L. (1987). *Plans and situated actions: The problem of human–machine communication*. Cambridge, England: Cambridge University Press.

Traweek, S. (1988). *Beamtimes and lifetimes: The world of high energy physicists*. Cambridge, MA: MIT Press.

Wenger, E. (1998). *Communities of practice: Learning, meaning, and identity*. Cambridge, England: Cambridge University Press.

CHAPTER FOUR METALOGUE

Authentic Science Education: Contradictions and Possibilities

G. Michael Bowen
Wolff-Michael Roth
Randy Yerrick

Randy: Mike, you have described scientists teaching college science by orchestrating their courses in ways that lack the attributes of actual scientific work. Why do you think that scientists, who are so closely connected to the context of science research, do not offer experiences to novices that are more similar to their own?

Michael B.: This occurs for several reasons. The first I can think of is that many scientists view courses as a "filter" presumably separating those who will be able to do research and those who will not be able to do field work. I don't mean to suggest they are uncaring about this—but there are many individuals who start science (especially biology) programs given the number of jobs available. So it makes sense to "filter out" people—whereas you can have 350 students in your first-year lectures, you cannot have that many in your fourth-year seminars for simple financial reasons. The second reason I can think of is that it is just a reconstruction of "what worked for me." The courses many professors present are of the same nature as those that had worked for them in their own training.

Michael R.: That is, in their teaching practices professors not only produce new scientists but also reproduce scientists who are similar to them. Undergraduate education is an obstacle course, and those who succeed well will eventually become professors, themselves presenting obstacle courses because they think (somewhat tautologically) that this leads to the selection of "good" scientists.

Randy: I have seen many teacher–scientist and student–scientist collaborations in which it boils down to the patronizing relationship that

Sharon Traweek (1988) described in *Beamtimes and Lifetimes*. The scientists see themselves as the "knowers" and their collaborators as subordinates and "receivers" of their crumbs that fall from the table of knowledge. I am certain I am overgeneralizing here and I have seen a few counterexamples. Does it have to do with their socialization? Is it something like, "I went through it and they should, too?"

Michael R.: A sort of rite of passage that is maintained over time....

Randy: Or do you suppose these scientists are not seeing the need to resolve the dichotomy between professed ideologies and actual teaching experiences with novices?

Michael B.: My experience both as a trained biologist and as an ethnographer of biology suggests that the enculturation process in biology is not as brutal and competitive as it is in physics. My impression from talking to fellow students majoring in physics is that Traweek presented a common experience in that field. Furthermore, I am not sure that professors see their role as preparing people who *want* to be biologists, but rather preparing those who *deserve* to be biologists—those who can best adapt, can best deal with the workloads and so on.

Michael R.: A little tongue in cheek I might say, according to the motto "survival of the fit."

Michael B.: Yes, but there is lots of room to talk about the aspect of "deserve" and whether the implicit and explicit filters enacted are actually the best ones.

Randy: Do you believe that most scientists are unaware of the differences in their walk and talk? That is, are they unable or unwilling to see the schism between opportunities available to their students and science experiences in the research laboratory?

Michael B.: They are unaware of it in many ways. Again, it is the old "what worked for me" thing. But, in addition, class sizes are often huge making any other teaching method than lectures impossible. I know professors who hate lecturing, who think that lectures do not work all that well, but who have little or no choice. They do the best they can in lecturing.

Michael R.: But this, lecturing, nevertheless, is an anachronism. They worked at a time when books and the Internet were not freely available, about 800 years ago. Professors lecture facts, tell students facts about concepts and theories, but these are available in many books and online.

Randy: Let me stop you and pursue the issue of facts, or as Mike suggested, claims that are presented. Students get to hear about facts and claims but very little about the actual practices that lead to those facts and claims, which have been simplified for them. I agree with the point that scientists' experiences need not be so separate from the students' experiences in the ontological and epistemological component of scientific knowledge production. Teachers and researchers need to know about science from scientists, but rarely is *this* question asked directly: "What do scientists need to know about teaching from teachers?"

Michael B.: My answer is both "lots" and "nothing." Fourth-year students, at least in the different Canadian universities with which I am familiar, have a lot more contact with professors and their storytelling than do first-year students. That is a consequence of several factors, an important one of which is that fourth-year classes typically comprise between 20 to 25 students.

Michael R.: Of course, administrators tell us that we cannot afford to operate teaching classes that have such low average sizes.

Michael B.: I have always suspected that the only way universities could afford the fourth-year courses is by having the first-year ones with 350 students; many of these students tend to drop out by the fourth year. So, if the question is, "Could first-year courses be dramatically re-tooled to do a better job knowing what we understand in our discipline?" The answer is of course, "Yes." However, given current funding priorities, I cannot see that it would make sense to do so. Most universities have fourth-year science courses that offer the kind of "authentic" experience we talk about. In most universities, fourth-year students also have the option to do an honors thesis project in which they learn how to do "real" research. Most students, however, choose not to take that course.

Michael R.: I know that one of your major gripes is with what you conceive of as a poor preparation of science teachers to teach inquiry, especially inquiry in which future generations of university students, in the sciences or humanities, could get a taste of how research actually gets done.

Michael B.: My main concern in writing about the discontinuities in undergraduate education is not to suggest that it needs a reform in how it produces biology undergraduates. Rather, my concern is to try and understand the insufficiencies of these programs with respect to science teacher preparation. We have defined our needs in very specific ways with respect to the different forms of knowledge and competencies we would like to have our teachers possess.

Michael R.: So there are really two issues that we are concerned with here. One concerns science education, whether for future scientists or science teachers, which provides some experiences of what everyday scientific praxis is like. The other concerns how teacher education programs need to be structured to provide necessary experiences more specific to science teachers.

Randy: Pertaining to the first point, Mike showed that the learning of science cannot be divorced from the context in which it is taught. Let me twist this a bit. Students often engage as students with an implicit contract (syllabus) and teachers/professors are having a temporal, truncated, immersive, and somewhat estranged engagement with their students. There are real constraints in knowledge and resources, common experiences, and interests on both parts. As professors we have all seen the minimalist, negotiating nature many students of science bring to college from their public school socialization. How are we to reconcile the two distinctly different timelines, expectations, rewards, support, and other attributes that are clearly contrasted by classrooms and well-funded laboratories? What specific advice can be given to scientists as baby steps toward more authentic engagement within such parameters?

Michael B.: I'm not sure I have "specific advice" in the way you mean. I think there are many different ways of getting there, and the way I get there with my own classes varies substantially by class, topic, and so forth.

Michael R.: What about my second point?

Michael B.: One notable thing is that faculties of education are in a position to drive the system somewhat. My experience in Canadian universities is that those going into education do not take the courses that have independent inquiry components. Faculties of education could require this, which would lead to more science teachers having some preparation in doing independent work, a basis for teaching independent inquiry work. In my secondary science methods classes this year I had several students who talked about having participated in relevant activities for becoming and thinking like a biologist—doing an honors thesis project, volunteering as research assistants. They also were usually my best students in all aspects of the methods course. I think that this is in part because they had many relevant experiences that they could take as objects of professional reflection.

REFERENCE

Traweek, S. (1988). *Beamtimes and lifetimes: The world of high energy physicists.* Cambridge, MA: MIT Press.

CHAPTER FIVE

Dialogic Inquiry in an Urban Second-Grade Classroom: How Intertextuality Shapes and Is Shaped by Social Interactions and Scientific Understandings

Maria Varelas
Christine C. Pappas
University of Illinois at Chicago

Amy Rife
Chicago Public Schools

In this chapter, we explore the nature of dialogic inquiry that takes place in Amy Rife's second-grade urban class as children and teacher engaged in integrated science–literacy experiences during a unit on states of matter and changes of states associated with the water cycle. We pay particular attention to the concept of intertextuality—the juxtaposition or reference to other texts. Elsewhere, we have described a typology of intertextuality, providing examples of the types that are possible in such units to support children's scientific understandings (Pappas & Varelas, with Barry & Rife, 2003). Here, we focus on intertextuality in more depth. Using an excerpt of a read-aloud session—the third in the unit—we examine the ways in which instances of intertextuality contribute to the acquisition of science and literacy. The classroom talk, as dialogic inquiry, enabled this urban class to appropriate ideas, extend them further, question their validity, and even generate new ones.

BRIEF THEORETICAL BACKGROUND
OF THE LARGER PROJECT

In a larger collaborative university-school action research project, we study how culturally and linguistically diverse, low-socioeconomic-status (SES) first- and second-grade children make sense of scientific ideas in the context of lessons that include read-aloud sessions of information books on a science topic, related hands-on explorations, and a variety of writing opportunities. The major theoretical framework of our work is a sociocultural, constructivist one derived from Vygotsky (1934/1978, 1987) that argues that children do not reinvent scientific inquiry and literacy understandings all by themselves, but instead learn by encountering concepts, ideas, procedures, and strategies that have been established by others over the course of time. In the best of circumstances, children use their own reasoning to make sense of these sociocultural achievements, and at the same time they are influenced by them to reorganize their understandings.

An important characteristic of scientific practice is the interplay between theory and data—between developing a network of concepts and processes that are logically linked and have explanatory power, and examining empirical evidence collected through observations and experiments (Varelas, 1996). Literacy is integral in this enterprise. In the ongoing process of scientific inquiry, written texts often serve as important cultural tools as scientists grapple with the ideas, thoughts, and reasoning of others (Goldman & Bisanz, 2002). An important feature of literacy is the expression of these scientific concepts and reasoning via particular linguistic registers or genre (Halliday & Martin, 1993; Lemke, 1990, 1998). That is, various activities of "doing science" are realized by certain discourse forms (Halliday, 1998). For example, concepts are expressed by generic nouns; processes are typically realized by timeless, present tense verbs that are frequently found in relational clauses; both concepts and processes are articulated by technical vocabulary (as opposed to everyday wordings); and so forth (Halliday, 1998; Martin, 1990; Pappas, 1993). Thus, creating scientific understandings entails constructing new conceptual entities and the wordings to express those entities (Ogborn, Kress, Martins, & McGullicuddy, 1996). As Sutton (1992) argued, learning science is based on the linkage between a "new way of seeing" any science topic and a "new way of talking" about it. Our chapter provides insights into how young urban children might take on this new seeing and talking.

Teachers have an important role in facilitating children's understandings as they engage in scientific/literacy practices (Becker & Varelas, 1995; Driver, Asoko, Leach, Mortimer, & Scott, 1994; Pappas & Zecker, 2001; Varelas, Luster, & Wenzel, 1999; Wells, 1993, 1999). A critical part of their role is to nurture, support, and orchestrate the classroom discourse in curricular activities—the being, doing, and saying in this community (Gee & Green, 1998; Green & Dixon, 1993). Thus, language is seen as a major semiotic tool to mediate intellectual activity and knowledge building (Halliday, 1993; Varelas & Pineda, 1999; Vygotsky, 1934/1987; Wells, 1999; Wertsch, 1991).

Bakhtin's (1981, 1986) ideas on dialogism especially inform our framework. For him, discourse is a social activity within which participants take turns offering utterances that are responsive to each other. Because discourse is a continual weaving and reweaving of responsive utterances, the meaning of any one utterance is unstable—each depends on the discussion in which it emerged. Moreover, there is conflict in this endeavor (Cazden, 1992). As Bakhtin (1981) noted: "Language is not a neutral medium that passes freely and easily into the private property of the speaker's intentions; it is populated—overpopulated—with the intentions of others. Expropriating it, forcing it to submit to one's own intentions and accents, is a difficult and complicated process" (p. 294). Dialogically oriented instruction, such as the one that teacher researchers in our collaboration bring to their classes, moves away from the traditional teacher-controlled three-part IRE pattern of classroom discourse—Initiation/Response/Evaluation (Cazden, 2001). It makes possible "the joint construction of a new sociocultural terrain, creating spaces for shifts in what counts as knowledge and knowledge representation" (Gutierrez, Rymes, & Larson, 1995, p. 445). Thus, both teacher and child voices are privileged in the classroom talk—as children grapple with new scientific ideas and try out how to express them, teachers work to contingently respond to, or take up, children's efforts to sustain and further extend them (Nystrand, 1997; Pappas & Zecker, 2001; Wells & Chang-Wells, 1992).

THE ROLE OF INTERTEXTUALITY IN DIALOGIC INQUIRY

As teacher and students attempt to build common knowledge as a community, they capitalize on connections they make between texts and their lives, among texts, between texts and classroom activities, and so forth. Therefore, intertextuality (Bloome & Bailey, 1992; Bloome & Egan-Robertson, 1993; Lemke, 1985) becomes an essential process that allows children and the teacher to make sense of ideas, develop understandings, and further expand and elaborate on them. Intertextuality means juxtaposing of texts. In our work, it means making sense of "texts" from other contexts that children and teacher bring to and instantiate in classroom discourse. Using ideas from Wells (1990; Wells & Chang-Wells, 1992), "text" is conceived of in an expansive way—it is more than another book that a child or teacher might refer to. As Wells (1990) argued, "it is heuristically worthwhile to extend the notion of text to any artifact that is constructed as a representation of meaning using a conventional symbolic system" (p. 378). Thus, texts may be algebraic equations, scientific formulae, diagrams, charts, musical notes, movies, TV shows, and a range of oral texts created in various speech contexts, including speakers' "recounts" of previous events or experiences. Furthermore, usually occasions of intertextuality occur by participants juxtaposing a present text with a *prior* text. Our framework further extends what others have considered as intertextual links by including the possibility of intertextuality as a future phenomenon, where a speaker refers to a future "text."

As already noted, in prior writings (Pappas et al., 2003), we have identified the various forms and functions that intertextuality can take. Our typology is quite robust. Category I includes references to various more traditional texts; other texts orally shared, such as poems, rhymes, sayings, songs; other media, such as TV/radio shows or movies; and prior classroom discourse. Category II involves connections to hands-on explorations. Category III covers the recounting of events—specific events that participants relate, as well as generalized ones that participants report on as habitually occurring. The last Category IV is what we call "implicit" generalized events where there is no explicit personal involvement or recounting, but participants implicitly refer to events that could or should be habitually experienced. In Table 5.1 we describe these categories and offer examples from first- and second-grade classes (see pp. 166–168). Shortly, in the excerpt of a read-aloud session in Amy's second-grade class, we illustrate many of these intertextuality types and show how they, along with other participant contributions, create a dynamic webbing of ideas. These types of intertextual links are vivid instantiations of life for the children and the teachers.

From a socioconstructivist perspective, to establish intertextuality, a juxtaposition or reference to another text must be proposed, interactionally recognized, acknowledged, and have social significance (Bloome & Egan-Robertson, 1993). However, sometimes this last criterion, the social consequence of intertextuality, might not manifest itself until later in the discourse. At particular times, the meanings created, the social identities formed, the ideas given significance, and so forth, influence the direction and shape of discourse. As participants communicate with each other, they act and react on ideas and meanings. These actions and reactions are not necessarily taken up immediately and in a linear fashion.

Examining intertextuality allows us to appreciate the funds of knowledge (Moll, 1992) that young urban children bring to the class along with the teacher's role in legitimizing and using these funds to facilitate the building of new understandings or elaborate prior understandings. In dialogically oriented instruction, intertextuality takes place as a negotiated dance among teacher, children, and texts in the construction of knowledge (Oyler, 1996). It is a complicated endeavor.

In this chapter, we illustrate this complexity—we explore which texts were juxtaposed, at what levels, for what purposes, and by whom. Implicated in this analysis are the ways in which teacher and children shared authority for directing and constructing the knowledge of the classroom, how they made meaning and deepened their knowledge of each other and of science. In other words, we examine how intertextuality allowed both children and teacher to contribute to the knowledge building as knowers of and participants in science. In dialogic inquiry, though, teacher and students offer many comments and initiations, and not all of them are intertextual links. Some participant contributions do not mark the connections between their understandings and personal experiences, thus we do not label them as intertextual links. However, they may be important in understanding the context and flow of the language and thinking that unfolds in classroom discourse, and they may be "critical moves" that spur or are spurred by intertextual links. Thus, we do not examine intertextuality in and of itself but rather how instances of intertextuality are embedded in the classroom discourse as a whole.

CONTEXTUAL BACKGROUND

Rife's second grade class is in an urban elementary school with a diverse ethnolinguistic student population, including African American, Latino/a, and Anglo children, where most of the children are eligible for federally sponsored food programs. There were 28 children in the class at the year of this study.

As part of the larger project, Rife taught three integrated units. During each unit, children engaged in science activities and whole-class discussions of them; participated in read-aloud sessions in which a range of children's illustrated information books on the topic being investigated were shared; explored and shared the findings of small-group literature studies on information books; participated in at-home explorations that they reported on in class; wrote and drew in science journals their observations and findings of experiments and other unit activities; and made their own illustrated book at the end of the unit. This chapter focuses on the middle unit in the year, the unit "States of Matter and the Water Cycle." The theme of the unit centers on characteristics of matter in different states, changes of states of matter, and how they take place, and how these changes are related to how rain is produced. Discussions on weather provided one of the contexts for thinking about the latter.

As the year was unfolding we collected various types of data: audio- and videotapes of the lessons (science activities and their whole-class debriefings; and whole-class read-aloud sessions of informational books); written artifacts created by the teachers and students; various children's drawings and writings related to the integrated inquiries; audiotapes of project staff meetings; teachers' research journals; and classroom field notes taken by university-based researchers. We use a qualitative, interpretive method with a range of ethnographic techniques (Lincoln & Guba, 1985; Wolcott, 1994) to look across all these data. Denzin and Lincoln (1994) noted that qualitative research and analysis is especially well suited to "seek answers to questions that stress how social experience is created and given meaning" (p. 4). For the excerpt of classroom discourse that we examine below, we used detailed discourse analyses drawing on approaches from the systemic functional model (e.g., Halliday & Hasan, 1985) and ideas on intertextuality (Bloome & Egan-Robertson, 1993; Fairclough, 1992; Lemke, 1992). Using an iterative constant comparative method (Glaser & Strauss, 1967), we compared our analyses to create a unified set of interpretations about the discourse.

ANALYSIS OF CLASSROOM DISCOURSE: INTERTEXTUALITY IN ACTION

We present our analysis of a relatively lengthy excerpt from the third lesson of the unit. The excerpt lasted 16 minutes and took place about 30 minutes into the lesson. The teacher was reading to the class the book *What Do You See in a Cloud?* (Fowler, 1996). She was already 9 minutes into the read-aloud of the book when the discourse excerpt took place. (Transcription conventions are explained in the Appendix.) We italicize intertextual links in speakers' turns, and we double-underline utterances—critical moves—that are related to or are significant relative to the intertextual links.

Water Vapor and How Clouds Are Formed:
Making Sense of Book Ideas

1 Amy: Alright, let's keep going. I want to talk about this page [p. 12] a little bit. Look carefully at this picture. It says, IT RISES IN THE FORM OF VERY TINY DROPS // IT RISES IN THE FORM OF VERY TINY DROPS CALLED WATER VAPOR.

2 Cs: What's water vapor?

3 Amy: Water vapor, that's a new one, huh. A new one we haven't really talked about. Water vapor, tiny drops. *Remember what um Timothy was saying about the water going up? And then Raoul told us that it's called evaporation.* How did you learn that word evaporation?

4 Raoul: *From the Magic School Bus.*

In unit 3, Amy initiated an intertextual link to prior classroom discourse as a response to several children's question (unit 2) that was triggered by the written text. Amy acknowledged the question, positioned it as a legitimate one, and proceeded to engage the children in thinking about it. She reminded children of two of their peers' earlier contributions hinting

5. INTERTEXTUALITY, SOCIAL INTERACTIONS, AND SCIENCE

their importance and foregrounding these children's voices. As Raoul answered Amy's question about how he knew the word evaporation, he framed an intertextual link to a different semiotic medium—a TV program.

5	Amy:	From the Magic School Bus. You heard Ms. Frizzle talking about evaporation. And when that water goes up tiny little drops called water vapor. YOU CAN'T SEE WATER VAPOR // Shh // YOU CAN'T SEE WATER VAPOR, BUT IT'S ALL AROUND YOU. Even right now. THE WATER VAPOR COOLS OFF // THE WATER VAPOR COOLS OFF AS IT RISES, OR MEETS COLD AIR, OR PASSES OVER COLD LAND OR WATER. WHEN THE WATER VAPOR HAS COOLED ENOUGH, THE DROPLETS COME TOGETHER {TO} MAKE CLOUDS. Wait a minute, Elena said something. Elena, what were you going to say?
6	Elena:	<It's> *like a little bit of what Timothy said.*

Elena initiated another intertextual link to prior classroom discourse—Timothy's contribution, a contribution that was just noted by the teacher. Elena's intertextual link contributed to the social significance and recognition of the intertextual link brought up previously by the teacher. Elena's contribution would be acknowledged by Amy and elaborated.

7	Amy:	A little bit of what Timothy said. He said they come together. Right. He said they make bubbles. But we said they kind of come together. It says THE DROPLETS COME TOGETHER AND {THEN THEY} MAKE {THE} CLOUDS. Okay, so what are clouds made of?
8	Cs:	Water!
9	Amy:	How does that water get up there?
10	Cs:	(*** ***)
11	Amy:	Shh, one at a time. . . . Shh. What?
12	C1:	<It evaporates>.
13	Amy:	It evaporates. And we call that water vapor, those little tiny drops that come up [...m...]. Alright, let's talk about it a little more. {SO IT RISES UP} TINY DROPS CALLED WATER VAPOR. Alright, and those little tiny drops come together and they make clouds. YOU CAN'T SEE WATER VAPOR, BUT IT'S ALL AROUND YOU. {AND IT} COOLS OFF AS IT RISES. Okay, and those little drops come together. Yes, Diego.
14	Diego:	<How come> // how does the water get out // get out of the <houses when it's full>?

As Amy and the children were engaged in reading the book, Diego offered a question, a critical move that was triggered by the written text. The book has a picture of a winding river and land, with clouds in a blue sky.

So, Diego asked about a different kind of situation—how does water (water vapor) get out of presumably closed-in houses? His question will be acknowledged by Amy and Roberto, and will spur Roberto's intertextual link in unit 18.

15	Amy:	Ahh.
16	Roberto:	<That's> was my other idea.
17	Amy:	Your other idea.
18	Roberto:	If you take a leaf // the idea that (*** ***) // <if> *the water* // *if you take* // *pull off a leaf off of a plant**...
19	Amy:	Uh huh.
20	Roberto:	*And it has water on it...if you keep it somewhere where the water can stay on in the house how would it come up?*
21	Amy:	How does it get out of the houses? Alright, so he wants to know how water gets out of the houses. What do you think? He said // we'll think // like the water that's in your houses, how does it get out of the houses?
22	Elena:	Steam.
23	Brittany:	If you open the window.
24	Amy:	Hmm?
25	Brittany:	If you open the window.
26	Amy:	Oh, you think opening the window lets it out. Pamela, what do you think?
27	Pamela:	Huh?
28	Amy:	How does the water get out of houses?
29	Pamela:	The water get out of houses. Maybe um the people when they open the windows maybe it comes out.
30	Amy:	Okay, maybe when we open our windows it comes out. What else? Elena.

Roberto, claiming that he shared the same idea as Diego, contributed in units 18 and 20 an intertextual link in the form of an implicit generalized event, namely, referring to a leaf having water on it in closed-in houses from which water may not be able to escape. Roberto used the general "you" structure to implicitly indicate a possible experience participants might have had. Category III of intertextual links includes two major categories of recounting of events—specific and generalized. Intertextual specific recounts include instances where speakers relate short narratives about their own (or related-to-them persons') experiences; generalized recounts are hybrid accounts (which include both narrative and informational linguistic features) where they tell of typical, habitual behaviors, actions, or experiences. Roberto's intertextual link is an example of Cate-

5. INTERTEXTUALITY, SOCIAL INTERACTIONS, AND SCIENCE 147

gory IV and it is "fuzzier" than the recounting of events. Moreover, this implicit generalized connection moves away from either narrative or hybrid registers to be expressed more fully via scientific language. In terms of content, Roberto's contribution gave further significance to Diego's question in unit 14. Amy recognized Roberto's intertextual link by revoicing it and involving the rest of the class in answering his version of the question. This is one form of sharing power in Amy's classroom community—a combination of the teacher using her power and status to foreground children's contributions and acquisition of their power by shaping the meanings negotiated and discussed.

How Is Steam Made? Beginning to Answer Elena's Question

31	Elena:	How does water make steam?
32	Amy:	How does what?
33	Elena:	Water make steam?
34	Amy:	How does water make steam? Good question. What do you think? Who has an idea? How does water make steam? [...m...]
35	Pamela:	I um // *my mom she always makes tea early in the morning.*

As the class was pursuing Diego and Roberto's question about how water gets out of houses, Elena's contribution "steam" (unit 22) was not taken up by the group at the moment. She came back to her proposition attempting to develop meaning—"how does water make steam?" (unit 31). Elena's critical move—in the form of a question—will spur several intertextual links as we will see below. First, as Amy encouraged the class to take up Elena's question, Pamela contributed her intertextual link in the form of a recounting of a generalized event—an everyday experience, a habitual action, where another person related to her—her mom—was involved. Pamela attempted to make meaning in the context of that habitual action, namely making tea involves water making steam (as will be elaborated upon below).

36	Amy:	She makes tea.
37	Pamela:	Yeah.
38	Amy:	Okay, tell us about it.
39	Pamela:	*So um she has this pot and she cooks it over the stove and she um ha // has like cold water*...*
40	Amy:	Uh huh.
41	Pamela:	*And she um has // she boils it up with fire*...*
42	Amy:	With fire. So she puts the water over fire from the stove and she boils it. What does that mean when water boils just in case somebody doesn't know?

43 Pamela: Um... ...
44 C1:   <u>It means it's hot.</u>

Pamela's intertextual link in unit 35 continued to be extended and recognized. Pamela offered elaborations in units 39 and 41, bringing up scientific and colloquial registers that would help the class elaborate on how water turns to steam—"cooks it over the stove," "boils it up with fire." Amy kept probing the class for linking scientific registers, such as, boiling, with these evolving ideas about steam. Children eventually explicitly associated boiling with higher temperature (unit 44).

45 Amy: Okay, it gets hot. Keep going tell me more about it.
46 Pamela: *Ah // so um // it's a // she always // today she like burned the tea...so it // it um // it has like bubbles <going> everywhere and I was talking to her and then she had the tea it // it was popping out (*** ***) and then she had to hold it this way [Pamela gestures as if she is pulling a tea kettle off the stove] and it was coming on the stove and it looked orange <#and there# was steam> so*...*
47 C1:   #Oooh!#
48 Amy: You saw steam. Describe the steam to us. Tell us about the steam.

Pamela's intertextual link that started in unit 35 continued to acquire recognition and significance. Amy encouraged the class to "keep going" and Pamela continued her intertextual link, but shaped it now to a recounting of a more specific event—today her mom burned the tea (unit 46). It was in the context of this specific event that Pamela made the connection back to steam—the concept that Elena's question pushed the class to consider. And it was in the context of the burning of the tea that Pamela used scientific registers ("bubbles going everywhere," "popping out") that she associated with the process of water making "steam." We see in this exchange the underlying assumption that water needs to reach high temperatures in order to make steam. Maybe Pamela thought that this was not what necessarily happened when her mom just made tea without burning it. Of course, Pamela used the verb "boil" in unit 41 before she brought up the specific case of burning the tea. Boiling is a technical register that as another child said (unit 44) means hot, but maybe that wasn't the meaning that Pamela intended when she used it.

49 Pamela: It had bubbles on the water.
50 Amy: Bubbles on the water. And how would // how did you // how would you describe the steam?
51 Pamela: Um it was hot and, I almost burned myself.

52 Amy: Yeah, you have to be careful with that, right. *[Amy points to Elena.]* So you're asking how does water turn to steam? Elena, is that your question?

As Amy continued to probe the children to articulate their thinking (unit 48), Pamela articulates that bubbles were on the water, and that steam is hot and can burn somebody (units 49 and 51). Another intertextual link was initiated by Amy in the form of referencing prior discourse and a past discourse participant (unit 52) as Amy turned back to Elena's question—a contribution continued to be recognized. Amy is the first to introduce a new register in the discourse—she referred to water turning to steam rather than making steam that was used so far as introduced by Elena's question. Necessary differentiations were being implicitly introduced as technical, scientific language was replacing colloquial language.

Considering Condensation Situations

53 Elena: Yeah.
54 Amy: How does water turn to steam? What do you think? [...m...]
55 Elena: I think since it's hot // since it's boiling and...it gets real hot and while the bubbles are coming up // while the bubbles are coming up *like when I go to the shower and I have the water real hot the mirrors they have like little spots on them.*

As Amy explicitly asked for Elena's way of thinking of water turning to steam, Elena attempted to put together ideas from Pamela's intertextual link (it's hot . . . it's boiling . . . bubbles are coming up), but very quickly turned to another intertextual link that she initiated—a recounting of another generalized event, the fogging of bathroom mirrors after a hot shower. Giving children the space and time they need to make sense of scientific ideas contributes to establishing a classroom community where teacher and children share the floor and the meaning making process, where voices of different participants dominate the discourse at different times, and where the children's own experiences become valuable funds of knowledge that shape the discourse and the learning.

56 Amy: Oh, that's interesting. Elena says when she takes a shower and it's really hot in the room the mirror gets little bubbles on them // on it. How many have ever*...
57 Cs: (*** ***)!
58 Amy: It has what on it? Let's talk about this. Shh. Let's talk about this. *Like the car window sometimes you see this.*

As Amy acknowledged and gave significance to Elena's intertextuality, she (possibly echoing a child's or children's contribution in the inaudible unit 57) initiated yet another intertextual link, recounting a very similar but different generalized event, the fogging up of car windows (unit 58).

59	Cs:	(*** ***)
60	Pamela:	<And you can't see in the mirror>.
61	Amy:	And then you can't see in the mirror. What is that on the mirror?
62	Pamela:	It's…kind of like part of clouds or something.
63	Amy:	Kind of like what? Part of clouds.
64	C1:	Water.
65	Amy:	It's water.
66	Pamela:	When like // it's like a baby cloud.

Pamela (in unit 60) returned to Elena's intertextual link—the fogging up of the bathroom mirror, thus contributing to its social significance. As Amy pushed for further elaboration of what fogs up the mirror, Pamela described the mist on a mirror as a cloud, echoing ideas discussed in the read-aloud book (i.e., the idea of droplets making clouds), although she did not explicitly make an intertextual link. Pamela was holding on to her metaphor—as another child offered that the mist was water, Pamela described it as a "baby cloud" (unit 66). This is another example of how children in Amy's class negotiated power and voice.

67	Amy:	It's like a baby cloud on your mirror, okay.
68	Pamela:	Yeah, baby clouds but not // but not really and…it's not hot but um it just comes on your mirror. *You have to like wipe it with something or just leave it, and then it'll get dry.*
69	Amy:	Or if you leave it, it will dry up?
70	Pamela:	Yeah.
71	Amy:	Where do you think it goes if you leave it? […m…]
72	Pamela:	Um I think it goes…
73	Amy:	So we're talking about that // the water that gets on your mirror when you take a hot shower. Where does that go if you just leave it, if you don't wipe it off?
74	Pamela:	Maybe it like goes everywhere trying to get out of the house.
75	Amy:	Trying to get out of the house, okay.
76	Pamela:	Yeah, but like nobody opens the door that much.
77	Amy:	Okay.
78	Pamela:	So maybe it like goes up through the housetop through the roof and it comes up in the air.
79	Amy:	Comes up in the air. Okay, so you think that water kind of evaporates like the water that's going up. Okay, Martin.

5. INTERTEXTUALITY, SOCIAL INTERACTIONS, AND SCIENCE   151

As Pamela continued to think aloud about the fogging up of the mirror, the class continued to give recognition to Elena's intertextual link. In unit 68, Pamela contributed yet another link to an implicit generalized event—drying off of the mist. The idea of drying off was picked up by Amy increasing its social significance as the class was building some common knowledge. As Amy kept pushing for more thinking, Pamela (unit 74) echoed earlier discussions that stemmed from Diego's question about how the water got out of houses—a critical move that contributed to the significance of Diego's contribution. As Pamela came up with her own colloquial description, Amy (unit 79) brought up the technical register that the class had encountered earlier ("so you think that water kind of evaporates like the water that's going up").

80	Martin:	I have a question!
81	Amy:	Okay, Martin.
82	Martin:	*I think that when I'm taking a bath and // and with hot water // I think when it goes down and when it comes on the mirror.*
83	Amy:	What comes on the mirror?
84	Martin:	The wind.
85	Amy:	The wind?
86	Martin:	*Yeah // what // like whe // the water // when the water comes down and then the wind comes up. If it go down the wind comes up.*
87	Amy:	The wind comes up?
88	Martin:	Yeah.
89	Amy:	Okay.
90	Martin:	*Like if it's <dropping> down because it's like this [Martin gestures up and down with his hand], it comes da da da da da da.*
91	Amy:	What do you mean, out of the faucet?
92	Martin:	Yeah, it's like [Martin continues gesturing up and down] da da da da.
93	Amy:	Okay, and it splashes down.
94	Martin:	[Now he gestures in an upward motion indicating water coming up.]
95	Amy:	And you think some of it comes up? You think that's how it gets on the mirror?
96	Martin:	[Martin nods.]

As the class discussion was unfolding, Martin, with Amy's help, contributed an alternative explanation about how mirrors get "water" on them. Martin used his prior experience with water coming out of a faucet splashing on a mirror to explain what Pamela had called "baby cloud." His intertextual link in the form of recounting a personal generalized event offered the context for his theorizing. Multiple meanings had been created so far that Amy was leaving as tentative and open to revision—another characteristic of dialogically-oriented instruction that encourages children to build their own identity as science learners.

Reexamining Steam Formation

97	Amy:	Okay. Have we answered Elena's question yet? She said how does water turn to steam? Have we answered that question yet? Adam what do you think?
98	Adam:	<Cause the>*...
99	Amy:	You guys are working hard. I can tell everyone's thinking so hard. Some of these are tricky. What do you think Adam?...Shh. We have to be patient and give people a chance to think it through.
100	Adam:	[Adam's hand shoots up then he struggles as he tries to answer scratching his head.] <Sometimes...when um>...something's really hot you can see the steam coming up.
101	Amy:	Okay, when something's really hot you can see steam coming up from it. You mean // what kind of things? Give me an example. ... ... [No response from Adam.]
102	Amy:	Where have you seen that Adam?... ...[No response from Adam.]
103	Amy:	Where have you ever seen steam coming up from something?
104	C1:	Coffee.

In unit 97, Amy initiated another intertextual link to prior classroom discourse, asking whether the class had answered Elena's earlier question, thus continuing to give significance to Elena's contribution. Amy invited Adam to offer his ideas. Adam, offering an intertextual link in the form of an implicit generalized event, shared that steam comes from something hot. Amy, with her critical move in unit 101, probed Adam to give an example, thus pursuing his link further. As another child came in the discourse when Adam was not answering the teacher's request, a different liquid (coffee) was brought up, obviously one that children would have seen and smelled in their everyday life—an "object" of generalized action.

105	C2:	#Water.#
106	Adam:	#Coffee.#
107	Amy:	Coffee, water...on the stove. Okay, hot water, how does that happen? Martin. How does that happen? I've heard a lot of people say coffee. I've seen the steam coming up from coffee, I've seen the steam coming up from water on the stove. How does it hap-

pen? How does that happen? What do you think Brittany? Please, one at a time.

108 Brittany: *<This one time when my cousin was visiting> he asked me how // how come that steam was coming out of the water // how come that steam was coming out of the hot water because um and I told him because um my mom // my mom was making something // I forgot.*

109 Amy: You forgot. Okay. Does anybody know? Does anyone have an idea? Why does that happen? Roberto.

Amy accepted the children's bringing up coffee as an example of where they see steam coming up from, and with her usual teacher moves, she pushed the class for further elaboration. Brittany (unit 108) tried to think in terms of yet a different intertextual link that took the form of a recounting of a specific event where she and her mom and cousin were involved. The link did not go much further as Brittany "forgot" her point.

110 Roberto: *My mom she lets me cook (***). When I cook them I just // I wait right there and I // I always // I stand on the chair and look at the water*...*

111 Amy: Uh huh.

112 Roberto: I just look at the water and then I see all these bubbles come up. I see bubbles just popping*...

113 Amy: You see bubbles popping in the water?

114 Roberto: *But when I cook hot dogs // when I cook hot dogs...when I first made em it got burnt...so then // so then um I didn't see bubbles floating up or anything like Pamela said like the tea um...I didn't see bubbles coming up*...*

Roberto offered his own intertextual link in the form of recounting a generalized event. As the discourse unfolded it became clearer that Roberto was talking about cooking hot dogs in boiling water. Eventually, in part of unit 114, Roberto switched to recounting a specific event—his first time when he burnt hot dogs and he could see bubbles. That was the place that another link to prior classroom discourse occurred—Roberto referred to Pamela's recount about her mom making tea. Roberto signaled a difference between the two cases—his burning of hot dogs had not produced "bubbles floating up" (i.e., steam) which is different from Pamela's burning tea that involved "orange" steam, although Roberto did not make this difference explicit.

115 Amy: Okay.

116 Roberto: *But I keeped on seeing bubbles in the water*...*

117 Amy: Uh huh.
118 Roberto: *But now <when I make> // it was that...when // when it burned it wasn't...// it wasn't // it didn't have steam...<in them or> // now // now I know how to cook hot dogs better um it has steam in it.*
119 Amy: Where do you think that steam comes from? What do you think? <u>We have some great ideas here.</u> I // <u>you're really doing a great job of thinking it through.</u> [...m...] Where do you think that steam comes from? What do you think?
120 Roberto: *It's like // it's like the bubbles that Pamela said when it comes up like you know it's like the mirror.*

In unit 118, Roberto further explicated his contrast with Pamela's contribution. At first, there was no steam when burning hot dogs, but now he cooks hot dogs better so there is steam (like Pamela's example). Amy then offered explicit social recognition and acknowledgment to the ideas that the children were bringing up, most of which had been intertextual links. As Roberto continued to contribute so did the intertextual links. Roberto (unit 120) kept referring to Pamela's contribution about bubbles, but also implicitly to Elena's contribution about the mirror mist. It appears that these two situations were similar somehow in Roberto's mind. Roberto was empowered in this classroom community to share his ideas and connect them with other children's contributions. Still, Amy was facilitating the children's sense-making by pushing them to think more deeply about the phenomena that were brought up.

121 Amy: Okay.
122 Roberto: *The mirror when it's cold it turns // it's like // it turns // you know it turns into a cloud // the water's turning into a cloud and it turns // and it gets like stuck on the mirror and then*...*
123 Amy: When it's cold.
124 Roberto: *Except when you open the door it goes outside and then it's like a // if you're cooking something the bubbles...// there's like the water and the water it's // it'll // it'll // like it makes the steam and like // it's like it gets smooshed up like the clouds and then #it makes steam.#*
125 Amy: #Smooshed up like the clouds#...and it makes steam, okay. So you think that's the bubbles coming up...and it gets smooshed together like a cloud?

As Roberto continued to think about the fogged-up mirror (unit 122), this became the first time that it was explicitly stated that the mirror was cold. Amy echoed "when it's cold" (unit 123), thus marking the significance of an important condition for condensation that was brought up in the midst of so many intertextual links that the children and Amy were bringing up. The information book that the teacher, Amy, was reading had linked being cold with clouds being formed. Roberto quickly moved from talking about condensation on the mirror (unit 122)—without using this register—to boiling (unit 124)—again without using that register. The two phenomena seem related to Roberto but the mostly colloquial registers he used were not capturing or indicating differentiated understandings.

126	Roberto:	Just the bubbles that come up.
127	Amy:	Just the bubbles.
128	Roberto:	*And they pop. So they pop because it's too hot...and they // when they pop the bubbles get // the bubbles get burnt and it's like they disappear because it's like they're like drowning but then they come back up it's like somebody's hitting them and then there's steam flying out.*

As Amy and Roberto kept talking about these ideas, Roberto stayed with the recounting of the generalized event of boiling and bubbles being formed. In unit 128, an extension of his intertextual link in unit 124, Roberto used two metaphors ("like drowning," and "like somebody's hitting them") to describe and explain how steam forms during boiling. Roberto's thinking included that steam need to be released from the bubbles in order to "fly out." Both these metaphors indicate implicit generalized events that Roberto had in mind—drowning and hitting bubbles. Multiple levels of intertextuality were coming up as a way of making meaning.

Turning to the Book: Considering Further Condensation Situations

130	Amy:	And then there's steam flying out. Okay, that's interesting. Alright let's keep reading. <u>Let's look at these children.</u> [Amy shows them the book. It is open to a picture of children breathing. Their breath is condensing.] What are they doing?
131	Cs:	Breathing!
132	Amy:	Ahh. What kind of day do you think it is outside?
133	Cs:	Cold! Winter!
134	Amy:	How can you tell?
135	Cs:	(*** ***)!
136	Amy:	How can you tell?
137	C1:	Cause there's a little bit of whiteness.
138	Amy:	Cause there's // shh. One at a time.
139	Cs:	(*** ***)
140	C2:	<You can see> yourself breathing.

141	Amy:	When it's cold.
142	Brittany:	There's // there's this um // there's um // because there's <a> breath // when they talk there's <a> breath coming out of their mouth.
143	Amy:	Alright, let's read it. HAVE YOU EVER SEEN YOUR BREATH IN FRONT OF YOUR FACE ON A WINTER DAY? THEN YOU ACTUALLY SAW A SMALL CLOUD BEING FORMED AS YOUR WARM, MOIST BREATH REACHED THE COLD AIR. Okay. Remember what we read about earlier that that water that goes up when it reaches cold air turns into a cloud? Right, now…the breath that's coming out of your mouth, is it warm or cold?
144	C1:	Warm!
145	Amy:	Yeah, warm because the inside of your body's warm, right. But when it comes out and reaches that cold air it turns into a little baby cloud right in front of your mouth. So you're blowing real clouds. You can make your own clouds right outside.
146	Jewel:	And then it (*** ***) into a big cloud.
147	Amy:	Into a big cloud in the air. Roberto.

At the beginning of unit 130, Amy revoiced, and acknowledged as interesting, part of Roberto's contribution. Continuing in unit 130, Amy and the children returned to the read-aloud book—a critical move that gave them the opportunity to consider the illustrations in the book that has pictures of two boys' breath during a cold, winter day. Amy and the children discussed that it needed to be cold for this to happen and that the kids' breath was warm. The book and Amy established a connection between formation of clouds and "seeing" breath on a cold day ("it turns into a little baby cloud right in front of your mouth" unit 145). This echoed the way that the children were talking about baby clouds earlier in this discourse excerpt—the baby clouds on mirrors in the bathroom during a hot shower or bath that Pamela initiated in unit 66 and Roberto talked about in unit 122. Both these earlier intertextual links actually foregrounded this idea in the book that was reinforced further as Amy and the children referred to the book. Furthermore, in units 150 and 152 that follow, Roberto brought the conversation back to the mirror clouds. All these point to the richness and complexity of dialogic discourse.

148	Roberto:	I think I know why // why*…
149	Amy:	Can we please make sure we're listening?
150	Roberto:	Why the mirror // why the mirror gets that // like that steam <it does>*…
151	Amy:	Why do you think the mirror gets the steam on it now? Explain it to us.

5. INTERTEXTUALITY, SOCIAL INTERACTIONS, AND SCIENCE 157

152	Roberto:	Because...// sometimes you know *like some people use cold water <for their babies>* // *they use cold water <for their babies>* and when they take a bath all of them are // you know like when you get bigger and you breathe in the shower and then it's // then um // it makes like your own cloud and then it gets // the air in your mouth is like steam*...
153	Amy:	Uh huh.
154	Roberto:	And it's like steam spreads // you could spread it // it goes like this [starting with his hands together he spreads them apart in the air to demonstrate] on the mirror and then it starts spreading everywhere.
155	Amy:	And it spreads out. So, it's like a cloud on the mirror.

As noted, Roberto returned to the earlier discussion on the fogged-up mirror. Roberto kept using colloquial registers as he was attempting to make sense of several ideas. In unit 152, Roberto made significant deviations from the ways he and the class were talking earlier. He was now talking about cold water, whereas earlier he had talked about the mirror being cold (unit 122). He was now talking about breathing causing the fogging up of the mirror instead of the hot water from the shower (as Elena contributed in unit 55). Making sense involves coordinating ways of thinking about multiple scenarios, contexts, and ideas that come up. Although Roberto did not explicitly elaborate his thinking, his language seemed to suggest that he may have thought of the phenomenon of the shower mirror fogging up as similar to seeing the breath on a cold day in the following way: If there is cold water running in the shower it makes the room cold which then it makes "the air in your mouth [that] is like steam" (unit 152) "make like your own cloud" (unit 152).

156	Roberto:	<Because you're breathing>.
157	Amy:	Because you're breathing. Do you always get a cloud on the mirror when you walk in the bathroom and breathe though?
158	Cs:	No.
159	C1:	Yes, yes, yes!
160	Roberto:	*Sometimes when me and my little brother and my mom when we're on a train he breathes when it's cold um // he breathes on the window to practice his ABC's #(***) <on the mirror cause the mirror has this (***)>#*
161	Amy:	#Ahhh.# So, if you breathe on a window you're saying, you might get that same little cloud like what you get in the bath // ahhh, so we're starting to see these things happen over and over again, right. If you breathe on a cold window you're saying and he could practice his ABC's drawing them.

162	Roberto:	(***) practices.
163	Amy:	What is that that goes onto the window?
164	C1:	#Your breath.#
165	Cs:	#(*** ***)#
166	Amy:	But what is it?...Shh. But what is it?
167	Roberto:	If you're in the shower the air is cold but if you could breathe the air and you feel it's warm air.
168	Amy:	It's warm air, okay.
169	Cs:	(*** ***) It's a cloud.
170	Amy:	It's what?
171	Cs:	(*** ***) It's a cloud.
172	Amy:	It's a cloud from inside. Alright let's keep going and finish up here.

Another intertextual link was contributed by Roberto as a reference to a generalized event, a habitual action, that involves the same object—a cold glass surface—but in a different context (unit 160). Roberto brought mom and little brother in the picture. Roberto's link foregrounded the concept of coldness although in unit 160 he didn't specify what exactly was cold. Amy recognized the intertextual link and further promoted its significance by acknowledging that the class was coming to appreciate the different contexts, circumstances, and times that similar phenomena happen. Amy's teacher-move legitimizes the class discussion so far. Amy specified what was cold ("If you breathe on a cold window" unit 161), and pushed the children to think about and express what it is that gets on the window, what breath contains. As Roberto and other children repeated ideas about cold air, warm breath, and clouds, Amy encouraged the class to move on.

INTERTEXTUALITY AS A MAJOR MEANS OF MEANING MAKING

As we hear children and teacher talk around scientific ideas in the foregoing excerpt of classroom discourse, we see the many forms and shapes that intertextuality can take and the opportunities it offers for developing understandings. We see how individuals create and recreate contexts, and how they interpret and give meaning to texts in a classroom community that values and respects both collaboration and individuality. We see how this second-grade class created the space, time, and context to bring texts together and juxtapose them. We also see how learning takes place not within individuals, but in transactions between them, how the voices of the teacher and the students come together as they co-construct understandings. Dialogic inquiry is manifested in many ways, in a range of inquiry acts (Lindfors, 1999). We get a glimpse of the "collaborative microinteractions" (Cummins, 1994), interactions that allow for joint participation among students and the teacher. In these interactions meaning was treated as tentative, provisional, open to alternative interpretations and revision. In this way we get a taste of how scientific literacy was emerging in this class. Figure 5.1 captures in a schematic form (to be read row by row from left to right and from top to bottom like regular text) all the intertextual links provided, their order, the person who initiated them, their type, and how they are linked with prior links in the midst of other critical moves. The arrows indicate relationships between contributions, which the discourse participants sometimes explicitly indicated. For example, Amy, with her intertextual link to prior classroom discourse (PCD) in unit 97, went explicitly back to Elena's critical move (CM) in unit 31. Other times these relationships were obvious by the content of the contributions. For example, Roberto, with his last intertextual link in the form of recounting a personal generalized event (PGE) in units 160–161 referred to a situation (fogging up of windows) similar to the intertextual link that children and teacher had contributed—again in PGE form—in unit 58. Roberto did not explicitly mark this relationship. Figure 5.1 relates in a powerful and succinct way the richness and complexity of the discourse that unfolded in Amy's class. We see a multifaceted network of intertextual links and critical moves that several children and the teacher contributed and how these contributions built on each other. The figure also depicts the variety of intertextuality that children and teacher initiated to make sense and develop further scientific understandings and linguistic registers.

Furthermore, in the excerpt of classroom discourse we analyzed we see how registers of scientific discourse were being used and created in the knowledge-building process. Although children offered personal examples, many were expressed as generalized actions that are distinctive in scientific talk (Kamberelis, 1999; Pappas, 1993; Wollman-Bonilla, 2000). They used present verbs, generic nouns, and tried to provide explanation structures.

Amy, as the teacher, was a frequent contributor to the classroom discourse. We heard her voice in approximately every other turn. However, her contributions did not fit at all the traditional IRE (Initiation, Response, Evaluation) form. At times she echoed a child's contribution or part of it, at times she asked probing questions, at times she celebrated the variety of ideas and thinking that children were contributing. And of course, she intermixed the reading of the information book as she allowed children to initiate intertextual links (and other comments and questions) and talk science in their own "ways with words" (Heath, 1983). This was her way of negotiating between authority and autonomy, and between her responsi-

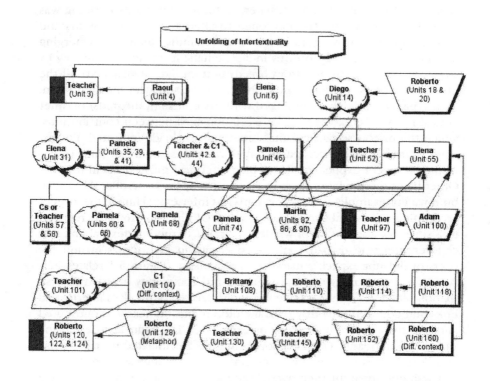

FIG. 5.1.

bilities of bringing scientific ideas to the children and letting them imbue ideas with meaning. These interactions and collaborative talk represent critical examples of sociocultural theory in action/practice.

One of the tensions we all feel as teachers as we go through this analysis is the sense of rich ambiguity versus narrow clarity. Connections were brought up, acknowledged, and taken up further in most cases. The class was heading toward "creating a sense of commonality, not of unity, . . . [a sense of] mutualism . . . to weave into one carpet the threads of hundred viewpoints" (Barber, 1984, p. 184). Discourse was not linear—there were plenty of diversions, but, at the same time, plenty of opportunities for connections. The children were clearly engaged, but the discourse was fragmented at times. This leaves us, teachers, sometimes uneasy (see also Gee, chap. 1, this volume). How do we make sure that all these intertextual links lead to some coherent ideas, to a rich but also coordinated network that brings together the variety of ideas, experiences, understandings that are brought up in a community of learners? In the particular discourse excerpt we analyzed we saw teacher and student moves that facilitated the integration of some fragmented contributions. For example, Amy made sure she got back to Elena's question after many turns where other students offered their sense and brought up several intertextual links. Additionally, students themselves brought up prior discourse contributions as they tried to make sense of their own ideas. Such moves attempt to bring integration of understandings that get expressed in fragmented speech. Furthermore, such moves give social significance to the discourse participants' contributions. Thus, we see through the prior analysis how social significance of intertextuality was at times gained later in the discourse and was not apparent right away.

We note that, as the unit unfolded, there were no specific moments when Amy and the children tied all their ideas together toward a common "scientific" understanding of the water cycle and changes of states of matter. As the class engaged in hands-on explorations and discussed their observations and findings, and as they continued to read children's informational books, ideas continued to be explored and the children were scaffolded to consider "scientifically accepted" ways of thinking about the various phenomena. And as the unit progressed, recounting of specific events was reduced while generalized-event links increased, indicating that the hybrid discourse that incorporated the mini-narratives of the children's everyday lives was transformed to classroom discourse that was more scientific in nature (Varelas & Pappas, 2002).

One of the findings of the present analysis is the recognition of students as knowers and active participants in their learning, who bring into the classroom important and various funds of knowledge. This analysis challenges the deficit view of urban school children still quite prevalent within the general public and schoolteachers. The examples here show urban children as individuals whose life experiences allow them not only to juxtapose ideas and develop scientific understandings, but also to share authority with the teacher (Oyler, 1996; Pappas & Zecker, 2001) and co-construct scientific entities. As teachers we need to view our students' prior experiences and understandings as the "capital" that we need to be investing on. And as we saw this capital is not insignificant. Furthermore, we need to create spaces so that this capital is revealed and worked upon. One way to view intertextuality is as a means for uncovering and foregrounding this capital. Through intertextuality children's tacit understandings become overt ways of meaning making. These intertextual connections allow children to play with ideas in a public way as their voices and minds mingle with those of their peers and their teacher.

REFERENCES

Bakhtin, M. M. (1981). *The dialogic imagination: Four essays by M. M. Bakhtin* (M. Holquist, Ed.; M. Holquist & C. Emerson, Trans.). Austin: University of Texas Press.

Bakhtin, M. M. (1986). *Speech genres and other later essays*. Austin, TX: University of Texas Press.

Barber, B. (1984). *Strong democracy: Participatory politics for a new age*. Berkeley: University of California Press.

Becker, J., & Varelas, M. (1995). Assisting construction: The role of the teacher in assisting the learner's construction of pre-existing cultural knowledge. In L. Steffe (Ed.), *Constructivism in education* (pp. 433–446). Hillsdale, NJ: Lawrence Erlbaum Associates.

Bloome, D., & Bailey, F. (1992). Studying language and literacy through events, particularity, and intertextuality. In R. Beach, J. L. Green, M. L. Kamil, & T. Shanahan (Eds.), *Multidisciplinary perspectives on literacy research* (pp. 181–210). Urbana, IL: National Conference on Research in English.

Bloome, D., & Egan-Robertson, A. (1993). The social construction of intertextuality in classroom reading and writing lessons. *Reading Research Quarterly, 28*, 305–333.

Cazden, C. B. (1992). *Whole language plus: Essays on literacy in the United States and New Zealand*. New York: Teachers College Press.

Cazden, C. B. (2001). *Classroom discourse: The language of teaching and learning*. Portsmouth, NH: Heinemann.

Cummins, J. (1994). From coercive to collaborative relations of power in teaching of literacy. In B. M. Ferdman, R.- M. Weber, & A. G. Ramierz (Eds.), *Literacy across languages and cultures* (pp. 295–331). Albany: State University of New York Press.

Denzin, N. K., & Lincoln, Y. S. (1994). Introduction. In N. D. Denzin & Y. S. Lincoln (Eds.), *Handbook of qualitative research* (pp. 1–17). Thousand Oaks, CA: Sage.

Driver, R., Asoko, H., Leach, J., Mortimer, E., & Scott P. (1994). Constructing scientific knowledge in the classroom. *Educational Researcher, 23*(7), 5–12.

Fairclough, N. (1992). *Discourse and social change*. Cambridge, England: Polity Press.

Fowler, A. (1996). *What do you see in a cloud?* New York: Children's Press.

Gee, J. P., & Green, J. L. (1998). Discourse analysis, learning, and social practice: A methodological study. In P. D. Pearson & A. Iran-Nejad (Eds.), *Review of research in education, Vol. 23* (pp. 119–169). Washington, DC: AERA.

Glaser, B. G., & Strauss, A. L. (1967). *The discovery of grounded theory: Strategies for qualitative research*. New York: Aldine de Gruyter.
Goldman, S. R., & Bisanz, G. L. (2002). Toward a functional analysis of scientific genres: Implications for understanding and learning processes. In J. Otero, J. A. Leon, & A. C. Graesser (Eds.), *The psychology of science text comprehension* (pp. 19–50). Mahwah, NJ: Lawrence Erlbaum Associates.
Green, J. L., & Dixon, C. N. (1993). Talking knowledge into being: Discursive practices in classrooms. *Linguistics and Education, 5*, 231–239.
Gutierrez, K., Rymes, B., & Larson, J. (1995). Script, counterscript, and underlife in the classroom: James Brown versus Brown v. Board of Education. *Harvard Educational Review, 65*, 445–471.
Halliday, M. A. K. (1993). Towards a language-based theory of learning. *Linguistics and Education, 5*, 93–116.
Halliday, M. A. K. (1998). Things and relations: Regrammaticising experience as technical knowledge. In J. R. Martin & R. Veel (Eds.), *Reading science: Critical and functional perspectives on the discourses of science* (pp. 185–235). London: Routledge.
Halliday, M. A. K., & Hasan, R. (1985). *Language, context, and text: Aspects of language in a social-semiotic perspective*. Victoria, Australia: Deakin University Press.
Halliday, M. A. K., & Martin, J. R. (1993). *Writing science: Literacy and discursive power*. Pittsburgh, PA: University of Pittsburgh Press.
Heath, S. B. (1983). *Ways with words: Language, life, and work in communities and classrooms*. Cambridge, England: Cambridge University Press.
Kamberelis, G. (1999). Genre development and learning: Children writing stories, science reports, and poems. *Research in the Teaching of English, 33*, 403–460.
Lemke, J. L. (1985). Ideology, intertextuality, and the notion of register. In J. D. Benson & W. S. Greaves (Eds.), *Systemic perspectives on discourse: Selected theoretical papers from the 9th International Systemic Workshop* (Vol. 1, pp. 275–294). Norwood, NJ: Ablex.
Lemke, J. L. (1990). *Talking science: Language, learning, and values*. Norwood, NJ: Ablex.
Lemke, J. L. (1992). Intertextuality and educational research. *Linguistics and Education, 4*, 257–267.
Lemke, J. (1998). Multiplying meaning: Visual and verbal semiotics in scientific text. In J. R. Martin & R. Veel (Eds.), *Reading science: Critical and functional perspectives on the discourses of science* (pp. 87–113). London: Routledge.
Lincoln, Y., & Guba, E. (1985). *Naturalistic inquiry*. Beverly Hills, CA: Sage.
Lindfors, J. W. (1999). *Children's inquiry: Using language to make sense of the world*. New York: Teachers College Press.
Martin, J. R. (1990). Literacy in science: Learning to handle text as technology. In F. Christie (Ed.), *Literacy for a changing world* (pp. 79–117). Victoria: Australian Council for Educational Research.
Moll, L. C. (1992). Literacy research in community and classrooms: A sociocultural approach. In R. Beach, J. L. Green, M. L. Kamil, & T. Shanahan (Eds.), *Multidisciplinary perspectives on literacy research* (pp. 211–244). Urbana, IL: National Conference on Research in English.
Nystrand, M. (1997). *Opening dialogue: Understanding the dynamics of language and learning in the English classroom*. New York: Teachers College Press.
Ogborn, J., Kress, G., Martins, I., & McGullicuddy, K. (1996). *Explaining science in the classroom*. Buckingham, UK: Open University Press.
Oyler, C. (1996). Sharing authority: Student initiations during teacher-led read-alouds of information books. *Teaching & Teacher Education, 12*, 149–160.
Pappas, C. C. (1993). Is narrative "primary"? Some insights from kindergarteners' pretend readings of stories and information books. *Journal of Reading Behavior, 25*, 97–129.

Pappas, C. C., Varelas, M., with Barry, A., & Rife, A. (2003). Dialogic inquiry around information texts: The role of intertextuality in constructing scientific understandings in urban primary classrooms. *Linguistics and Education, 13*(4), 435–482.

Pappas, C. C., & Zecker, L. B. (Eds.). (2001). *Transforming literacy curriculum genres: Working with teacher researchers in urban classrooms*. Mahwah, NJ: Lawrence Erlbaum Associates.

Sutton, C. (1992). *Words, science and learning*. Buckingham, UK: Open University Press.

Varelas, M. (1996). Between theory and data in a 7th grade science class. *Journal of Research in Science Teaching, 33*, 229–263.

Varelas, M., Luster, B., & Wenzel, S. (1999). Meaning making in a community of learners: Struggles and possibilities in an urban science class. *Research in Science Education, 29*, 227–245.

Varelas, M., & Pappas, C. C. (2002, June). *Exploring meaning-making in integrated primary science-literacy units: The nature of intertextuality*. Paper presented at the symposium "Instruction in Reading Comprehension," at the joint conference of the Society for Text and Discourse, and the Society for the Scientific Study of Reading, Chicago.

Varelas, M., & Pineda, E. (1999). Intermingling and bumpiness: Exploring meaning making in the discourse of a science classroom. *Research in Science Education, 29*, 25–49.

Vygotsky, L. S. (1978). *Mind in society*. Cambridge, MA: Harvard University Press.

Vygotsky, L. S. (1987). Thinking and speech. In R. W. Rieber & A. S. Carton (Eds.), *The collected works of L. S. Vygotsky (Vol. 1): Problems of general psychology* (N. Minick, Trans.). New York: Plenum Press. (Original work published 1934)

Wells, G. (1990). Talk about text: Where literacy is learned and taught. *Curriculum Inquiry, 20*, 369–405.

Wells, G. (1993). Reevaluating the IRF sequence: A proposal for the articulation of theories of activity and discourse for the analysis of teaching and learning in the classroom. *Linguistics and Education, 5*, 1–37.

Wells, G. (1999). *Dialogic inquiry: Towards a sociocultural practice and theory of education*. Cambridge, England: Cambridge University Press.

Wells, G., & Chang-Wells, G. L. (1992). *Constructing knowledge together: Classrooms as centers of inquiry and literacy*. Portsmouth, NH: Heinemann.

Wertsch, J. V. (1991). *Voices of the mind: A sociocultural approach to mediated action*. Cambridge, MA: Harvard University Press.

Wolcott, H. F. (1994). *Transforming qualitative data: Description, analysis, and interpretation*. Thousand Oaks, CA: Sage.

Wollman-Bonilla, J. E. (2000). Teaching science writing to first graders: Genre learning and recontextualization. *Research in the Teaching of English, 35*, 35–65.

APPENDIX

Keys to symbols used in transcripts

//	false starts or abandoned language replaced by new language structures
...	short pause within language unit
... ...	longer pause within language unit
*...	breaking off of a speaker's turn due to the next speaker's turn

(***)	one word that is inaudible or impossible to transcribe
# #	overlapping language spoken by two or more speakers at a time
CAPS	actual reading of a book
{ }	teacher's miscue or modification of a text read
[ ]	identifies what is being referred to or gestured and other nonverbal contextual information
. . . .	part of transcript has been omitted
< >	uncertain words
( )	transcriber's notes
[...m...]	omission of teacher's speech concerning class management
(*** ***)	longer stretches of language that are inaudible and impossible to transcribe
Italics	emphasis

TABLE 5.1
Categories of Intertextuality Identified in "States of Matter and the Water Cycle" Unit Read-Alouds

Type of Intertextual Connection	Definition of Connection	Example
CATEGORY 1		
1) Written texts		
a) information books in unit	Refers to a particular information book by title, or by noting other information books in the unit.	Ooh, lightning. We're going to read a book called *Flash, Crash, Rumble, and Roll* and that one has some stuff about lightning... [G2, WTWT, TI]
b) text around classroom (on charts, board, etc.)	Refers to a written text found on charts, the board, etc.	Remember how I did that on the board to show. That's kind of just to show you something that's invisible, okay? [G2, DCTR, TI]
c) other books available (in or outside of classroom)	Refers to other books in or outside the classroom.	Now, I'm going to go over to get a book. In fact, Alejandro, no, Manuel, you go over and get the Emperor penguin book. [G1, WSCU, TI]
d) children's own writing (and/or drawings)	Refers to children's own writing (and/or drawings).	We're going to talk about it [today's weather] on a piece of paper. We're not only going to talk about it, we're going to draw a picture and we're going to write. [G1, WTWT, TI]
2) Other texts (*orally shared*)		
Poems, rhymes, sayings, songs	Refers to a poem, rhyme, saying, or song, by orally sharing some part or all of it.	It's raining, it's pouring, the old man is snoring. [G1, WTWT, CI]
3) Other media		
TV/radio shows or movies	Refers to a TV/radio show or a movie.	I was watching Ms. Frizzle it was like...it was raining and the...wind was blowing in the water and it was like the wind...flew up and made the clouds. [G2, WTWT, CI]
4) Prior classroom discourse		
a) in current Read-Aloud session	Refers to prior discourse in the current Read-Aloud session.	You know, probably in the month of March...just like Alexandra said. [G1, WSCU, two TIs]

b) in unit, but outside present session	Refers to prior discourse in the unit, but not in the present session.	**The other day Julio was kind of describing them [tornados] as having hands that can pick things up, right? Cause they're so strong.** [G2, WSCU, TI]
c) outside unit	Refers to prior discourse related to previous units or other curriculum outside unit.	**Remember we talked about the equator...and people who live around the middle part of the earth are always warm.** [G1, WTWT, TI]

CATEGORY II

Hands-on explorations

1) within unit in classroom	Refers to classroom explorations within the unit.	**Now...one half of the class yesterday was up here in front with me and we were heating up the teapot and we were seeing the exact thing, right?** [G2, WIWMO, TI] *[Students had done evaporation experiments using water.]* **If you left like juice on the table would it evaporate?** [G2, DCTR, CI]
2) other explorations	Refers to other explorations (at-home explorations developed for other units, other explorations conducted at home or at other settings).	

CATEGORY III

Recounting events

1) specific events

a) personal specific events	Refers to a personal, specific event.	**Last time I poured cold water in...my plate...cause...I was gonna use my mom's water and I seen (***) and I seen air coming up.** [G1, AIAAY, CI]
b) specific events—personally-related others involved	Refers to specific events in which speakers are not personally involved, but others who they know are.	**One day when...I was in my uncle's car and he was...picking me up from school because it was...time to go cause we had half a day and then when he...started to go back cause he wanted to go front but it went back cause all the snow...was on the ground...And there was ice on the ground and then he started to go...back.** [G2, WSCU, CI]

(Continued)

TABLE 5.1
(Continued)

Type of Intertextual Connection	Definition of Connection	Example
2) generalized events		
a) personal generalized events	Refers to personal, generalized events that are habitual actions.	**This is what I do in the bathtub.** *[He takes the cup, pushes it in the water upside down, and then lets go of it.]* [G2, AIAAY, CI]
b) generalized events—personally-related others involved	Refers to generalized, habitual events in which speakers are not personally involved, but others they know are.	Like my brother, he goes downstairs, that's his room, got to go downstairs. [G1, WSCU, CI]

CATEGORY IV

"Implicit" generalized events	Refers to generalized events in which speakers do not indicate any **explicit** personal involvement. However, they seem to implicitly refer to events they *could/should have* habitually experienced.	**Sometimes water isn't always thin when it comes out...the faucet...it's always thick when it comes out the faucet but when it lands in the...sink...it's always thin.** [G2, WIWMO, CI]

KEY: G1 = First grade; G2 = Second grade

Read-Aloud Book:
- WTWT What Is the Weather Today?
- DCTR Down Comes the Rain
- WSCU When the Storm Comes Up
- WIWMO What Is the World Made Of?
- AIAAY Air Is All Around You

CI = Child-initiated; TI = Teacher-initiated

CHAPTER FIVE METALOGUE

Lifeworld Language, Scientific Literacy, and Intertextuality

G. Michael Bowen
Maria Varelas
Wolff-Michael Roth
Randy Yerrick

Michael B.: In reading this chapter I am struck by the contrast between it and my own. In the setting described in this chapter, it is possible to examine the intertextuality of discourse, which is generated among the many participants in the setting, many of whom are peers. Even in many of the formal settings of science (such as conferences and reports) in which findings are reported and questions asked there is such a normative structure to it all that there is little resemblance in the discourse to Amy's class. I find myself thinking about what discourse there is in the community of scientists I have studied and that when it occurs it is of two types. On one level they talk about things they have done, things they have observed. On another level, they talk about why they think things happened the way they did. The kids are talking about things they have seen, experiences they have had, and trying to integrate that with new terms and ideas. Their experiences may not be as immediate, as experience-near, but they were obviously salient enough to be brought to the explanatory task.

Michael R.: I think that Jim Gee makes the point (see the metalogue to chapter 1) that you cannot compare the two forms of discourse, especially if you take the sense-making talk of children that come from poor neighborhoods.

Maria: We agree with Jim Gee's overall argument that lifeworld language can be very different from academic social languages, but we find Michael Bowen's comment on the similarities between the two

169

forms of discourse quite interesting and important. Maybe it is these similarities that could allow us, teachers, to use the children's spontaneous (in a Vygotskian sense) discourse "as a way to bridge to and enhance the acquisition of one or more academic forms of language" that Jim Gee in his chapter metalogue argues, "is a good thing." And as we discussed in our chapter, Amy was helping her students in several ways to become more specific and explicit, to distinguish between examples, to provide and think about distinctive details and differentiations—all important elements of more academic forms of language—as, at the same time, she was accepting and validating their lifeworld language.

Randy: Is it really necessary that *all* of the intertextual links lead to coherent ideas? Perhaps my question is, should we assume that they always will? Why should we expect all those connections would lead to fruitful observable outcomes? Perhaps you meant it more collaboratively, that each of these different aspects come together to contribute to a part of what the community represents and not the other meaning I drew from it. If this is your question, what are the key components of learning that forward contexts like these into powerful learning moments with a long-term impact on individual or collective meaning?

Maria: You talk about an "observable" outcome. The question we posed (i.e., How do we make sure that all these intertextual links lead to some coherent ideas, to a rich but also coordinated network that brings together the variety of ideas, experience, understandings that are brought up in a community of learners?) does not necessitate that the coordinated network is observable. I think you might argue that many times it is very difficult and maybe impossible to capture all the understandings that an individual holds, and I wholeheartedly agree with such a position. I also think that the coherence and coordination may take very different shapes and forms in individuals' minds and words. The question we raise has also the collaborative flavor you offered—different contributions coming together to shape the thinking and discourse in which a particular community is engaged. I would add, though, that we need to be concerned with, and explore, not only how individual contributions come to shape collective thinking and talk, but also how the classroom discourse and collective thinking come to shape the individual's meaning making and learning. Although assessment is not a topic that we undertake in this chapter, I consider both these points as important dimensions of assessment practice and theory.

Michael R.: How individual contributions relate to the overall shape of a conversation is probably much more complex than we think—we may actually need to draw on theoretical models grounded in a dialectical logic, whereby overall topic and individual contributions are but two different parts of the same unit. The present state of the conversation and the history of topic development shape what participants consider as good continuations. At the same time, each individual contribution shapes topic development. If teachers are truly committed to a place for lifeworld language in the classroom and what other (inter) texts are brought in, it will be impossible to predict where a specific conversation will end up and what course its topical development takes.

Randy: I want to come back to these intertextual links. What is it that contrasts with regular school so much that when teachers try to make such intertextual connections they are drawn back into a pattern that minimizes the creation of substantive discourse? I think you have made a strong case that intertextuality is central to the process of acquiring science discourse. Is there evidence to suggest which components are most central or in what order they must be emphasized for such acquisition to occur?

Maria: I see two different questions here. It seems to me that the first question relates to the tensions we have written about elsewhere (Varelas, Luster, & Wenzel, 1999) where we explored two dimensions of a community of learners—a social–organizational and an intellectual–thematic dimension—and their interplay. Trying to offer all students opportunities to participate in classroom discussion may interfere with sustaining substantive discussion with a limited number of students and pursuing their ideas further. Trying to develop connections between participants' contributions and meaning making may impose limitations on how much individual ways of thinking and talking can be enhanced. The second question relates to research we have been pursuing (Varelas & Pappas, 2002) where we studied intertextuality across seven read-aloud sessions in two classrooms. This study shows that recounting events (specific and generalized) plays a central role in both classrooms, especially at the earlier read-aloud sessions of the unit. This recounting eventually diminishes in later read-aloud sessions to give place to the connections with hands-on explorations that the classes were pursuing or about which the children were wondering about. Furthermore, an interesting dynamic between recounting of specific versus generalized events unfolded in the second-grade classroom. There was relatively more recounting of specific versus generalized events earlier in the unit that became about equal in the middle of the unit and then reversed direction in the later read-aloud sessions of the unit.

Michael B.: Maria, you and your collaborators write about seeing the foundations of scientific literacy in the discourse of the students and teacher. I wonder more and more what that means. Not just in this chapter, but overall, what is scientific literacy? But does it mean bringing an interpretive framework to an experience similar to that of scientists? Does it mean being able to use scientific tools, like graphs and empirical arguments, to everyday common concerns and interests? Or to manipulate the real tools that assist in the observing and measuring like rulers and microscopes (to name two)? Does it mean being able to construct arguments that are internally consistent and coherent (without contradictions)? Does it mean being able to relate five important things about a magnet? Or, alternatively, pick up a magnet and figure out those five important things? Is it the participation in a community that is collectively scientifically literate, where those within it have different strengths and perspectives in many different areas that overall contribute to the "collective" literacy but which are not completely found in any member?

Maria: I guess I constantly struggle with this question. An important aspect of the science genre we espouse is what we call the "theory–data dance"—the going back and forth between empirical evidence and ways to explain it and theorize it. In this process, we may use empirical evidence that we have already available to us or we collect for a particular purpose. We may use certain tools, for example, equipment or symbolic tools, such as data tables and graphs. We may observe or measure. Somehow we draw on data that we try to make sense of and in the process we use, form, and reform our theories. And we use our networks of ideas, concepts, processes, our theories to ponder about existing data and data we need to collect. And we draw on or acquire certain data that are strongly influenced by our own conceptions, beliefs, lifestyles, and sociopolitical agendas. This endeavor has an individual and a collective aspect. Somehow scientific literacy is, I think, "collective" literacy that is not completely found in every member, but every member needs to have enough of the fundamental feel for and understanding of both concepts, processes, language that allow her or him to contribute to the "collective" literacy.

Michael R.: This is just what I have argued in various places—scientific literacy is not something that individuals hold in their minds but rather is the outcome of the dialectical unit in a conversation that I referred to earlier.

Maria: Awareness of these dynamics is an important aspect of such scientific literacy. The construct of intertextuality that we use in our research captures in a certain way the "dance," if you like, between togetherness and individuality in the class's pursuit of scientific literacy. We wrote about this elsewhere (Pappas & Varelas, in press) where we discuss how, for example, connections to prior discourse foreground togetherness, and connections to recounting of events foreground individuality. In addition, scientific literacy involves being able to express scientific ideas or explanations in terms of typical linguistic registers. Recounts of specific events are narrative in nature (e.g., include personal pronouns, past tense verbs, temporal conjunctions, and sometimes even dialogue); recounts of generalized events are realized by more typical scientific registers, (personal pronouns are present but generic nouns and timeless present-tense verbs are found). The wordings of recounts of implicit generalized events are fully scientific—timeless present tense verbs, generic nouns, third-person (rather than personal) or the general "you" pronouns, if- or when-then and because constructions, and so forth, are found. As I indicated earlier, in our analysis of seven read-aloud sessions, the use of specific events—narrative talk—was high in the beginning and then decreased across the unit; whereas the use of generalized events—scientific talk—was low in the beginning and then increased over the read-aloud sessions. We see this cline from narrative to scientific linguistic registers as a critical aspect of scientific literacy.

Michael B.: Perhaps there is more scientific literacy in this class than I can actually sense. Maybe it is not just "emerging" but actually there. At least as "there" as is possible for 7 year olds. These students adopted new terms, integrated them with their (observed) experiences, drew in related experiences and examples to those initially discussed to broaden their understanding of the issues that arose in the discourse. They tried out ideas, cast aside some, accepted others. Brought in new examples to clarify old ones. These are the practices of science communities, although they're not foregrounded as such in the particular classroom episode that was described. So the science literacy is not just around the students integrating examples with concepts but also in the practices in which they engaged in while conducting that integration. Does engaging in some scientific practices mean that one is on the trajectory to being "scientifically literate"? Or does one need germinal forms of all the parts to be on that trajectory?

Maria: Neat question! We find ourselves leaning toward the latter. But, of course, it is debatable what we mean by "all parts." Science educators, scientists, science education policymakers have not reached a consensus in what these parts are—maybe they agree in a vague, general level, but there is significant disagreement in a more specific and articulated level. As I wrote on the previous point, for me an important element of scientific literacy is awareness, consciousness of the practices in which an individual and a community of practitioners engage. These second graders were not necessarily aware of their engaging in scientific practice. Not that we would strive for this at this grade, but this is one of the reasons why we would claim that scientific literacy was emerging in their class. In addition, the teacher's scaffolding was critical in sustaining the children's engagement in such practices. We do not know whether the children could self-sustain their engagement in such practices.

Furthermore, when we talk about emergent understandings we refer to the ill-defined, "messy" understandings that some children held and brought out in the discourse that do not conform to the scientifically accepted ways of thinking about a particular phenomenon. These were emergent in two ways: (a) not fully fleshed out and articulated with each other and (b) not being validated by the community of practitioners—both reasons that we would use to call "emergent" understandings of practicing scientists, not only children.

REFERENCES

Pappas, C. C., & Varelas, M. (in press). Dialogic inquiry around information texts: The role of intertextuality in constructing scientific understandings in urban primary classrooms. *Linguistics and Education*.

Varelas, M., Luster, B., & Wenzel, S. (1999). Meaning making in a community of learners: Struggles and possibilities in an urban science class. *Research in Science Education, 29*, 227–245.

Varelas, M., & Pappas, C. C. (2002, June). *Exploring meaning-making in integrated primary science-literacy units: The nature of intertextuality*. Paper presented at the symposium "Instruction in Reading Comprehension," at the joint conference of the Society for Text and Discourse, and the Society for the Scientific Study of Reading, Chicago.

CHAPTER SIX

Meaning and Context: Studying Words in Motion

Cynthia Ballenger
Chèche Konnen Center
TERC
Cambridge, MA

Teachers engaged in inquiry-based science attempt to build on children's ideas and prior knowledge toward mature concepts and practices representative of the discipline (American Association for the Advancement of Science, 1993; National Research Council, 1996). Guided by basic constructivist principles, they are committed to immersing children in a community of practice rich with experiences, ideas, and scientific tools. And yet, educators have been led to ask in recent years how children will learn about the formalities of language meaning and use associated with academic success when it is the children's own ideas and questions and their everyday language and experience that are central in this view of teaching (Cope & Kalantzis, 1993; Delpit, 1995). This is of particular concern in relation to children who do not speak English as a native language and children from families that are not immersed in school-based ways of talking and acting. How do children learn school-based and discipline-based ways of talking in school? What do educators need to do to insure that this happens?

In the field of literacy, discussion on these issues has addressed questions of narrative and expository language, standard and nonstandard dialects, and various forms of participation in discussion, among other issues (e.g., Ballenger, 1999; Delpit, 1995; Gallas, 1994; Heath, 1983; Hymes, 1972). The analogous concerns in science and mathematics education have generally focused on what we might call everyday and scientific ways of using language (e.g., Ballenger, 1997, 2000; Clement, Brown, & Zeitsman, 1989; Conant, Rosebery, Warren, & Hudicourt-Barnes, 2001; diSessa, Hammer, Sherin, & Kolpakowski, 1991; Gee & Clinton, in press; Hammer, 1997, 2000; Lee & Fradd, 1996; Lehrer & Schauble, in press; McDermott, Rosenquist, & van Zee, 1987; Michaels & Sohmer, 2000; Minstrell, 1989; Nemirovsky, Tierney, & Wright, 1998; Viennot, 1979), seeing these two arenas of sense making variously as supports for each other, largely separate, or in some kind of conflict. Visions of what counts as scientific language and scientific practice have been elaborated and deepened in these discussions as well as in related work in the sociology of science (e.g., Biagioli, 1999; Fox-Keller, 1983; Gooding, Pinch, & Schaffer, 1989; Goodwin, 1997; Jacob, 1987/1995; Latour & Woolgar, 1986; Lynch, 1985; Ochs, Jacoby, & Gonzales, 1996; Rheinberger, 1997; Root-Bernstein, 1989.)

Among the important things we have learned through these conversations has been the value, for teachers and researchers alike, of regularly consulting detailed accounts of classroom practice, what children have said and done and what they have learned from this. Doing so regularly complicates dichotomies and challenges our categories and assumptions. Furthermore, among some segments of the science education community, there has been increasing recognition of the value of exploring in particular the ideas and talk of those children whose learning is somewhat less transparent to the educators involved, children whose ideas are puzzling (Ballenger, 2000; Gee & Clinton, in press; Michaels & Sohmer, 2000; Warren, Ballenger, Ogonowski, Rosebery, & Hudicourt-Barnes, 2001; Warren, Pothier, Ogonowski, Noble, & Goldstein, 2000).

In this chapter I provide a detailed account of two immigrant fourth-grade boys and the work they did to understand the language being used in their class's study of motion. This chapter has two purposes. First, I believe that these boys and the specifics of their case have something to offer us in our concern about how one talks in school, about formality and vernacular and why we move from one to the other. I hope that studying the boys' remarks will help us to consider, in a specific case, how school-based usages become evident and become useful and to see the kinds of classroom activities that support attention to language in meaningful contexts. Second, I hope to demonstrate the value for an understanding of language use in context of a focus on what we might call "puzzling children."

SETTING

Mary DiSchino is a third–fourth-grade teacher in a diverse urban school. She has been teaching for 25 years. Ms. DiSchino has explored her own teaching and learning and has written about this (DiSchino, 1987, 1998).

She has also participated in the Chèche Konnen Center (CKC) seminars for 8 years, learning science there with other teachers and researchers, and exploring her teaching.

The Chèche Konnen Center is organized around the principle that good science teaching, whatever else it may include, involves continual inquiry into the science itself and into the diversity of children's ways of talking and knowing. Teachers in the Chèche Konnen Center routinely as part of their everyday professional practice come together to explore their questions about the science that they are teaching and about the ways in which their students understand and explain this science. Teachers and researchers do science together, and, as they learn, they uncover new questions. At the same time, participants explore tapes and transcripts of the work the students are doing in the classroom, with a particular focus on talk and discussion. Through this experience, teachers come to see themselves as participants in the discipline of science and to recognize a broad range of ways of connecting with science on the part of their students (for more, see Rosebery, 1998). In 1999–2000 I was involved, as a member of the Chèche Konnen staff, in Ms. DiSchino's classroom while she taught science. The ideas presented here were developed in close collaboration with other members of the CKC.

The children were engaged in a science unit on motion that Ms. DiSchino developed with the help of conversations at CKC, her own experience learning motion at CKC, and some pieces from the Investigations curriculum on speed and the mathematics of change (see Rosebery, 2000). She began with the idea of representing motion as a means of communication. The children made pictorial representations of trips they took down a 9-meter paper strip (e.g., skip for three steps, fall down, twirl to the end). These representations, which developed and changed as the class tried them out, were intended to be read and then followed by other children so that they took the same trip and ended up at the same place (see Monk, 2000, for an account of these trips and their development). Throughout this unit the children found themselves concerned with questions such as: What are the different ways a child might move to get from one end to the other of this strip? How can we show this in a chart? How can we insure that each traveler ends up at the same place? How can we insure that each traveler goes at the same speed as the original traveler did? These concerns led to talk about standard measures, about speed, about systems of representations, about distance, and about how a chart functions, among other issues. Ms. DiSchino did not plan the curriculum in advance although she had goals in mind; rather she videotaped class discussions and then watched moments from them with others, CKC staff and sometimes other CKC teachers, in order to investigate what and how the children were thinking and, in light of this, to plan the following classroom activities. This involved one 2- to 2½-hour meeting per week after school.

METHODS AND DATA COLLECTION

The Chèche Konnen Seminar, following and modifying a tradition of the Brookline Teacher Researcher Seminar, has developed a practice of focusing teaching and research attention on children considered "puzzling," children about whose academic progress there is concern, children we are not always sure how to teach (Brookline Teacher Researcher Seminar, 2004). We do this in order to help these children and to find out what they are thinking and how they are approaching topics. But more than that, we do this because it has been our experience that these children have the most to teach us. Their assumptions, their habits of thought and their ways with words are often not what we expect; investigating what they are engaged in often reveals implicit and powerful aspects of the learning and associated practices that we would otherwise miss. Thus it was a normal, if still emerging, part of CKC practice for Ms. DiSchino to ask me, as a part of the CKC staff, to pay particular attention to two Haitian-American boys in her class, Saintis and Roland. I followed these boys at her request because she found them puzzling and was concerned about their academic progress, Saintis more than Roland. The boys were both fourth graders. Saintis was newly literate, not having attended much school in Haiti, and both were newly integrated into mainstream classes. I became, on occasion, a part of their cooperative groups in science and paid particular attention to them in the group discussions. I brought their ideas to the planning sessions with Ms. DiSchino. I videotaped and transcribed what they said in discussion, questioned them and took notes on their ideas in informal contexts, collected their science writing, and interviewed them twice, once after a few sessions of science and once at the end of the unit. We did these interviews because we wanted them to have time to elaborate on their thinking, time not always available within the classroom schedule. The interviews were very open-ended. They began with the question, "What have you learned?"

There is, of course, in this corpus a great deal of information on what and how these boys were thinking. In this chapter, I chose to pursue what the boys themselves felt they had learned. After hearing their responses to this question in the earlier interview, I then looked through all instances in the class discussions where they talked and chose for further analysis those instances that seemed to bear on what they said they had learned.

THE STUDY

What is most noticeable in what the boys have to say about their learning is the way they were struck by the chart that their class developed and the way it, in concert with the activities to which it referred, affected for them the meaning of familiar words. The chart, which was intended to represent the trips children took down a 9-meter strip of paper and to communicate these trips to others, eventually contained the headings *speed, movement,* and *distance.* The boys referred to these familiar words, words that they knew well and had used in the past, as concepts they had learned. These children were especially intrigued by the way this tool and the context in which it was used affected the meanings and uses of these familiar words; they were intrigued by the way words came to take on new aspects of meaning in new intellectual and social contexts. Their attention was directed to language by their engagement with the chart and the trips.

Although I am seeing these boys as individuals who seem to have done at least some of the thinking discussed here somewhat silently while the class went on, it is nevertheless very important to note that they were students in a classroom that encouraged thinking out loud and restating significant ideas in various forms, and in which words, particularly big impressive scientific words, were regularly questioned. Attention to language and alternative points of view were thus a part of the content of science discussion on a regular basis.

I begin with Saintis and explore some examples from his class participation and then from his interview.

Classroom Conversations: Saintis. In the course of the first classroom conversation in which children were beginning to develop the chart (2/11/98), they were exploring their sense of the categories they would need to communicate their trips (see Monk, 2000). The early stories they wrote included such directions as "walk three steps, stop to pick up a dog, twirl three more steps." Ms. DiSchino had proposed that one heading should be *kind of movement.* She asked the children, after proposing this, "What are some kinds of movement?" For most children the question prompted them to list such activities as skipping, running, walking. But some children brought up movement that did not cover space in the same way, like blinking. These suggestions were written down on chart paper for everyone to consider. Saintis, when asked for his contribution, suggested *playing,* or at another point, he suggested the word *moving* itself. When he suggested *moving* his teacher tried to help him understand that she wanted examples of the category, not the name of the category itself.

Puzzled by his contributions, I asked him later for an example of movement. He suggested *cooking* and acted out moving up and down a counter while doing various things with his hands, suggesting, to me, a restaurant chef dealing with multiple dishes. *Playing*, he told me again later, "like in playing volleyball," he explained. Here is the puzzling child. Although earlier we thought he was unable to understand the category of movement as it was being used, further reflection revealed that everything he proposed certainly did involve movement. He was not choosing anything that was actually still. However, while the other children suggested movements that either covered linear distance or contained fairly discernible rates of movement (like *blinking*), Saintis, by suggesting *cooking* and *playing*, even *moving* itself, was offering unexpected ideas. We stored our puzzlement.

In the next class session the sense of appropriate movements had narrowed. Ms. DiSchino had proposed headings for the chart: *kind of movement, distance, speed,* and *pause* at this point in the study. The children were trying to use the chart to see how these headings worked. They were charting the following trip that they had written together: Walk two steps. Run three steps. Powerwalk to the end. As they charted, it turned out that many students felt that *run* belonged in the *speed* category, not the *kind of movement* box, although they did classify *walk* as a kind of movement. Other students argued that, if *walk* was a kind of movement, *run* must be too. These children eventually proposed words like *slowly* or *fast* for the *speed* category.

Kind of movement	*Distance*	*Speed*	*Pause*
Walk	2		
(Run)	3	**(Run)**	

In the course of an animated discussion on whether *run* was a speed, Saintis, who said he believed that *run* was a speed, told the class, as he moved his hand up and down slowly, "That's a speed. . . . And you can feel the air." The other students waved their arms slowly like he did. Gail responded, "And you can stay still but you're still breathing; that's a speed." Saintis had broadened the discussion with his sense of *speed* and at the same time introduced a physical sense of *speed*.

Saintis and some of the others were interested in a kind of speed that you could feel, but one that did not necessarily involve distance in the same way that running three steps does. The activities the children were engaged in, traveling down a 9-meter strip of paper and making a chart of this movement, were planned to move them toward a view of speed that was based on linear distance and time, even if the actual mathematical calculations were not done. But eye blinking does have a speed as does arm waving, and cooking and playing do require movement in various senses.

6. MEANING AND CONTEXT: STUDYING WORDS IN MOTION

It is worth noting that Ms. DiSchino returned to discuss the chart and where *run* belonged a number of times during this unit. She regarded the children's various ideas on this subject as worthy of considerable time. Still we were not able to tie Saintis' interests into the class's direction in a way that both was useful to him and developed the curriculum further. He participated less in later discussions in the unit.

The Interview: Saintis. Saintis, fortunately, did not give up on his view of movement. In his interview he told me, when I ask him what he had learned:

Saintis: I learned that anything you do is a movement, like when you walk, anything that you do, like when you are sleeping and you turn.

I followed up by asking about speed, and he tied his answer back to movement, of the body and more cosmically.

CB: What have you learned about speed?

Saintis: Speed is a movement too because when you go like this [waves arm slowly], you go slowly, it's fast. you think it's going slow, but it's fast, cuz the world is turning, is turning very fast, so the body of someone is turning very fast too. They think they're going slow, they're going very fast even if you're going so slowly, like if someone is calling you slowpoke you're going mad slow, you're like like this [moves arm slowly] you're like going fast.

Saintis first said that *speed* is movement. Equating *speed* and *movement* in this way recalled the idea that *run* might be a speed. Saintis had offered his demonstration of speed during that discussion by moving his hand slowly and feeling the air—*speed is* movement. Since he purposely moved his hand slowly, he clearly did not feel that speed must be fast movement. Rather it is something you feel when you move.

He perhaps was also distinguishing a sense of movement that entails a change in position "like when you walk," from a kind of movement in which you end up where you started, "like when you are sleeping and you turn." He noted the differences, but he was interested in the way in which they are all kinds of movement—"anything that you do."

As he went on, Saintis tied *speed* and *movement* together in a different way, a way that problematized what you can directly feel. For example, in the second utterance Saintis seemed to be viewing *speed* and *movement* from outside his body: If you're in space and watching the earth turn, people who are barely moving must seem to be going quickly. In his view of movement he seemed to be considering the idea that it is impossible to be still: Even when you're asleep, you are moving from the perspective of outer space. Although he was interested in the way you can feel the air when you move, he was also exploring a perspective on movement in which you would not be able to sense your own speed.

At the same time as he was making these rather cosmic or philosophical points—a search maybe for the limiting case of *movement* or for the way the term can have unexpected meanings in new contexts—he was also invoking a very down-to-earth scenario of some other child calling him slowpoke while he was going "mad slow." He used his view of speed to argue with this hypothetical other who only sees him moving in relation to distance on earth, who does not realize how fast he is going from a different perspective. He tricked this hypothetical other by this shift of perspective on determining speed. This suggests perhaps a riddle—How can you move slowly and fast at the same time?—a practice well developed in Haitian culture.

The interview suggests that when Saintis offered *playing* as an example of a kind of movement in the class discussion, it truly might not have been because he did not understand the category system the others were using; in fact other activities within this unit demonstrate that he could on occasion categorize these actions conventionally. Rather, he seemed to be interested in extreme cases, perhaps in order to figure out what aspects of meaning were central or what could count in one situation and not in another—a trick upon which riddles often depend. What is wrong with *playing* or *cooking* as examples of movement? Or *turning over in your sleep*? How could something be fast and slow at the same time? How could something move and yet not cover distance?

The process of filling in the chart had in some ways for the entire class been one of questioning the meanings of words in this new context, for example, Does spin have a distance? Is run a speed? These are not questions one normally asks about these terms. They were prompted by the chart and its structure and by the trips that the children were trying to fit into the chart's structure. For all the children, the chart and the interaction with it as they jointly constructed it directed attention to the language used in it.

For Saintis, because he had only recently learned to read, charts may have been particularly unfamiliar. The head words in a chart are undoubtedly very striking to someone who has not always been familiar with this kind of literacy tool. Because they are outside of a sentence, the constraints that the other words in a normal sentence offer, by filling in time or place or other features, by being part of narrative structures or structures of explanation, etc., may not seem to exist at first in a chart. "Sam walks slowly to school" constrains the imaginative possibilities for the scene in a different way than the stark sequence "walk" in one box and then "slowly" in the next does in the chart. Saintis, however, appeared to

6. MEANING AND CONTEXT: STUDYING WORDS IN MOTION **183**

take this to be a particularly hypothetical space. I see him as asking of *speed* and *movement* What possibilities are there for their meaning? What limits if any are there on what they can mean?

From Saintis we gain a sense of the range of ways students might consider *movement* and *speed*. He also leads us to recognize more distinctly the role of new contexts in influencing word meaning. Saintis in his riddle-like formulations seems to be focused himself on a view of this.

Unfortunately we were not able to see Saintis's thinking very clearly at the time. His interview was a help, but it took place after the class had moved on from the issues about which he remained concerned. It is certainly worth asking whether, if we had been better able to address Saintis's view of *movement* or *speed*, he would have moved further with his class in addition to pursuing his own view.

Classroom Conversations: Roland. Let us turn to Roland, who took a somewhat different approach. We see him first in the classroom talking about a hypothetical walking trip down the piece of adding-machine tape laid on the floor and marked into 9 meters. Ms. DiSchino (3/12/99) asked him what the numbers (meter marks) on the tape were doing.

Roland: The numbers are doing/see how far you go
like, say, like anybody in the classroom that's walking is/lands on like 5
and you see how far you are
and you look back
and you're like **fa:r**.

Ms. DiSchino asked a question that in its form could be a request for one particular answer. The answer to "What are the numbers of the tape doing?" would be, "They mark the meters." In fact she may have been looking for that answer. And yet, categorizing a teacher's questions as open or closed often misses the role of classroom culture in how children respond and how their responses are taken up. In this classroom, Roland felt he could answer this question with a very different sort of answer. His teacher could have corrected him and might have in another instance. Here, she gave him the time to offer his thoughts, different from hers, on the role of the numbers.

Roland responded by telling a kind of a story and at the same time offered something that sounds like a mathematical rule. Say you or I or "anybody in the classroom that's walking" took this walk, he said, thus claiming that what he was going to say was true in general and had the status of a rule, at least in his classroom. But then he moved to a more expressive kind of talk, reminiscent of oral stories, "You look back and you're like fa:r." He emphasized *far* and elongated it. Let me consider each aspect a little further.

His use of this rule-like beginning, "Like anybody in the classroom that's walking is/lands on like 5," reminded his listeners that this experience of *far* had a standardized aspect. This was an issue that the class had taken up: Step size can vary; 20 steps might be far for a short child, while 5 steps might be far for a taller child with larger steps, but if you attend to the numbers on the line, 5 is far. This, he claimed, was true for all people in the classroom.

Then, when he said, "And you look back and you're like **fa:r**," he seemed to be giving some felt sense of the distance traveled. His use of the indeterminate pronoun *you*, no particular "you" but "you" in general, seems, at one and the same time, to offer the listener the opportunity to put herself into the situation and to make reference again to the standardizing role of numbers. You will go to a particular place designated by 5, and you will feel "you're like far." Using *like* as he does allows a hesitation before the next word, which then receives an emphatic stress, a real down beat and one held for an extra length of time; Roland did that with the word *far*. *Like* is also often used to introduce direct discourse in everyday speech. When he said, "I was like far," it feels rather as if Roland had walked the 5 meters in front of the listener, looked back, and then exclaimed, "Far." All of this serves to give his listener the experience of how very far Roland had come.

And yet, he is only 5 meters from the start. One might ask, "5 meters is far in what world?" Roland, I think, was also realizing that 5 is far within this activity but not necessarily elsewhere. Five meters is not far in every world, the world of airplane travel for example. There seems to have been a sense of irony in that "far." Someone outside the classroom, who had not shared their experiences, might not recognize 5 meters on the tape as *far*.

Of course we have no way of knowing for certain how much of this Roland was aware of in his use of language in that instance. However, when I asked him in his interview what he was learning, it was language and particularly distance that he mentioned.

The Interview: Roland. I began, as I did with Saintis, by asking Roland what he had been learning.

> Roland: I'm learning how, like, I never knew what distance were and some stuff from speed.
> I never knew what that like, hopping, I never knew that was [a] speed.
> Me, I used to hop in one spot, so now I know, what hopping means.
> I can hop here to there and hop anywhere.

6. MEANING AND CONTEXT: STUDYING WORDS IN MOTION   185

>And now I think what speed is, like moving your body, like speed.
>I used to think it was just walking or running or just racing or crab walking.

Roland claimed to have learned what *distance* meant and "some stuff from speed." He was struck that hopping could cover distance. The children also discussed this in relation to spinning. In this context, for example, Roland knew that hopping had to cover distance because of the necessity of getting down the meter tape and the demands of the chart, where the next heading is *distance*. The inferences he made came from a very different place than in the language of sentences. This felt like a new way of understanding the word and it was.

He agreed with Saintis that *speed* is moving your body, a much larger category, he told me, than he had previously, "just walking or running or just racing or crab walking." I asked him to tell me more about what he thought about *distance* now.

>Roland: now I know what distance means.
>Because distance, I think it means the numbers and movement.
>It means like when you move, like movement, like crab walking, like running.
>And distance, I used to never know what that means.
>I didn't even know what distance means.
>Now I found out, it means numbers like distance between 5.
>Now I know what the distance [of] 5 means.

When I asked him what he now understood about *distance*, he addressed the question in terms of movement. As he said, "I think it means the numbers and movement. It means like when you move." He was referring to the chart here, I believe, and the way that the trips had come to be charted. Roland was defining these words in terms of each other: "Distance . . . means the numbers and movement." Saintis too said, "Speed is movement." It appears that the structure of the chart led them to see the words *speed*, *movement*, and *distance* in relationship to each other and to the particular imagined world in which they were engaging.

The "distance [of] 5" he said he now understood. If *skip* is the movement, it is possible to label the distance as 5, meaning 5 skips. For a period of time the children did exactly this, but they noted that if you calculated distance in this way, 5 skips might not get you to the same place as 5 runs would, nor were every child's skips the same size. However if by 5 you mean 5 meters, then it did not matter what the movement is, or who the person taking the movement was, you would end up in the same place. It seems possible that it was this understanding that Roland was claiming when he said, "Now I know what the distance [of] 5 means."

From Roland, like Saintis, we gain an increased sense of how context influences word meaning. Unlike Saintis, I think that Roland did not see the chart's headings so much as spaces for cosmic or riddling questions about word meaning. Rather, he seemed to see the headings as constraining meanings, but in new ways that intrigued him. In the chart, Roland realized that movement and distance and speed all must remain in some relationship. *Hopping* must have a distance as well as a speed. *Distance* must have movement as well as numbers. He was trying to do this. And he was very struck by the way meanings shift in this context.

And yet it is not exactly true that he learned what *distance* meant or at least that "he never knew." When I asked him the distance to his house, he said "like 5 miles" without hesitation, which is about right. I do not think that *distance* was a new word to him that he never understood before. I believe it was an old word that he was seeing anew in the context of the chart.

CONCLUSION

What have Saintis and Roland learned? All the children in this classroom were developing "technical" meanings for words, meanings that are necessary for particular disciplinary activities such as the ones in which they were engaged, meanings that are a part of this particular set of problems and tools. In this example the terms that were developing particular disciplinary meanings were long-familiar words. The meanings that they were developing were the result of a perspective that the students took, with their teacher, on *movement, distance*, and *speed*, as they tried to communicate successfully in a world of charted and actual trips. These meanings were also the result of the possibilities contained and suggested by the structure of the type of chart they had chosen.

The overall classroom inquiry was made up of activities that the children found intellectually challenging and engaging. The teacher did not have a full plan worked out in advance but rather observed what the children did regularly on video and planned the next weeks' activities with this in mind. The children worked often in small groups in which each child's participation was necessary. Perhaps most importantly, a classroom culture in which the conversation on "is *run* a speed?" was held multiple times is a culture that promotes an attitude of awareness of and inquiry toward language use. Saintis and Roland, supported by the habits of mind cultivated in their classroom, made their own inquiries. Roland explored the way the activity of walking down the 9-meter strips defined what was *far* in general terms for anybody in his classroom. He went on to redefine for himself *distance, speed*, and *movement* in this context.

Saintis used the chart and some of the classroom discussions to further his own thinking about speed and movement, somewhat outside of the conventions being developed among his classmates and yet in ways that were very engaging and productive for him and certainly concerned with uses of language in school and outside.

The students here were accomplishing something perhaps analogous to what Goodwin (1997) called developing "a professional vision." He discussed, for example, a group of graduate students working with a chemical reaction that they must stop at just the right moment. This moment is signaled when they observe just the right color of *black*. *Black* is unproblematic in many contexts, hardly considered sophisticated vocabulary. It is something we know, but these students, in order to stop the reaction at the right moment, discover qualities of *black*, ways of discussing what *black* means and of seeing the complexities within *black* that are important for them. They refer, for example, to the "gorilla fur" aspect of the right kind of *black*.

Saintis and Roland were highly aware of the necessity to uncover the meaning—in this new context and with these intentions—of these formerly unproblematic terms. However, their discovery is more than just the actual new sense of *motion* and *distance* and *speed*. The excitement that they communicate about what they now know results from their broader discovery, that is, that words change with context. A good part of what we term *academic language* is exactly this—not new words but old ones with new resonances, new connections, new commitments. To speak the language of power in our schools, a knowledge of how words can be bent, shaped, expanded, and interrogated is a crucial component.

What have we learned from Saintis and Roland? Puzzling students like these regularly present us with moments when at first glance they seem to be missing the point or to be confused, and yet upon further reflection it appears that they are seeing something in a way we cannot because we are constrained by our own goals and assumptions. Close investigation of what these students are saying rewards us with a better sense of the complexity beneath terms and usages that otherwise we take for granted. The chart, the terms it contained, the sense of *movement* that Saintis was considering, the sense of *far* with which Roland was concerned, help us to see what it means to talk in disciplinary ways on this topic and in this classroom. In a rich context such as the one provided in this classroom, where there was regular attention to language and interpretation as an integral part of doing science and where the curriculum was responsive to children's ideas and to the discipline, closely watching children such as these will deepen our own understanding of what constitutes school language and disciplinary language itself and how it functions in various contexts. We can see when they are taking on new meanings and purposes for words and why. We can see where they put their energy and what they find challenging. There will always be surprises. And then planning what to do next in the classroom can be based on our own fresh thinking and vital curiosity about the issues involved in what is being taught and what is being learned.

The question "What children do you learn from?" is not often asked in teacher preparation. And yet I think it might be a question that, as educators, we ought to ask routinely, picking out our puzzling children in various ways and at different points for interview, for further probing and study. When we are using what we think are the best practices and they do not work for all children, we generally blame the child's background, sometimes without even realizing it. We say, "The parents were too busy working to take him to museums," or, "Her mother was too overwhelmed with the younger children," or, "It's difficult when they don't yet speak English well." We are trying to be kind. We are sorry he or she did not have the advantages others have. We try to provide them. What I think this attitude does however, despite what are good intentions, is take our attention away from the process of reflective practice. It takes us away from a stance of inquiry toward the child's ideas and toward what we are teaching. It prevents us from learning from these children.

Although we did focus on children from whom we expected to learn, we were not able to explore as deeply as we would have liked what they had to say while the curriculum itself was being formed. Because we knew they had something to teach us, we did give them extra time to explain themselves, stopped ourselves from correcting them in many cases until they made us understand their meaning. However, next time I believe there would be additional ways to bring some of what children like Roland and Saintis were interested in to the attention of the entire class, to everyone's benefit. Learning from and with these children not only is a powerful way to bring them into the curriculum, it also challenges our own sense, and that of many other students, of what we know and how we know it. I do not mean to convey that puzzling children are only minority and bilingual children or that exploring the ideas of the children one finds puzzling is a way to help them only. Rather the goal in listening to children like Roland and Saintis is to keep both the teacher and all the students in touch with the complexity and the mystery of what they are learning.

ACKNOWLEDGMENTS

The research reported herein was supported by the National Science Foundation, REC-0106194 and ESI 9555712, the Spencer Foundation, and the U.S. Department of Education, Office of Educational Research and Improvement, Cooperative Agreement No. R305A60007-98 to the National Center for Improving Student Learning and Achievement in Mathematics and Science, University of Wisconsin, Madison. The data presented, statements made, and views expressed are solely the responsibility of the authors. No endorsement by the funding agencies or foundations should be inferred.

REFERENCES

American Association for the Advancement of Science. (1993). *Benchmarks for science literacy*. New York: Oxford University Press.
Ballenger, C. (1997). Social identities, moral narratives, scientific argumentation: Science talk in a bilingual classroom. *Language and Education, 11*(1), 1–14.
Ballenger, C. (1999). *Teaching other people's children*. New York: Teachers College Press.
Ballenger, C. (2000). Bilingual in two senses. In Z. Beykont (Ed.), *Lifting every voice: Pedagogy and the politics of bilingualism* (pp. 95–112). Cambridge, MA: Harvard Education Publishing.
Biagioli, M. (Ed.). (1999). *The science studies reader*. New York: Routledge.
Brookline Teacher Researcher Seminar. (2004). *Regarding children's words. Teacher research in language and literacy*. New York: Teachers College Press.
Clement, J., Brown, D., & Zeitsman, A. (1989). Not all preconceptions are misconceptions: Finding anchoring conceptions for grounding instruction on students intuitions. *International Journal of Science Education, 11*, 555–565.
Conant, F., Rosebery, A., Warren, B., & Hudicourt-Barnes, J. (2001). The sound of drums. In E. McIntyre, A. Rosebery, & N. González (Eds.), *Classroom diversity: Connecting curriculum to students' lives* (pp. 51–60). Portsmouth, NH: Heinemann.
Cope B., & Kalantzis, M. (1993). *The powers of literacy: A genre approach to teaching writing*. Pittsburgh, PA: University of Pittsburgh Press.
Delpit, L. (1995). *Other people's children: Cultural conflict in the classroom*. New York: New Press.
DiSchino, M. (1987). The many phases of growth. *Journal of Teaching and Learning 1*(3), 12–28.
DiSchino, M. (1998). "Why do bees sting and why do they die afterward?" In A. Rosebery & B. Warren (Eds.), *Boats, balloons, and classroom video: Science teaching as inquiry* (pp. 109–133). Portsmouth, NH: Heinemann.
diSessa, A., Hammer, D., Sherin, B., & Kolpakowski, T. (1991). Inventing graphing: Metarepresentational expertise in children. *Journal of Mathematical Behavior, 10*, 117–160.
Fox-Keller, E. (1983). *A feeling for the organism: The life and work of Barbara McClintock*. New York: W. H. Freeman.
Gallas, K. (1994). *The languages of learning*. New York: Teachers College Press.
Gee, J. P., & Clinton, K. (in press). An African-American child's "science talk": Co-construction of meaning from the perspective of multiple discourses. In M. Gallego & S. Hollingsworth (Eds.), *Challenging a single standard: Multiple perspectives on literacy*. New York: Teachers College Press.
Gooding, D., Pinch, T., & Schaffer, S. (Eds.). (1989). *The uses of experiment: Studies in the natural sciences*. Cambridge, England: Cambridge University Press.
Goodwin, C. (1997). The blackness of black: Color categories as situated practice. In L. B. Resnick, R. Saljo, C. Pontecorvo, & B. Burge (Eds.), *Discourse, tools, and reasoning: Essays on situated cognition* (pp. 111–140). Berlin: Springer-Verlag.

Hammer, D. (1997). Discovery learning and discovery teaching. *Cognition and Instruction*, *15*, 485–529.
Hammer, D. (2000). Student resources for learning introductory physics. *American Journal of Physics* (Physics Education Supplement), *68*(S1), S52–S59.
Heath, S. B. (1983). *Ways with words: Language, life, and work in communities and classrooms*. Cambridge, England: Cambridge University Press.
Hymes, D. (1972). Introduction. In C. B. Cazden, V. P. John, & D. Hymes (Eds.), *Functions of language in the classroom* (pp. xi–lvii). New York: Teachers College Press.
Jacob, F. (1995/1987). *The statue within*. Plainview, NY: Cold Spring Harbor Laboratory Press.
Latour, B., & Woolgar, S. (1986). *Laboratory life: The social construction of scientific facts*. Princeton, NJ: Princeton University Press.
Lee, O., & Fradd, S. (1996). Interactional patterns of linguistically diverse students and teachers: Insights for promoting science learning. *Linguistics and Education*, *8*, 269–297.
Lehrer, R., & Schauble, L. (in press). Modeling in mathematics and science. In R. Glaser (Ed.), *Advances in instructional psychology* (Vol. 5). Mahwah, NJ: Lawrence Erlbaum Associates.
Lynch, M. (1985). *Art and artifact in laboratory science: A study of shop work and shop talk in a research laboratory*. Boston: Routledge and Kegan Paul.
McDermott, L., Rosenquist, M., & van Zee, E. (1987). Student difficulties in connecting graphs and physics: Examples from kinematics. *American Journal of Physics*, *55*, 503–513.
Michaels, S., & Sohmer, R. (2000). Narratives and inscriptions: Cultural tools, power and powerful sensemaking. In B. Cope & M. Kalantzis (Eds.), *Multiliteracies* (pp. 267–288). London: Routledge.
Minstrell, J. (1989). Teaching science for understanding. In L. Resnick & L. Klopfer (Eds.), *Toward the thinking curriculum: Current cognitive research* (pp. 131–149). Alexandria, VA: Association for Supervision and Curriculum Development.
Monk, S. (2000). *Designing representations of motion*. Report to National Center for Improving Student Learning and Achievement in Mathematics and Science, University of Wisconsin.
Monk, G. S. (in press). "Why would run be in speed?" Artifacts and situated actions in a curricular plan. In R. Nemirovsky, A. Rosebery, B. Warren, & J. Solomon (Eds.), *Learning environments: The encounter of everyday and disciplinary experiences*. Mahwah, NJ: Lawrence Erlbaum Associates.
National Research Council. (1996). *National science education standards*. Washington, DC: National Academy Press.
Nemirovsky, R., Tierney, C., & Wright, T. (1998). Body motion and graphing. *Cognition & Instruction*, *16*, 119–172.
Ochs, E., Jacoby, S., & Gonzales, P. (1996). "When I come down I'm in the domain state": Grammar and graphic representation in the interpretive activity of physicists. In E. Ochs, E. A. Schegloff, & S. Thompson (Eds.), *Interaction and grammar* (pp. 328–369). New York: Cambridge University Press.
Rheinberger, H. (1997). *Toward a history of epistemic things: Synthesizing proteins in the test tube*. Stanford, CA: Stanford University Press.
Root-Bernstein, R. S. (1989). How scientists really think. *Perspectives in Biology and Medicine*, *32*, 472–488.
Rosebery, A. S. (1998). Investigating a teacher's questions through video. In A. Rosebery & B. Warren (Eds.), *Boats, balloons and classroom video: Science teaching as inquiry* (pp. 73–80). Portsmouth, NJ: Heinemann.
Rosebery, A. (2000). *"What are we going to do next?" A case study of lesson planning*. Manuscript submitted for publication.

Warren, B., Ballenger, C., Ogonowski, M., Rosebery, A., & Hudicourt-Barnes, J. (2001). Rethinking diversity in learning science: The logic of everyday languages. *Journal of Research in Science Teaching, 38*, 1–24.

Warren, B., Pothier, S., Ogonowski, M., Noble, T., & Goldstein, G. (2000, November). *Children's accounts of motion: Re-thinking the role of everyday experience.* Presentation at the NCISLA Seminar on Case Studies and Instructional Design, Ashland, MA.

Viennot, L. (1979). Spontaneous reasoning in elementary dynamics. *European Journal of Science Education, 1*, 205–221.

CHAPTER SIX METALOGUE

Balancing Our View of Science With Language Usage and Culture

Cynthia Ballenger
Randy Yerrick
Wolff-Michael Roth

Randy: I am drawn back to Greg Kelly's chapter to make reference to how and why we must look at things in this way to solve your problem that seems to be, "How are progressive teaching strategies interpreted by children not indoctrinated with American school culture?" It is always these children who teach me the most about my teaching—or my assumptions about teaching. I think teachers (less experienced ones especially) are prone to think of lessons that are not perfectly executed or interpreted as containing mistakes. What are the most direct lessons to be learned from this chapter for teachers as they consider teaching children like Saintis and Roland? It is clear that Ms. DiSchino is a thoughtful and reflective teacher. How were these discussions central in shaping her next choices as a teacher? How were these discussions central in shaping your next choices as a researcher? How does this inform the wider educational reform audience about what we are missing from the ongoing conversation concerning students' learning?

Michael: Randy, your questions are valid, but we also must not forget that her competencies become salient because of the very type of analyses that Cindy conducted. Others in our field not attuned to the discourse in the same way may look at the same classroom interactions and ask the question, "So where is the science?" and not recognize the value of knowing that Cindy allows us to see in Saintis's and Roland's contributions.

Cindy: This approach has developed at Chèche Konnen through collaboration with teachers and researchers. The approach involves decisions made as we teach based on what we think the children are saying, and some later level of analysis of what went on that usually is useful later on, in our overall understanding of what we are teaching and how children take it up. I guess it's true that the type of analyses we defer to in the Chèche Konnen project and other work contrasts much of the current work of the field, but it is not just my own invention. It is one that is shared and has emerged from and been informed by a broad and diverse number of perspectives from Haitian and American teachers, to parents and community members, to researchers and linguists, and many others. It has been a long, fruitful, and rewarding project that has helped us to understand the role of culture and language in learning science and the intricacies of making science accessible to all. I offer this chapter as one that explicates inherent methodological decisions in an instance of learning to effectively teach two Haitian children who otherwise may be interpreted as "off task."

Randy: These children demonstrate the vast array of interpretations that are present in any given class. It is this kind of work coming out of TERC that has helped teachers and science education researchers come to understand that culture, voice, and content are not easily separated. Anyone familiar with the work of CKC will come to understand that it is one of the exemplary programs engaging parents, insiders, and outsiders into the voices and lives of children to make more equitable conditions for all learners of science. The issues Saintis and Roland face in engaging in the science lesson are not so unlike students who are fluent English speakers but lie outside the mainstream of successful students in science. In what ways does your approach inform researchers about what they may have missed from examining this classroom from a more traditional research perspective? How might this research approach help us to guide future research and reform in other classrooms?

Cindy: I am hoping that we realize how we can become a culture that promotes an attitude of awareness of and inquiry towards language use. Saintis and Roland are just two examples of children who, when given the made their own inquiries, brought to bear their own knowledge and experiences and explored the way the activity and representative model helped them co-construct some meaning of classroom constructs like *far* and *fast*. They are children, like others, who are continually engaged in a process of redefining for themselves meanings, uses of charts and other artifacts, and classroom discussions to further their own thinking—though admittedly it is sometimes outside of the conventions being developed by other classmates. This is not unique to this classroom. I believe children in every classroom are actively engaged in this process as well but in different ways that reflect their cultures, their previous experiences and their individual interests. This classroom was unusual in the way the language was often publicly interrogated by the teacher and her students. I see that if we are able to see what concerns and interests students arrive with earlier in the learning process, we can better facilitate such habits of mind for all children to make their own inquiries. But I wouldn't limit the value of this kind of research to methodological advice for researchers. I see this chapter as having an impact on other areas of reform besides educational researchers. Don't you?

Randy: Yes, I think your approach can have a major impact on all science teachers and potentially the way we prepare them for such work. The conscientious teacher is one who welcomes such questions, challenges, and perspectives, for the teachers themselves must come to grips with what we mean by *speed* and *motion*. The reader would be wrong to assume that this chapter is written exclusively for the teachers operating in multicultural contexts. It is for all teachers who dare to create such discourse so that divergent child-centered views, regardless of their cultures are welcomed into the conversation. The child is usually asking quite sincerely, "Why isn't 'eyes blinking' a movement?" These questions have lasting effects and move us (in the moment and far beyond) closer to our own deeper understanding of what we hope children understand and how to get them there.

Michael: Your point is well taken. As longtime teacher and researcher I have encountered few children that were not sincere. Many science educators fail to recognize this, which I think arises from the rationalist approach they take to knowing and learning. For example, you suggested that there is a "vast array of interpretations that are present in any given class." Such expression put emphasis on rational acts of interpretation. I read the chapters by Cindy, Jim Gee, as taking a lifeworld perspective, whereby human beings are viewed as taking the world at face value. In our everyday lifeworlds, we do not interpret traffic lights, "If the light is red, then stop" but rather we simply stop; we do not plan shopping trips but simply shop. Even the "understanding" scientists develop is intimately tied to their lifeworld experiences—if these experiences are missing, they cannot even interpret the simplest graphs from undergraduate courses in their own domain (Roth, Bowen, & Masciotra, 2002).

Randy: But I wonder how we can move greater numbers of teachers in this direction or how we can get teacher standards to reflect such an open and evolving definition of teacher knowledge. Current standards do not yet clearly demarcate such thoughtful activity and even proven National Board Certified teachers have quite a range of this kind of knowledge. With the context of teacher testing, electronic portfolios, and other kinds of focus on "performance" I wonder if there is a way to move the demonstration of expertise to one that exposes teachers who are not extending themselves to understand school outside of their own personal experiences and which allow children unlike themselves to flounder. Have you found any clearly defining attributes that demarcate teachers like Ms DiSchino from teachers who are not making such strides as a way to reward those teachers making such cultural and language connections?

Cindy: I am skeptical of any generalized approach to change science teaching. However I do think Mary DiSchino represents some useful practices—she gave the children a great deal of room to express their own ideas, while carefully planning the activities about which they would talk. She then, with colleagues, took the time to look carefully at what they had said, not assuming that the meaning of the children's ideas was either obvious or the same as her meaning. She expects to learn about the science itself from the ideas of her children, and she welcomes this when it happens. There are some common attributes of teachers who effectively use linguistic analyses to better engage children and we must continually ask how progressive teaching practices are taken up by children unfamiliar with American school culture because the assumption is that this is automatic. I think where we often fail in reaching all children is in not recognizing the difficulty for children, particularly children from families with little education, to learn about the formalities of language meaning and use associated with mature practice in academic disciplines. Exploring what these children are learning then opens for us a view of what we are teaching. These are children from whom we can learn about the language and about the science. I have hope that, when children's own ideas and questions and their everyday language and experience are given a central place, further analyses of other contexts will build a knowledge base to assist teachers in connecting science concepts with children's experiences. Like Saintis and his use of extreme contrasts of cooking as motion, perhaps we should be concerned with extreme cases of interpreting the failures as well as success of teachers trying to make such cultural and language connections to better understand the challenges of such works (Ballenger, 1995).

Michael: Cindy, your analysis goes in a direction that is similar to what I have tried in my chapter, emphasizing that we do not encounter an abstract world but always a lifeworld, including the utterances of and social relations with others. This lifeworld, we do not interpret but because it is so familiar take it at face value. For Saintis and Roland, words accrued to existing meaning, the lifeworld, rather than becoming meaningful (receiving meaning). I think that this turns the traditional problematic in science education around that assumed words get meaning; rather, it recognizes that words come to meaning (Heidegger, 1996). If this is the case, then the teaching problem also reverses in the sense that science educators have to wonder how to allow students to evolve lifeworlds and meanings; this is a different project than saying, "Here are some words (speed, distance); learn them, make them meaningful!"

REFERENCES

Ballenger, C. (1995). Because you like us: the language of control. *Harvard Educational Review, 62,* 199–208.
Heidegger, M. (1996). *Being and time* (J. Stambaugh, Trans.). Albany: State University of New York Press.
Roth, W. -M., Bowen, G. M., & Masciotra, D. (2002). From thing to sign and 'natural object': Toward a genetic phenomenology of graph interpretation. *Science, Technology, & Human Values, 27,* 327–356.

CHAPTER SEVEN

Science for All: A Discursive Analysis Examining Teacher Support of Student Thinking in Inclusive Classrooms

Kathleen M. Collins
University of San Diego

Annemarie Sullivan Palincsar
Shirley J. Magnusson
University of Michigan

> *All students, regardless of sex, cultural or ethnic background, physical or learning disabilities, future aspirations, or interest in science, should have the opportunity to attain high levels of scientific literacy. By adopting this principle of equity and excellence, the* Standards *prescribe the inclusion of all students in challenging science learning opportunities and define a high level of understanding that all students should achieve. In particular, the commitment to science for all implies inclusion of those who traditionally have not received encouragement and opportunity to pursue science—women and girls, students of color, students with disabilities, and students with limited English proficiency. It implies attention to various styles of learning, adaptations to meet the needs of special students and differing sources of motivation.*
>
> —National Science Education Standards
> (National Research Council, 1996, p. 221)

Although educational equity is a concern of those who support standards based reform efforts, the only national content standards to include students with specific disabilities are the *National Science Education Standards*, as noted above (McDonnell, McLaughlin, & Morison, 1997). Even so, the emphasis is on adapting instructional approaches to meet the needs of "special students." This discourse suggests there is some static quality located within certain students (those identified as "special" or as having "disabilities") that impedes them from learning in instructional science contexts. The implication is that it is the students who are deficient, and who therefore require adaptations of classroom learning contexts in order to overcome their deficiencies.

This approach to addressing inequities in educational opportunities and differential academic achievement has become known as the "deficit model." In various forms, deficit discourses have been used to explain the low school achievement of students of color, poor and working-class students, students whose first language is one other than English, and students identified as having learning disabilities (see discussions in Valencia, 1997). One common form that deficit discourses take is the assertion by (often well-meaning) teachers that some students (due to differences in ethnicity, class, language, or their identification as students with special needs) bring less to school-learning contexts in terms of background knowledge and experience than their peers. For example, Gersten and Baker (1998) argued that students with disabilities must be taught theoretical and factual knowledge *prior* to engaging in inquiry instruction, rather than advocating that teachers design inquiry instruction to capitalize on the knowledge that such students bring to the classroom.

Elsewhere we have presented research that challenges this deficit perspective by illustrating the efficacy of guided inquiry instruction in tapping into the rich science-related background knowledge of students with special needs (e.g., Palincsar, Collins, Marano, & Magnusson, 2000; Palincsar, Magnusson, Collins, & Cutter, 2001). Our program of research was informed by an emerging perspective that, in contrast to a deficit model, is designed to consider the ways in which context shapes student performance. Informed by social constructivist and sociocultural theories, this perspective locates ability and disability not within the individual learner but rather as co-constructed within situated activity (see discussions in Trent, Artiles, & Englert, 1998; Collins, in press). The assumption is not that some learners are deficient, but that specific and identifiable features of instructional activity contexts, such as forms of symbolic mediation and patterns of discursive interaction shape learners' forms of participation, and hence shape the abilities they are able to show. This perspective thus points to the importance of examining instructional conversations as "construction zones" in which knowledge building may occur (Newman, Griffin, & Cole, 1989).

In our previous work, we identified the range of challenges faced by students with special needs who engage in a particular form of inquiry science instruction (Palincsar et al., 2000; Palincsar et al., 2001). These challenges ranged from the representational (i.e., challenges involved in representing one's understanding to others through written documentation) to the social-interactional (i.e., challenges related to being heard or responded to by one's peers and gaining access to materials). In each of the classrooms in which we worked, gaining access to instructional conversations was a challenge for those students identified as having special needs (Palincsar et al., 2001). Furthermore, the teachers we worked with emphasized that the most powerful "intervention" they implemented was the intentional monitoring and responding to students' thinking, strategies that were actualized through their instructional conversations and specific discursive choices.

In an effort to make these strategies visible, in this chapter we focus closely on the instruction of one of these teachers and examine the conversations that took place during her inquiry teaching. Our aim is to identify the discursive features that supported students' construction of specific forms of scientific understanding, particularly those that seemed to facilitate the participation of those students who had been identified as having special needs. In the "construction zones" of these instructional conversations, we explored what the discursive strategies were that seemed to help fulfill the promise of "science for all."

THE GUIDED INQUIRY ORIENTATION TO TEACHING

The inclusive classroom of fifth-grade teacher Laura Bozek provided the context for our study.[1] During the course of this study, Laura's class was engaged in investigating the behavior of matter, specifically with respect to the phenomena of floating and sinking. This occurred as part of Laura's participation in the Guided Inquiry supporting Multiple Literacies (GIsML) Community of Teacher Practice (Palincsar, Magnusson, Marano, Ford, & Brown, 1998). At the time of this study, the community had been meeting regularly for approximately 2 years in an effort to co-construct understanding of effective practice relative to the GIsML orientation to teaching science.

GIsML is a particular form of inquiry-based instruction that draws on the knowledge-building practices of the scientific community (Magnusson, Palincsar, & Templin, in press; Magnusson & Templin, 1995). The GIsML orientation to teaching science was developed to support the engagement of individuals in sustained inquiry about the physical world and to provide opportunities for social interaction in ways that mirror the knowledge-building practices of the scientific community. GIsML assumes a classroom learning community (Rogoff, 1994) and is an orientation that supports the development of such a culture. The goal of this instruction is to construct "a context in which learners can practice the language and

---

[1] Portions of the transcripts presented here first appeared in *Ability Profiling and School Failure: One Child's Struggle to Be Seen as Competent* (Collins, K., in press).

tools of scientific problem solving in socially situated activity" (Magnusson & Palincsar, 1995, p. 43).

The GIsML orientation has three main phases: engage, investigate, and reporting, with two supporting phases—preparing to investigate and preparing to report. Students participate in the same phases of instruction recursively, either in the same context or in a different context, to ensure multiple opportunities to construct and deepen understanding. New definitions challenge and extend static, elitist, and functional and traditional definitions of literacy that argue for consideration of representational practices as socially and historically situated (Gee, 1991, 1996; Michaels & O'Connor, 1990). The "Multiple Literacies" aspect of the GIsML orientation refers to its potential to support diverse forms of representation and ways of meaning making (Gee, 1991; see discussions in Au, 1998; Keller-Cohen, 1994; Sleeter, 1986; Trent et al., 1998; see also Anderson, Holland, & Palincsar, 1997, re science literacy). For example, within a single cycle of investigation students may employ graphic documentation, informal writing in lab notebooks, formal writing of presentation materials, reading of second-hand investigation materials, individual and small-group oral presentations, and whole-group discussions (Collins, 1997; Collins, MacLean, Palincsar, & Magnusson, 1998).

RESEARCH METHODS

Theoretical Underpinnings

As described earlier, the questions, methods, and analysis of this study were informed by social constructivist theory. Social constructivism, drawing primarily from the work of Lev Vygotsky, calls our attention to the dynamic interplay between individual and community ways of knowing and of representing knowledge (e.g., John-Steiner & Mahn, 1996; Forman, Minick, & Stone, 1993; Moll, 1990; Wertsch, 1991). Our analysis of the instructional conversations was also informed by the work of Joseph Schwab, who explored the nature of knowledge construction specific to scientific inquiry.

Schwab (1963, 1964) defined *substantive knowledge* as referring to the major products of science, that is, the principles, laws, and theories of science. These substantive, or conceptual, structures inform our understanding of phenomena by telling us what to look for in our examination of data, as well as how to interpret our findings. Schwab argued (1963, 1964) that substantive structures undergo a constant process of revision as more is learned and as inconsistencies between our current substantive knowledge and our new data are confronted. *Syntactic knowledge* was defined by Schwab (1963, 1964) as the structures and conventions for generating knowledge within the scientific community. His reasoning was that in order to fully understand the products of science, the substantive knowledge, students must understand the *process of inquiry* that created these products (1963, 1964).

In this study we employ Schwab's (1963, 1964) substantive and syntactic knowledge as analytical constructs in examining the moment-to-moment discursive interactions within Laura's classroom community. Although Schwab's description of substantive knowledge seemed to connect in a straightforward manner to the theories, laws, and principles inherent in Laura's context of inquiry (i.e., floating and sinking), we found that our consideration of syntactic knowledge needed a bit of unpacking. We therefore identified three dimensions of syntactic knowledge, based on the purposes they served in supporting students' development of understanding of the knowledge-building practices of scientists:

1. *Investigative:* This form of syntactic knowledge maximized the learning opportunities inherent in the investigation at hand. Investigative syntactic knowledge was knowledge about how to engage in and conduct the first-hand investigations; that is, direct study of the physical world.
2. *Communicative:* This form of syntactic knowledge maximized the opportunities for communication through supporting both the use of precise language and the development of conventions for representation. Communicative syntactic knowledge was thus knowledge about how to "talk science" (Lemke, 1990).
3. *Explanatory:* This form of syntactic knowledge supported the development and refinement of explanations of phenomena. It includes application of the "3 Cs": consistency, coherence, and completeness (Smith, diSessa, & Roschelle, 1993, describing the assertions of Einstein, 1950, cited in Magnusson, Templin, & Boyle, 1997)[2]

Whereas this identification of different dimensions of syntactic knowledge was useful in helping us understand the different types of knowledge-building opportunities available, it must be noted that any one contribution to the discourse could be evaluated from more than one perspective. This is illustrated through the analysis of transcripts that follow (and see further discussion in Collins, Palincsar, & Magnusson, 1998).

---

[2]As Magnusson and colleagues elaborate on this concept, "scientific explanations are expected to be able to be usefully applied in multiple contexts (consistency), they are expected to fit together with respect to one another (coherence), and they are ultimately expected to contribute to a complete explanation of a particular aspect of the physical world (completeness)" (Magnusson et al., 1997).

Data Collection

The current study draws primarily from four data sets. First, ethnographic field notes were taken in Laura's classroom for the purposes of documenting the development of her classroom culture, beginning with the first day of school and collected periodically thereafter (approximately once every 2 weeks). In December, when Laura began her Guided Inquiry teaching, we began collecting our second set of data: videotapes and ethnographic field notes of Laura's daily instruction. This set was later expanded by the addition of transcripts of all videotapes, expanded field notes, and student artifacts (writing, documentation, etc.). The third set of data we drew on for this chapter is comprised of transcripts of a videotaped preteaching interview and transcripts of videotaped debriefing sessions conducted after each day's Guided Inquiry teaching. In addition, we solicited feedback from Laura Bozek about our interpretation and representation of her teaching. Laura supported this effort by reading and responding to excerpts of transcripts and drafts-in-progress. Transcripts of her comments and responses relative to our representation and interpretation of her instruction constitute a fourth data set.

Data Analysis. For the purposes of this study we drew from data collected during the first two cycles of investigation engaged in by Laura and her students, which took place over 6 days of instruction.[3] In preparing our analysis, we examined the transcripts and videos of Laura's Guided Inquiry instruction in a manner consistent with Erickson's (1992) description of ethnographic microanalysis of interaction. As Erickson described, this approach has roots in context analysis, ethnography of communication, Goffman's (1959, 1961, 1981) work regarding the presentation of self, conversation analysis, and continental philosophy regarding discourse and power relations. This approach involves a kind of tacking between individual segments of discourse and the whole text from which they were drawn. As Erickson concluded, ethnographic microanalysis of interaction illustrates the dialectical co-construction of activity. Ultimately, "such microanalysis provides an holistic perspective on the conduct of interaction and the processes by which human learning and change take place" (p. 222).

Drawing on Erickson's methodological orientation and informed by Vygotsky's construct of the dialectical nature of knowledge construction, we examined the transcripts and videotapes of the first 6 days of Laura's instruction. This instructional period consisted of two cycles of first-hand

---

[3]The students and Laura later conducted a second-hand investigation and one additional first-hand investigation.

investigations of a seemingly simple and yet complex phenomenon, known as the Cartesian Diver System (Penick, 1993; Roberts, 1982). Using Schwab's identification of substantive and syntactic knowledge to guide our analysis, we identified the types of knowledge-building opportunities present in these interactions and how these opportunities were utilized by participants. We then sought to identify the discursive strategies that Laura employed and, following Erickson, traced the students' responses over time to the particular discursive move. In doing so, we also noted the ways in which the interactions of the students identified by Laura as having special needs seemed to be supported or constrained by each strategy and examined their patterns of participation relative to those of the unidentified students. For the purposes of this chapter, our analysis was limited to the instructional conversations—for example, we did not fully address the role of graphic forms of knowledge representation in mediating students' understanding.

About the Identified Students. GIsML instruction seeks to engage all students, including those identified as having special needs, in science instruction that is challenging, productive, and supportive of individual approaches to meaning-making. In doing so, we are not taking a "one size fits all" approach to instruction that fails to adapt and support the divergent needs of students. Rather, we assert that good instruction, instruction that is characterized by an attention to developing and extending student thinking, holds the potential to benefit all students. In doing so we recognize the particular demands that this type of instruction places on teachers to attend closely to students' developing understanding, and that these demands may be increased when students' ways of interacting, constructing or expressing understanding, or cognitive styles may be divergent from those of the teacher or the majority of the students in the class.

Therefore, prior to beginning her guided inquiry investigation of floating and sinking, we asked Laura to describe five students whom she identified as having special needs. She described one child, a boy of European-American descent, Danny, as having Down's syndrome. Danny was assigned a one-on-one aide to help him participate in Laura's classroom community; he was out of the classroom receiving remedial services while she was engaged in science instruction. Laura identified two other boys, Jay, who was African American, and Ned, who was European American, as having learning disabilities that she anticipated would interfere with their ability to participate fully in the guided inquiry investigation. She described Jay as either "learning disabled or emotionally impaired," and explained that she had recently referred him for special education assessment. Laura described Ned as being very interested in science and as having Attention Deficit–Hyperactivity Disorder. At the time of this study

Ned was undergoing psychological counseling for diagnosis outside of the school day, but was not taking medication. Brian, a boy of European-American descent, was identified as having a written-language disability. Martin, a first-generation Korean American, was not identified as having a learning disability at the time of this study, although Laura referred him for testing later in the school year based on her assessment of his written-language abilities. With the exception of the full-time aide assigned to Danny, Laura did not have instructional assistance assigned to support these included students.

OVERVIEW OF THE INSTRUCTION THAT WAS STUDIED

Laura began instruction by engaging the students, and did this by introducing a functioning Cartesian Diver System (CDS) as a "puzzle." She showed the assembled system, as it appears in the following figure,[4] to the students. She then pressed on the rubber sheeting, which caused the diver to sink to the bottom of the large test tube. Laura then released her finger, which caused the diver to return to a floating position at the top of the test tube. The puzzle to solve was to determine how and why the diver behaved as it did.

The purpose behind Laura's introduction of this problem context was to spark students' interest in explaining the unexpected behavior of the diver. The CDS is a rich context because the same object floats *and* sinks, a

Cartesian Diver System

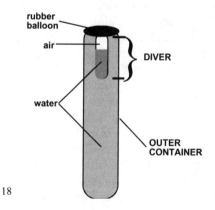

18

---

[4]This graphic was created by Shirley Magnusson and was used in several versions of text created by the GIsML research team to support students' second-hand investigations.

contrast that requires students to develop an explanation that can account for both states, and because additional contexts for investigation are easy using materials with different properties for the diver (e.g., heavier, smaller) and the outer container (less rigid, differently shaped). It is also a context that captivates students and has the potential to lead to many additional investigations of related concepts (e.g., characteristics of matter).

After Laura demonstrated the operation of the CDS, students made initial observations to the class about what they thought was happening in the system. Next, students worked in collaborative dyads at stations supplied with a container of water, two large test tubes, a small test tube, a small vial, an eyedropper, a beaker, two pieces of rubber sheeting, and two rubber bands. Each student constructed his or her own CDS, so that each pair had two CDSs with divers of different sizes. Before concluding for the day, Laura asked the students to respond to three writing prompts: (1) Describe how you constructed your diver, (2) describe how your diver behaved, and (3) explain your thinking about why the diver behaved the way it did. Students were told that they could use both words and illustrations in their responses.

On the second day of Guided Inquiry instruction, students regrouped with their partners to construct a poster to support reporting their findings from the previous day to the rest of the class. Students used the writing they had completed the day before to construct their posters. After the posters were finished, two pairs of students reported their observations and conclusions to their classmates.

The largest chunk of instructional time on the third day was devoted to continuing the reporting phase of Guided Inquiry, with two pairs of students reporting this day as well. Shortly before the end of class on the third day, Laura began the engagement for the second cycle of inquiry by introducing an "upside down" or "Australian" version of the Cartesian Diver System, which operates on the same principles as the original CDS but appears different because the external container is inverted. After viewing this system, students made written predictions about what they thought would happen when Laura pressed on the rubber sheeting, and they explained their reasons why this would happen.

The fourth day of instruction began with students investigating the CDS a second time, with many building Australian CDSs. Although they responded to the same writing prompts to which they had responded on the first day of instruction, this time students were encouraged to write while they were investigating rather than after the materials had been put away. The second round of first-hand investigation was again followed by 2 days of public sharing, or reporting, on Days 5 and 6 of the inquiry. Students again worked with their partners to construct a poster, and four pairs of students reported across the 2 days.

Analysis of Instruction: Identifying Supportive Strategies

In the following section we describe those discursive strategies that seemed to support the participation of all Laura's students, particularly those whom she had previously identified as having special needs, in the guided inquiry investigation. In identifying these moves as supportive, we considered issues of immediate access to the instruction conversation (i.e., whether the move supported students in attaining the floor, for example), response (such as whether the move resulted in uptake of the speaker's ideas by listeners), as well as the forms of substantive and syntactic understanding developed as a result of the move.

Explicit Introduction of Physical Tools of Inquiry and Their Labels in the Context of Inquiry.

One way in which Laura provided opportunities for all students to participate in the instructional conversations was through her explicit introduction of the physical tools they would be using. As one might expect, Laura's explicit introduction of the physical tools of the inquiry occurred predominately in the engagement phase of her GIsML instruction, as in the following example from the first day of the inquiry. Walking around her classroom, Laura held a Cartesian Diver System up at eye-level for her students (who were seated) as she explained:

Laura: I have another puzzle[5] that I was interested in having you work on. . . . This is called a Cartesian Diver. And that's the name that we have for this system. And I want to bring it around so you can look at it a little bit. If you'll notice there's some things inside. There's a rubber band around the top. This stuff on the top is called rubber sheeting. What I'm holding on to here, this is called a test tube.

In this example, Laura introduced tools in the form of language the students might use to support their construction of explanations at a later time, such as *test tube*, *rubber sheeting*, and *Cartesian Diver*. The introduction of these tools was an opportunity for building syntactic knowledge relative to the communicative aspects of scientific inquiry. This type of syntactic knowledge served to maximize the potential for dialogic knowledge building by providing students with a shared language with which to construct explanations. This shared language reflected the genre of science (Hicks, 1997) and thus supported students' understanding of

---

[5]Laura's reference to the CDS as "another puzzle" stemmed from her description of a previous investigation which involved identifying substances by their properties ("Grandma's Mystery Powder") as a "puzzle."

scientific ways communicating, that is, ways of "talking science" (Lemke, 1990).

There was no difference in the frequency or degree of uptake of these labels between those students Laura identified and others in the class. Each of the labels introduced in this manner was taken up by the students and used frequently during their classroom conversations to refer to the physical tools of their inquiry. This points to the importance of attending to students' perception of the materials with which they will be working during the planning of such instruction, and of attempting to predict which tools and labels they will need to take up in order to participate, physically and discursively, in the subsequent investigation. It is important to note that Laura's explicit introduction of these tools and labels took place as part of the active process of inquiry, as opposed to other forms of explicit instruction and labeling (such as vocabulary lists or worksheets) which may be introduced prior to or separate from students' active participation.

One label that was not introduced explicitly during the engagement phase of Laura's instruction, and which may have clarified students' subsequent instructional conversations, was the term *diver* to describe the small inverted test tube inside of the larger test tube. In the example discussed earlier, Laura made reference to the diver when she said, "There's some things inside." The students then went on to observe the changes in the system when pressure was applied to the rubber sheeting, and to build their own systems. It was not until they were actively engaged in building their own systems that Laura introduced a form of the graphic illustrated before that labeled the small test tube as the diver. Additionally, this introduction of the term *diver* was made only through writing on this graphic, not stated orally. Unsurprisingly, then, students' uptake of this term was extremely limited. The result of this limited uptake was some ambiguity in the instructional conversations that centered around the behavior of the diver, with students most commonly referring to the diver as "it," and, especially in the initial phases of the first round of inquiry, "the thing." Again, this underscores the importance of developing teachers' ability to predict which terms and tools need to be explicitly introduced in the initial phases of an inquiry in order to assure that the instructional conversations are as productive as possible for all students.

Introduction of Conceptual Tools and Their Labels in the Process of Inquiry

Another way that Laura's intentional monitoring and responding to students' thinking was actualized was through the explicit introduction of conceptual tools. Like the physical tools described earlier, each of the conceptual tools was also identified by Laura as having a label. Conceptual tools were most often introduced by Laura during the engagement and reporting phases of the inquiry, and then reinforced through repeated use in the other phases. There was no difference in the uptake of these tools between the students identified as having special needs and others in the class—all of the students tended to take up and apply the conceptual tools whose use was most often reinforced through Laura's discourse in the other phases.

The following examples of conceptual tools introduced explicitly by Laura are taken from the engagement phase of the first cycle of inquiry. At this point Laura was preparing the students to build their own CDSs. Before freeing them to do this, she spent a few minutes eliciting claims from them about the behavior of the system. After continuing to elicit students' assertions about the behavior of the diver for several minutes, Laura introduced the notion of a *claim*:

> Laura: Well, now you guys are really coming up with some interesting ideas. And I want to give these ideas a name. . . . These are called *claims*. A claim is something that you are going to give as an explanation for why something is happening.

Laura's instructional choice here was to introduce a semiotic tool reflective of the discourse of the scientific community. Laura not only introduced the tool but also explained its meaning in direct connection to the activity in which the students were already involved. In this way the students were introduced to the tool and its use relative to their own participation in classroom scientific inquiry. In her debriefing after this lesson, Laura reflected that her introduction of *claim* as a conceptual tool at this point in the instruction was deliberate and reflected her desire for the students to employ this term throughout the subsequent investigation.

Laura's introduction of *claim* illustrated the way in which the use of a single semiotic tool can support the construction of multiple dimensions of syntactic understanding. Laura's introduction of this tool, and her definition of *claim* as an "explanation," provided an opportunity for students to construct two forms of syntactic knowledge: syntactic knowledge relative to constructing and refining explanations and communicative syntactic knowledge. Explanatory syntactic knowledge was introduced here in the form of describing "why something is happening," not just describing what was happening. Laura's emphasis on including a reason for the students' observations was important because it encouraged students to go beyond making observations of phenomena and to begin to build explanations. The development of communicative syntactic understanding was supported here in that Laura introduced the *claim* as a conceptual tool with the intention that this "name" would become part of the students' shared language for talking about their investigations.

Several turns later Laura introduced the notion of a *sensitive system* as a conceptual tool in response to students' questions about the construction of the CDS. Just as she did with the introduction of *claim*, Laura connected the meaning of the new tool to the activity in which the students were engaged:

> Laura: The easier it is to . . . push down and have the diver move, that means that it's a very sensitive system. In other words it reacts very quickly to pressure on here *[indicates the rubber sheeting]*. What you guys want to try to make is a system that is very, very sensitive, so that when you push down it's not gonna take much pressure to get this to go down.

In introducing this tool and explicating its meaning relative to the students' activity, Laura provided the opportunity for students to construct syntactic knowledge about the assembly of their CDSs and hence maximize the learning opportunities available in their investigations. This was an important choice because it offered students the opportunity to be unencumbered by the complexities of building a CDS, freeing them to focus on the conceptual issues at hand. Laura's choice here thus illustrated an opportunity to build investigative syntactic understanding.

Furthermore, just as the conceptual tool *claim* supported the communicative dimension of syntactic understanding earlier, so did the tool *sensitive system* introduced here. This tool gave students a shared means with which to verbally communicate the behavior of their CDSs. This was further illustrated as students appropriated and applied the tool *sensitive system* during the investigation and reporting phases of guided inquiry.

Drawing Boundaries Around the Problem Space. In inquiry instruction there is a tension between "discovery teaching" and instructional strategies that are more explicit. One of the teachers in our community of practice, engaged in her own reflection about guided inquiry teaching, expressed this tension as "knowing when to hold and when to fold." Laura engaged this tension productively at several points during the two cycles of inquiry through decisions designed to identify and limit the problem space in which the children were working. During the engagement and investigation phases this most often took the form of providing direct information about the CDS when students asked questions about the system. For example, one student asked Laura, "Did you say it's just regular water?" and Laura confirmed, "Just regular water." Providing explicit information in this way served to free students from spending time and cognitive energy in unproductive ways, such as considering the effects of the properties of the liquid within the system. Further, it supported the students in planning for and conducting their own investigation. Responses to Laura's boundary-making discursive moves did not differ among different groups of students.

The following example is taken from the engagement phase of the first cycle of inquiry, just before the students were going to build their own CDSs:

Jason:	It goes down to a certain spot and then it stops and goes back up.	*Jason makes a claim about the behavior of the diver.*
Laura:	Well, it stops at the bottom 'cause it just hits the bottom. And then it comes back up. I mean, that's just as far as it can go. Larry?	*Laura provides explicit information about the behavior of the CDS.*

In this example, Jason asserted something he noticed about the system, "It goes down to a certain spot and then it stops and goes back up." Laura clarified Jason's understanding by telling the students why it stopped at a particular point, "Well, it stops at the bottom 'cause it just hits the bottom. And then it comes back up. I mean, that's just as far as it can go." Laura's response to Jason served to free him conceptually to focus on the investigation. A different CDS, with a longer outer tube, would have behaved differently. By providing this information at this time, Laura prevented students from spending time in their investigations exploring and developing reasons for the phenomenon described by Jason.

Laura's choice to provide direct information about the CDS reflects what we have termed *investigative syntactic understanding*. In both cases Laura's choice served to maximize the learning opportunities inherent in the investigation by bounding the problem space in which students were working. It is important to note that in both cases Laura could have made different choices, and the opportunity for building investigative syntactic knowledge would have been altered. For example, Laura could have responded to Jason by asking others what they thought about Jason's observation. This choice would have resulted in *opening* the problem space and potentially diverting children's investigative time and energy from the phenomena by encouraging them to focus on variables unrelated to the substantive issues represented in the behavior of their CDSs.

Another form of boundary-making that Laura enacted took place primarily during the investigation phases (Cycle one and Cycle two) of the inquiry. In these instances, Laura drew boundaries around the problem space by focusing students' attention on key aspects of the investigation. The following example of this form of boundary-making was drawn from

7. SCIENCE FOR ALL: THINKING IN INCLUSIVE CLASSROOMS

the investigative phase of the second cycle of inquiry. In this segment, Laura employed the "Australian" (inverted) CDS as a tool to elicit Jay's thinking. She held his system while she spoke, and used questions to focus his attention.

Instructional Conversation		*Analysis*
Laura:	What's different between now and now? Now and now? Now, what have I done? *[Laura is pressing and releasing the rubber sheeting while she is speaking]*	Laura uses her verbal questions and the Australian CDS to focus Jay's observation.
Jay:	You have pushed it [the rubber sheeting], and then probably pressure pushed it [the water] up, and pushed it in there. [pushed the water inside the diver]	Jay articulates a claim in response to Laura's scaffolding.
Laura:	OK, so when I pushed it [the rubber sheeting], you're calling that pressure?	Laura checks her understanding of Jay's use of pressure as a semiotic tool.
Jay:	Yeah.	Jay acknowledges that Laura's use of "pressure" agrees with his own.
Laura:	So where's that pressure go?	Laura uses Jay's semiotic tool "pressure" to prompt further explanation.
Jay:	It goes into the water, in there, and it, um, the water makes it, I mean the pressure makes the water go up [into the diver].	Jay extends and clarifies his original claim.
Laura:	And then because the water is going up?	Laura revoices Jay's claim "the water goes up" as a question to prompt further explanation.
Jay:	It gets heavier and it goes down.	In response to Laura's revoicing of his claim, Jay again extends his claim, for the first time giving a more complete reason ("it gets heavier") for the diver's sinking.

In response to Laura's manipulation of the diver system, Jay was able to focus on the changes within the system and then articulate a claim about the movement of the water, "You have pushed it, and then probably pressure pushed it [the water] up, and pushed it in there [inside the diver]."

Through her response, "OK, so when I pushed it, you're calling that *pressure*?" Laura clarified her understanding of Jay's use of the term *pressure*. She then appropriated this tool for use in her next question, "So where's that pressure go?" Laura's appropriation of Jay's term supported Jay in constructing the beginning of a claim about why the diver floats and sinks: "It goes into the water, in there, and it, um, the water makes it, I mean the pressure makes the water go up." Laura thus used Jay's own notion of *pressure*, first checking for intersubjectivity around the term, to scaffold further explanation. Laura's choices illustrated both the importance of a shared semiotic system in scaffolding interactions and the responsibility of the teacher to make sure that a system is in place by appropriating the students' tools when necessary.

Laura's scaffolding throughout this interaction, enacted through her use of questions designed to focus Jay's thinking on the critical aspects of the system, assisted him in building substantive knowledge of the system, particularly regarding the sinking of the diver due to its increased weight. This interaction also provided the opportunity for Jay to construct syntactic understanding relative to developing and refining explanations as he, with Laura's support, revised his earlier explanation and developed a more complete version.

Revoicing to Extend and Clarify Thinking. As illustrated in our discussion of the strategies Laura employed to define the problem space for the students, Laura used revoicing as a discursive strategy throughout the program of study (O'Connor & Michaels, 1993, 1996). In the previous example drawn from her work with Jay during the investigation phase, Laura used revoicing to scaffold Jay, responding to his claim, "the pressure makes the water go up" with "And then because the water is going up?" This use of revoicing elicited further clarification from Jay, and he finished the statement of his claim by adding, "It gets heavier and it goes down."

As in that example, Laura's use of revoicing most often took the form of repeating a word or phrase from each student's claim with the rising final intonation of a question. Laura's choice to revoice student claims as questions invited them to clarify their contributions or to extend them with the addition of new information. This resulted in the co-construction of more complete claims by Laura and her students. The co-construction of more conceptually and linguistically complete claims was the beginning of these students' construction of substantive knowledge. In these interactions, Laura's use of revoicing supported them in forming ideas about the diver's operation.

Laura used revoicing as a discursive strategy throughout each phase of both cycles of the inquiry. Laura's use of revoicing seemed to work equally well in helping all students to extend and clarify their thinking. The co-

7. SCIENCE FOR ALL: THINKING IN INCLUSIVE CLASSROOMS   215

construction of more complete claims that was achieved by Laura's revoicing of the student's initial contributions illustrated opportunities for the construction of two dimensions of syntactic knowledge. First, the explanatory dimension of syntactic knowledge was fostered in these exchanges as Laura communicated to the students that their initial contributions were incomplete, and she supported them in explaining themselves further.

Additionally, Laura also used revoicing throughout the inquiry to validate student assertions, as in the following example. Here Larry was responding to Laura's request for claims about the behavior of the diver, and he attempted to explain what made the diver sink:

Larry:	I think it's because when you push it down, like the water squeezes the air so it can't hold it up.	*Larry makes an initial claim.*
Laura:	The water squeezes the air?	*Laura revoices Larry's claim.*
Larry:	Yeah, it squeezes the air and it can't hold it up. Because my mother used to tell me that about holding my breath.	*Larry offers a revised claim, with his reasoning included as an indirect analogy.*
Laura:	Like you're holding your breath. Interesting.	*Laura revoices Larry's answer and makes his implied analogy more explicit.*

In this example the communicative dimension of syntactic knowledge was supported through Laura's interactions with Larry. Laura revoiced Larry's second contribution, "Like you're holding your breath. Interesting." This time Laura did not intone her revoicing move as a question, and therefore did not invite further response from Larry. Rather, this form of revoicing served the purpose of acknowledging and validating Larry's contribution while also changing it slightly to make his implied analogy more explicit. As in Laura's use of revoicing to help students construct more complete claims, there was no evidence that the identified students responded differently to this use of revoicing than the other students did.

Assigning Roles and Making Expectations for Appropriate Participation Explicit. Another strategy that Laura used to provided opportunities for all the students to participate in the instructional conversations was to explicitly describe the roles and ways of interacting she expected them to take up during the reporting phase of the inquiry. At this time Laura's students were already familiar with a form of sharing called "problem solving" that they used during math. Laura drew on that knowledge to support students' understanding of the appropriate behavior for reporting in Guided Inquiry science. In both problem solving and reporting, Laura provided students with what she termed a *format* for their sharing, a guide that told them what to include on their posters and how to organize their oral presentations. Laura reminded the students of this similarity between math and science, and then she elaborated on why this type of reporting was important to their science investigations in particular:

> Laura: One of the things that I want you to realize as to why we're doing this kind of reporting, instead of just having you fill out this paper and turn it in to me. When scientists do their investigations, many times they share their result with their colleagues or other people who are also scientists. So you guys, your colleagues are sitting right in this room. These are all people who shared the same experience yesterday. And, in reading through your things last night, it's amazing how many different ideas came out of that one experience. Now, if we get all those ideas out for everybody to think about and talk about, then the way you're thinking about something may get even bigger or grander than where you are right now. . . . That's part of the class this year, is that we all want to be able to learn from each other.

This excerpt illustrates how Laura used her guided inquiry investigation of floating and sinking to explicitly introduce the concept of dialogic learning into her classroom culture, stressing that, like scientists, the students had "colleagues" with whom they needed to share their ideas. The importance of creating instructional contexts that facilitated students' sharing with and learning from one other was an aspect of Laura's pedagogy that she developed through her membership in the GIsML Community of Practice. Here, Laura connected the notion of learning from one another with her interpretation of the practice of scientists, "When scientists do their investigations, many times they share their result with their colleagues or other people who are also scientists." Laura thus used the Guided Inquiry investigation to *build* aspects of her classroom culture.

Laura's explanation of why the students would be engaging in public sharing stressed the ways in which sharing would facilitate learning, "Now, if we get all those ideas out for everybody to think about and talk about, then the way you're thinking about something may get even bigger or grander than where you are right now." At this point one of the students questioned why they had to share their thinking with each other in class, stating that it sounded like "therapy." Laura extended her first explanation by connecting the notion of learning from one another to the way scientists develop and revise ideas.

> Laura: It seems a little odd to share your thinking with people doesn't it? I mean, that's not something that we usually do in class. But it's, you learn a lot from each other. You may not realize it but everything people say goes in your mind and rolls around in there and adds to your

thinking.... That's what scientists do. They just keep thinking about things and sharing their ideas with their colleagues and going back and thinking about it again.

Laura's own understanding of the way in which scientists develop and refine explanations, as a result of communicating their ideas to on another, directly informed her explanation of scientific knowledge building. This example illustrated an opportunity created by Laura for the students to gain a deeper understanding of syntactic knowledge relative to developing and refining explanations of phenomena. The recursive nature of building scientific explanations that was implicit in Laura's revoicing of students' initial claims described earlier was made explicit and connected to the practices of scientists here.

From here, Laura continued to emphasize the construction of syntactic knowledge as she described the desired behaviors of the listeners during reporting. She termed this set of behaviors the *role* of the listener.

Instructional Conversation	Analysis
Laura: You're taking on a role as a listener. And it's not just to sit politely and listen.... So one of the things I'd like you to listen for is, what is the claim? ... When someone gives their presentation, are they stating a reason for what happened with their diver yesterday? Is it a clear reason? Do you understand it? ... And if it's not clear to you, as a person in the audience, it's your job to raise your hand and say, "I'm not sure I know what you mean. Can you explain it a little bit further?" ... Another thing to look for is the evidence. Now, someone comes up and gives an explanation about why the diver works. Is that all they have? Or are they saying, "Well, I believe this, because this is what happened yesterday in my system." And evidence can be their drawing or a written explanation about things that happened yesterday.	Laura defines the role of the listener as active. Laura models questions listeners can use to identify and evaluate the presenters' claims for clarity and coherence. Laura models ways for listeners to elicit more information about the claim from the presenters. Laura states that evidence can be graphic or verbal in form and models ways to discern whether the presenters have evidence for their claim.

In describing the role of the listener as active and in explicating the responsibilities of this role in such detail, Laura created the opportunity for the construction of skeptical and communicative syntactic understanding. Communicative syntactic understanding was created by Laura's description of the role of the listener as active in helping the speaker construct an explanation that could be clearly understood. It was developed further as Laura reinforced the meaning of the tools she introduced earlier and clarified what counts as a claim, "a reason for what happened with their diver yesterday," and what evidence might include, "evidence can be their drawing or a written explanation about things that happened yesterday." The development of skeptical syntactic knowledge was supported by Laura's assertion that students should take an active part in evaluating claims and evidence of their colleagues. She modeled the skeptical habits of mind of investigators as she offered questions for listeners to ask themselves when identifying and evaluating the presenters' claims and supporting evidence.

Unlike the other strategies employed by Laura, this one did not seem to support all the students in participating in the reporting sessions. However, rather than a discrepancy between those students identified as having special needs and unidentified students, the differential uptake of the roles explicitly described by Laura was between a core group of seven students and the rest of class. This may have been at least partially due to a culture that was developing in Laura's classroom in which some students were responded to in ways that encouraged their increased participation, whereas others were not (see discussion in Collins, in press).

Jay, a student Laura had identified as having special needs, was one of the students who responded to Laura's explicit description of the appropriate ways of interacting by taking up the specific discursive tools she used in modeling the role of listener, as in the following example. Additionally, his uptake of those roles was more explicit, and he volunteered to speak more often than the other six students in the core group of seven.

In this segment of instructional conversation, Martin and Brad began their report to their classmates by reading the various parts of their poster aloud to their colleagues; Martin asserted, "When you push on the rubber we saw the water rise up a little and we saw bubbles coming out the bottom. They came out the bottom of the vial. But then when you let go the air pushes it up." Immediately after this Laura asked for questions about the group's presentation, and she called on Jay first:

Instructional Conversation	Analysis
Jay: What do you mean that the air started coming . . . out the bottom? Can you guys explain that?	*Jay appropriates the discursive tool introduced by Laura earlier, "Can you explain that?" and uses it to elicit further explanation.*

7. SCIENCE FOR ALL: THINKING IN INCLUSIVE CLASSROOMS

Martin: Well, what we meant this, by air going out of the bottom, maybe we could say water pushes air out. And then it makes the vial heavier so it goes down, and it wouldn't float. And like air comes back in and it floats up.

Martin extends his claim, reasserting that air came out of the vial but also noting that increased water made the diver heavier.

Jay: So are you saying the same thing as me and Brian, are you saying the same thing as us?

Jay attempts to build consensus around the claim by aligning it to his claim and to Brian's, which asserted that increased water made the vial heavier.

Martin: Same thing kinda like this.

Martin agrees with Jay interpretation of his claim, ignoring the differences.

Laura: Well, I heard two different things, so can I pick up on your question a little bit, Jay? 'Cause I thought it was a good question. . . . You're saying the air comes out of the vial and that's why it sinks? The water comes up [in the vial]?

Laura intervenes and revoices Martin and Brad's claim as a question, providing them with the opportunity to clarify or modify it.

Martin: Yeah, 'cause then the water, air comes out, water can go in, more water can go in so it makes, it kinda makes the vial heavier and it goes down.

Martin restates both aspects of his claim, that air leaves the vial and the vial gets heavier.

Laura [to Jay]: Now your explanation, did you think air was coming out of the vial?

Laura revoices the aspect of the claim that is in conflict with Jay's claim, and asks Jay whether his claim agrees with it or not.

Jay: Uh-huh <indicating No>

Jay acknowledges that the claims are not the same.

[. . .]

Laura: So this is a, this is a different claim that has been put out, at least from what I'm hearing. I don't know, maybe somebody else heard a different claim. Did you hear something different than what was already on our lists up there, with what was going on with the air?

Laura asserts that this claim is different and solicits students' interpretations.

Jay began the questioning session by using a tool introduced by Laura in her introduction to reporting. Laura had advised the students that, as listeners, it was their job to ask, "I'm not sure I know what you mean. Can you explain it a little bit further?" if they were unclear about the presenter's claim. Jay appropriated and applied this tool when he asked Martin and Brad, "What do you mean that the air started coming . . . out the bottom? Can you guys explain that?" Jay's transformation and use of the tool elicited further explanation from Martin, "Well, what we meant was this, by air going out of the bottom, maybe we could say water pushes air out. And then it makes the vial heavier so it goes down, and it wouldn't float. And like air comes back in and it floats up." This exchange illustrated the students' growing understanding of explanatory syntactic knowledge as Jay supported Martin in developing his explanation.

Next Jay attempted to build consensus around the claim by aligning it to his and Brian's, which had both asserted that increased water made the vial heavier, "So are you saying the same thing as me and Brian, are you saying the same thing as us?" When Martin agreed with Jay's interpretation of his claim, ignoring the differences, Laura chose to intervene, "Well, I heard two different things, so can I pick up on your question a little bit, Jay? 'Cause I thought it was a good question." Laura's intervention was done in such a way that Jay's attempt at consensus-building was favorably acknowledged but that also allowed her the opportunity to guide the students in exploring the difference in the claims. Laura's initial intervention when Jay attempted to build consensus created the opportunity for the development of the communicative dimension of syntactic understanding as she drew students' attention to the specificity of their descriptions.

After intervening and stopping the student discourse, Laura helped Martin re-state his claim clearly. Laura used revoicing to co-construct Martin's claim with him, "You're saying the air comes out of the vial and that's why it sinks? The water comes up?" Martin responded by re-stating his claim, "Yeah, 'cause then the water, air comes out, water can go in, more water can go in so it makes, it kinda makes the vial heavier and it goes down." Laura created the opportunity for the construction of the explanatory dimension of syntactic understanding.

After helping Martin construct his claim clearly, Laura allowed Jay to respond as to whether he still thought the claim reflected his own thinking. Laura revoiced the aspect of the claim that was in conflict with Jay's claim, "Now your explanation, did you think air was coming out of the vial?" Jay acknowledged that this was not in agreement with his claim.

Laura revoiced for the class that this was a different claim, both different from Jay's and different from the rest of those on the class list, "So this is a, this is a different claim that has been put out, at least from what I'm hearing. I don't know, maybe somebody else heard a different claim. Did you hear something different than what was already on our lists up there, with what was going on with the air?" She provided the opportunity for students to make the same comparison that she made between the list of class claims and Martin's claim. In asking the class to evaluate the difference between Martin's claim and those that were already posted, Laura invited them to participate in exercising skeptical syntactic understanding. Furthermore, this was an opportunity to build substantive knowledge as students discerned the differences in the claims and developed their thinking about the validity of the claims. She focused students' attention on the difference of air coming out of the vial which guided students' skeptical questioning of Martin's evidence.

DISCUSSION AND IMPLICATIONS

In this chapter we sought to identify the discursive features that supported students' construction of specific forms of scientific understanding in a guided inquiry context. In particular, we were interested in strategies that seemed to facilitate the participation of those students who had been identified by their teacher, Laura Bozek, as having special needs. We identified four discursive strategies employed by Laura that seemed to facilitate the participation of all of the students:

1. Explicit introduction of physical tools of inquiry and their labels in the context of inquiry
2. Introduction of conceptual tools and their labels in the process of inquiry
3. Drawing boundaries around the problem space
4. Revoicing to extend and clarify thinking

Additionally, we identified a fifth strategy, assigning roles and making expectations for appropriate participation explicit, that was taken up by some of her students, including one of the students identified by Laura as having special needs.

Across the strategies that were most successful in supporting all of the students in participating was an explicit approach to building communicative syntactic understanding. This seemed to be an important factor in providing all the students, including those identified as having learning disabilities, with an understanding of how to appropriately enter the discourse. One dimension of this included Laura's explicit explanation and demonstration of how to *talk science* (Lemke, 1990) while the students were actively engaged in the activity of doing science.

Another important theme running across each of the discursive strategies that Laura employed to support student thinking is that their effectiveness relied heavily on her own understanding of the substantive and syntactic dimensions of scientific understanding related to the inquiry context. This is consistent with other empirical research focusing on the role of the teacher in mediating the construction of scientific knowledge that has demonstrated the major impact that a teacher's knowledge (or lack thereof) can have on the learning opportunities created and on the nature of the knowledge constructed by students (Carlsen, 1992; Moje, 1995, 1997; Smith & Neale, 1989). For example, Carlsen found that novice biology teachers who were unfamiliar with the content area knowledge of biology (what Schwab would term substantive knowledge) curtailed student questioning and "artificially constrain[ed] science to an exploration of the known" (p. 15). Drawing on Schwab's constructs, Carlsen posited that the ways in which teachers controlled the discourse may have curtailed students' syntactic understanding although he did not explicate evidence for this assertion.

Although a teacher may never gain the facility with the tools of a community of practice that a full-fledged member would have, in order to guide students' inquiry the teacher must have an understanding of what those tools are, as well as an understanding of what the collected knowledge (substantive understanding) of that inquiry area is. As Dewey (1902/1990) asserted, the teacher must be able to see ahead to where the child's inquiry path may lead, and thus be able to guide him or her along the way. Guidance or scaffolding is not an external structure or order of thought imposed on a child but a sort of map that helps the child connect where he or she is on the inquiry path with where the child might be if he or she continues.

REFERENCES

Anderson, C., Holland, D, & Palincsar, A. (1997). Canonical and sociocultural approaches to research and reform in science education. *Elementary School Journal, 97,* 357–381.

Au, K. (1998). Social constructivism and the school literacy learning of students of diverse backgrounds. *Journal of Literacy Research, 30,* 297–319.

Carlsen, W. (1992). Closing down the conversation: Discouraging student talk on unfamiliar science content. *Journal of Classroom Interaction, 27*(2), 15–21.

Collins, K. M. (1997, December). *Talking about light: The construction of a shared discourse by first grade students in guided inquiry science.* Paper presented at the annual meeting of the National Reading Conference, Scottsdale, AZ.

Collins, K. M. (2003). *Ability profiling and school failure: One child's struggle to be seen as competent.* Mahwah, NJ: Lawrence Erlbaum Associates.

Collins, K. M. (in press). *Ability profiling and school failure: One child's struggle to be seen as competent.* Mahwah, NJ: Lawrence Erlbaum Associates.

Collins, K. M., MacLean, F., Palincsar, A. S., & Magnusson, S. (1998, November). *The case of Robert: Influences of one student's ability to negotiate verbal, print, and scientific discourses on the construction of shared knowledge in guided inquiry science.* Paper presented at the annual meeting of the National Council of Teachers of English, Nashville, TN.

Collins, K. M., Palincsar, A., & Magnusson, S. (1998, April). *Metaphor, mediation, and meaning: Dialectical knowledge construction in guided inquiry science.* Paper presented at the annual meeting of the American Educational Research Association, San Diego, CA.

Dewey, J. (1990). *The child and the curriculum.* Chicago: University of Chicago Press. (Original work published 1902)

Einstein, A. (1950). *Out of my later years.* New York: Philosophical Library.

Erickson, F. (1992). Ethnographic microanalysis of interaction. In M. D. LeCompte, W. L. Millroy, & J. Preissle (Eds.), *Handbook of qualitative research in education* (pp. 202–225). San Diego, CA: Academic Press.

Forman, E., Minick, N., & Stone, A. (Eds.). (1993). *Contexts for learning: Sociocultural dynamics in children's development.* New York: Oxford University Press.

Gee, J. P. (1991). What is literacy? In C. Mitchell & K. Weiler (Eds.), *Rewriting literacy: Culture and the discourse of the other.* Westport, CT: Bergen and Garvey.

Gee, J. P. (1996). *Social linguistics and literacies: Ideology in discourses* (2nd ed.). Bristol, England: Taylor and Francis.

Gersten, R., & Baker, S. (1998). Real world use of scientific concepts: Integrating situated cognition with explicit instruction. *Exceptional Children, 65*(1), 23–35.

Goffman, E. (1959). *The presentation of self in everyday life.* Indianapolis, IN: Bobbs-Merrill.

Goffman, E. (1961). *Encounters: Two studies in the sociology of interaction.* Indianapolis, IN: Bobbs-Merrill.

Goffman, E. (1981). *Forms of talk.* Philadelphia: University of Pennsylvania Press.

Hicks, D. (1997). Working *through* discourse genres in school. *Research in the Teaching of English, 31,* 459–485.

John-Steiner, V., & Mahn, H. (1996). Sociocultural approaches to learning and development. *Educational Psychology, 31,* 191–206.

Keller-Cohen, D. (1994). The web of literacy: Speaking, reading, and writing in 17th and 18th century America. In D. Keller-Cohen (Ed.), *Literacy: Interdisciplinary conversations* (pp. 155–176). Cresskill, NJ: Hampton Press.

Lemke, J. L. (1990). *Talking science: Language, learning, and values.* Norwood, NJ: Ablex.

Magnusson, S. J., & Palincsar, A. S. (1995). The learning environment as a site of science education reform. *Theory Into Practice, 34*(1), 43–50.

Magnusson, S. J., Palincsar, A. S., & Templin, M. (in press). Community, culture, and conversation in inquiry-based elementary science. In L. Flick & N. Lederman (Eds.), *Scientific inquiry and the nature of science: Implications for teaching, learning, and teacher education.* Dordrecht, The Netherlands: Kluwer Press.

Magnusson, S. J., & Templin, M. (1995). Scientific practice and science learning: The community basis of scientific literacy. In F. Finley (Ed.), *Proceedings of the 3rd International Conference on History, Philosophy, and Science Teaching.*

Magnusson, S. J., Templin, M., & Boyle, R. A. (1997). Dynamic science assessment: A new approach for investigating conceptual change. *Journal of the Learning Sciences, 6,* 91–142.

McDonnell, L. M., McLaughlin, M. J., & Morison, P. (Eds.). (1997). *Educating one and all: Students with disabilities and standards based reform.* Washington, DC: National Academy Press.

Michaels, S., & O'Connor, M. C. (1990). *Literacy as reasoning within multiple discourses: Implications for policy and educational reform.* Paper presented at the summer institute of the Council of Chief State School Officers.

Moje, E. (1995). Talking about science: An interpretation of the effects of teacher talk in a high school science classroom. *Journal of Research in Science Teaching, 32*, 349–371.

Moje, E. (1997). Exploring discourse, subjectivity, and knowledge in chemistry class. *Journal of Classroom Interaction, 32*(2), 35–44.

Moll, L. (Ed.). (1990). *Vygotsky and education: Instructional implications and applications of sociohistorical psychology.* New York: Cambridge University Press.

National Research Council. (1996). *National science education standards.* Washington, DC: National Academy Press.

Newman, D., Griffin, P., & Cole, M. (1989). *The construction zone: Working for cognitive change in school.* New York: Cambridge University Press.

O'Connor, M. C., & Michaels, S. (1993). Aligning academic task and participation status through revoicing: Analysis of a classroom discourse strategy. *Anthropology and Education Quarterly, 24*, 318–335.

O'Connor, M. C., & Michaels, S. (1996). Shifting participant frameworks: Orchestrating thinking practices in group discussion In D. Hicks (Ed.), *Discourse, learning, and schooling* (pp. 63–103). New York: Cambridge University Press.

Palincsar, A. S., Collins, K. M., Marano, N. L., & Magnusson, S. (2000). Investigating the engagement and learning of students with learning disabilities in guided inquiry science teaching. *Language, Speech, and Hearing Services in Schools, 31*, 240–251.

Palincsar, A. S., Magnusson, S. J., Collins, K. M., & Cutter, J. (2001). Making science accessible to all: Results of a design experiment in inclusive classrooms. *Learning Disabilities Quarterly, 24*(1), 15–32.

Palincsar, A. S., Magnusson, S., Marano, N., Ford, D., & Brown, N. (1998). *Design principles informing and emerging from a community of practice.* Paper prepared for a special issue of *Teaching and Teacher Education* focusing on Professional Development edited by R. Putnam and H. Borko.

Penick, J. E. (1993). The mysterious closed system. *The Science Teacher, 60*(2), 30–33.

Roberts, R. S. (1982). Teaching an old diver new tricks. *The Science Teacher, 49*(7), 25–27.

Rogoff, B. (1994). Developing understanding of the idea of communities of learners. *Mind, Culture, and Activity, 1*(4), 209–229.

Schwab, J. (1963). Problems, topics, and issues. In G. W. Ford & L. Pugno (Eds.), *Education and the structure of knowledge* (pp. 4–43). Chicago: Rand McNally.

Schwab, J. (1964). Structure of the disciplines: Meanings and significances. In G. W. Ford & L. Pugno (Eds.), *The structure of knowledge and the curriculum* (pp. 6–30). Chicago: Rand McNally.

Sleeter, C. (1986). Learning disabilities: The social construction of a special education category. *Exceptional Children, 53*(1), 46–54.

Smith, D., & Neale, D. (1989). The construction of subject matter knowledge in primary science teaching. *Teaching and Teacher Education, 5*(1), 1–20.

Smith, J. P., diSessa, A. A., & Roschelle, J. (1993). Misconceptions reconceived: A constructivist analysis of knowledge in transition. *The Journal of the Learning Sciences, 3*, 115–163.

Trent, S., Artiles, A., & Englert, C. S. (1998). From deficit thinking to social constructivism: A review of theory, research, and practice in special education. In P. D. Pearson & A. Iran-Nejad (Eds.), *Review of research in education, Vol. 23* (pp. 277–307). Washington, DC: American Educational Research Association.

Valencia, R. (1997). (Ed.). *The evolution of deficit thinking.* Bristol, PA: Falmer Press.

Wertsch, J. (1991). *Voices of the mind: A sociocultural approach to mediated action.* Cambridge, MA: Harvard University Press.

CHAPTER SEVEN METALOGUE

A Question of Perspective in Supporting the Learning of Students With Special Needs in Inquiry-Based Science Instruction

Margaret Gallego
Shirley J. Magnusson
Kathleen M. Collins
Annemarie Sullivan Palincsar
Wolff-Michael Roth
Randy Yerrick

Randy: I took this chapter to be one of exploring a theoretical framework for guiding research and teaching for equity among special-needs students. I agree that the *Standards* rhetoric is often over generalized and lacks substance regarding students with special needs. I think, however, that interpreting the *Standards* as built on a "deficit orientation" is perhaps rather restrictive.

Margaret: I, too, wonder if the authors could take the *Standards* and reinterpret the call to one that is clearly about modification of instruction. By this I mean, your chapter (and ultimately all educational reform) is advocating important changes in curriculum that is essential and called for by the Standards. Isn't there a great potential to reveal the way *Standards* for supporting teachers to modify curriculum and provide interesting, innovative ways of delivering instruction for those with special needs can then, in return, assist all children?

Shirley: Indeed there is potential to illustrate how modifications for students with special needs might benefit many more students. However, an important point of the chapter is to alert us to the deficit assumption that the language in the Standards reflects, which tends to be overgeneralized for students. That is, in our experience, special-needs students have much to bring to instruction, even when marginalized by their peers or teachers. More research is needed to determine what modification or accommodations, beyond the benefits of good instruction, are needed for students with special needs.

Michael: It would perhaps be more advantageous to take the notion of special needs as a temporary one, because while you are saying that there are some special needs you are also saying that the same children are also like the others in the classroom, having a lot to contribute to the emergence of learning communities.

Shirley: Rather than "temporary," we would encourage a conceptualization of special needs as a context-bound phenomenon. Our research indicates that good instruction (albeit with the employment of strategies to mediate the learning of individual students that many teachers do not routinely or systematically employ in their instruction) goes a long way in supporting special-needs students, along with other students who are typically low achieving. Thus, our point is as much about the need for good instruction, which requires high levels of knowledge and sophisticated skills on the part of the teacher if we're talking about inquiry-based instruction. We think there is still much to learn about how to effectively conduct inquiry-based instruction that leads to development of understanding consistent with the *Standards*.

Randy: You have combined several different scaffolding approaches to inform the teaching in the GIsML approach including revoicing, positioning, and explicit cognitive linking. How do these strategies assist in the development of your learning community and students' identities and how might this contrast other more traditional methods active in other learning communities at this school site?

Shirley: The issue of developing a learning community to support enculturation into the norms, conventions, values, and beliefs of science really is a chapter in and of itself, and the mediation of student learning is also a chapter in and of itself. Nevertheless, a few comments may be useful. First, we use the term *scaffolding* to refer to a particular type of interaction of teacher with a student, but which can have a variety of purposes. The different purposes we cited are important because they are meant to indicate that interacting with students in all these ways is part of good instruction. From a sociocultural standpoint that assumes that the "teacher" is always creating a zone of proximal development with a "student," . . .

Michael: In the presence of the teacher, students can always work beyond what they could do on their own.

Shirley: . . . scaffolding is an integral part of instruction. At the same time, the type of teacher move that is necessary in a particular context to support the student working in the zone varies. Determining when to employ particular moves is a function of the judgment of the teacher, guided by her knowledge of the targeted goals (Point A), her assessment of tools the student has to work with (knowledge- and skill-wise) to learn in the situation (Point B), and her knowledge of how tools that she has at her disposal might be employed to help the student bridge from Point B to Point A.

Michael:	Randy also raised the question about how your use of scaffolding contrasts with "more traditional methods active in other learning communities at this school site."
Shirley:	This raises many questions for us. What do you mean by "other learning communities at the school site?" Do you mean, "Contexts in other classrooms of which our students were not a part?" Or do you mean, "Contexts in other classrooms that some of our students may have been a part?" We would agree with the implication that there are many different communities in schools, but we would contend that most of those communities are communities via *social* relationships, and not relationships that have centrally to do with academic learning. For a context to constitute a learning community there has to be shared responsibility for learning and shared authority for knowing. In our experience there are few such situations in schools.
Michael:	I have suggested that "conversations," which are collective enterprises, inherently puts *all* participants in a zone of proximal development or, in other words, the conversational setting constituted by a collective "scaffolds" each individual (Roth & Lee, 2002).
Shirley:	We would agree that it has the potential to do so but only to the extent that any particular participant expends effort to achieve intersubjectivity with the others, which is not necessarily a reasonable assumption in many group settings. Furthermore, the conceptual focus of the zone of proximal development that is operating for any single individual relative to the group may be quite different. Thus, you describe quite a different perspective from that in the mind of a teacher who has particular conceptual goals for the scaffolding to support.
Margaret:	My understanding is that "talk" is a quite common literacy used within classrooms for judging students' understanding—even more common than writing pencil-and-paper examinations. I am interested in the role that the multiple literacies approaches provided prior to the talk performances. Without a more direct link between the multiple literacies available to the students, talk is the predominant focus. This could lead readers to believe that talk is the only variable that is necessary, which is to say, teaching these children to talk the way scientists sound. I don't believe that this is what you want to say, but I cannot see a complete connection with the multiple literacies you present in your data.
Shirley:	In our approach, the term multiple literacies applies to all aspects of the instruction. It signals that reading, writing, numeracy, and speaking are integral parts of the learning process in any situation. The different types of activities in which students participate—engagement with specific ideas or phenomena, preparing to investigate, investigation, preparing to report, reporting—have different demands with respect to each of these aspects of literacy. Hence,

	there is the need to more specifically examine the literacy demands for each type of activity.
Michael:	But in any event, talk, as Margaret suggested, plays an important role in your instructional approach.
Shirley:	Absolutely. If teachers begin to see their role as one of supporting and guiding conversation, then students will have many more opportunities to identify and examine their ideas, which is a critical part of the learning process, and teachers will have many more opportunities to learn of the types of conceptual tools students have to work with in building knowledge, which can stimulate the teacher to develop increasingly richer knowledge bases about how to help students construct the desired understanding. Having said that, it is important to remember that the talk is about what was investigated and what was found out. Thus, the talk is only meaningful because of the other types of activity in which students engaged.
Michael:	And others might say that *talking* is the thinking rather than expressing the thinking.
Shirley:	We would not make such a sharp distinction because we think both things are going on, but we would definitely agree that, just like any writing process creates new thinking, even when we think we know exactly what we want to write, the process of putting our ideas into words creates new thinking, and the process of framing words to speak to others, and then speaking to others, are not synonymous because we often change the words we speak from what we were thinking we would speak. From a larger perspective, scientific knowledge does not get produced because someone conducts an experiment, it gets produced first because investigation is conducted in particular ways due to that individual's enculturation in the scientific community, and second because it passes the scrutiny of the scientific community. As we have modeled after this in our instruction, it is important to recognize that the stakes of the conversation vary across phases of investigative activity in the instruction. When preparing to investigate, investigating, and analyzing data from investigation, multiple perspectives may be valued. When preparing to publicly report and reporting to the whole class, the situation is high-stakes. First, because it is public, and second, because what is presented will be evaluated according to the norms and conventions that reflect the particular values and beliefs of the scientific community.
Kathleen:	We also recognize that our work with students is embedded in a larger social and political system that imbues the discourses of science with more authority and power than it affords those discourses that students have already acquired from participating in their home and community groups. In designing instruction that combines explicit and implicit instruction in both the syntactic and

CHAPTER 7 METALOGUE: SPECIAL NEEDS AND INQUIRY

the substantive aspects of scientific knowledge building, we create avenues of access to the more privileged form of discourse, that is, the knowledge-building practices of scientists.

Randy: Let me return to the *Standards* issue. You argued that it embodies a deficit model. It is evident that your strategies allow students more access to oral participation in richer scientific discourse. I wonder if there is any evidence or reason to believe that these practices also impacted students' abilities to improve their performance or engagement on measures emphasized by the *Standards*? Do you suppose in the current standards educational context that it is important that they lead to such evidence?

Shirley: We think that good instruction should lead to understanding on the part of the students that can be reflected in students' responses to standardized assessments. However, it has not been our experience that either the *Standards* or associated assessments are written well enough so as to inspire the confidence that we think we should have in what student performances mean. Nevertheless, a regular feature of our research is pre- and postinstruction assessment of student learning. These assessments have either been constructed response or simply multiple-choice questions. These look very much like standardized state assessments, and some items are from existing standardized assessments (e.g., TIMSS). We saw impressive learning gains on the part of the special-needs students, along with the other students. I say impressive because the slope of the gains were similar or identical to the gains for normally achieving children.

Michael: But such tests and items often do not pick up what is really of interest to science educators, whether it is the use of language in a holistic way or whether someone wants to identify misconceptions, which requires considerable stretches of talk.

Shirley: We would agree that the assessments typically currently associated with the standards are limited in what they reveal. That is why our research has typically involved interpretive as well as analytic analyses of student performance in the classroom.

Margaret: I was wondering, where would the notion of misconceptions be placed within your descriptions and categories? That is to say where might theories and understandings representative of those within the group (immediate) and that run counter to larger group (scientific community) be situated in your analysis of student discourse?

Shirley: The issue of misconceptions, where that language has been used in science education research, falls within the purview of *substantive* knowledge as opposed to syntactic knowledge. We prefer the term *alternative conceptions* because it reflects that the ideas are simply alternative to how scientists have constructed understanding about the world.

Michael: So how do you use the idea of *misconceptions* constructively?

Shirley: We use the notion of *conceptual profile*. The notion of *conceptual profile* is that an individual can hold multiple ideas about a concept that may differ along both ontological and epistemological dimensions; hence, the use of the term *profile* to refer to the knowledge an individual holds about a concept. From this perspective, alternative conceptions in the area of science simply represent knowledge informed by different ontological and epistemological tenets, typically those that inform our everyday knowledge building as opposed to the tenets informing the development of scientific knowledge. As a result, these conceptions can be seen as knowledge held that may be useful in an everyday context but not with respect to doing science. Thus, they don't have to impinge upon new learning at all. It would be important, however, to make this explicit for students. When students are building scientific knowledge, we can then support them in keeping the values, beliefs, norms, and conventions that inform the production of knowledge at the scientific forefront.

REFERENCES

Roth, W. -M., & Lee, S. (2002). Breaking the spell: Science education for a free society. In W. -M. Roth & J. Désautels (Eds.), *Science education for/as socio-political action* (pp. 65–91). New York: Peter Lang.

CHAPTER EIGHT

Playing the Game of Science: Overcoming Obstacles in Re-negotiating Science Classroom Discourse

Randy Yerrick
San Diego State University

We will need to face the fact that methods are rarely neutral. On the contrary, the means we use color and modify the ends we actually achieve through them. How we teach will determine what our students learn. If a structure of teaching and learning is alien to the structure we propose to teach, the outcome will inevitably be a corruption of that content. And we will know that it is.
—Schwab (1962, p. 229)

For many years I have been engaged in an exploration of struggles in my own teaching as I seek to balance a desire to be open to children's thinking with a search for ways to connect the powerful and lasting ideas of the discipline I so enjoyed studying as a student. There was a marked departure in my teaching and research when I began to compare the ways children and scientists talk about the world in similar and often strikingly different ways. Although I had been well socialized to believe that teaching science was a venture to impart correct answers and scientific truth to high-school chemistry and physics students (Lortie, 1975), I became dissatisfied with those students who would helplessly defer, "Is this correct?" With each instance I became more committed to the quest to learn how to change students' thinking and dispositions toward the scientific endeavor I knew could be so much more. I was reminded that my students' questions about correct answers were results of my own hypocrisy. I was asking the wrong kinds of questions for my students to truly think, and it was the opening quote from Schwab (1962) that first challenged me to engage as a researcher of my own classroom. My teaching efforts were inadequate in getting students to act like scientists, and the fact that I was aware of it seemed worse than blissfully going about teaching the way it had always been done.

My understanding of the scientific discourse I sought to promote with my students evolved from my doctoral work where I had the luxury of examining the teaching of Deborah Ball, Magdalene Lampert, and Suzanne Wilson as they challenged traditional views of what children are able to accomplish. They challenged me as well to consider how Schwab, Popper, and Kuhn conceived of scientific endeavors. They also challenged me to consider what the core principles and dispositions of the members of a scientific community were and to reconsider in what aspects children could be invited to participate.

It was fortuitous that I turned the camera onto my own practices as Project 2061 was launching recommendations of science for "all Americans" (American Association for the Advancement of Science, 1989). I was disturbed by the lack of clarity proposed by this document and how such visions could be achieved in real science classrooms. There are a whole host of interpretations of this reform document, and in fairness to the reader, I must put forth my biases to provide a basis for understanding the changes I tried to invoke among my students. My interpretation of Project 2061's call for scientific literacy was framed in the hopes that my students would be able to

- create separate, visual, and detailed models that represent and predict everyday phenomena;
- design a variety of experimental tests and orchestrate a plan of inquiry for their own evidence in pursuit of determining which model best describes everyday events;
- decide on the rules of admissible evidence in constructing arguments and make group decisions regarding the importance of different types of evidence;
- critically analyze models on the basis of their ability to explain and predict all the evidence gathered;
- individually and collectively defend models and create new tests to resolve questions arising from discrepancies in existing models.

School science, which contrasts professional science in many ways, balances a tension between promoting accepted knowledge and engaging learners in constructing their own interpretations from the same data. The historical roots of this tension lie in the social context in which school science evolved as a subject (DeBoer, 1991; Layton, 1973). There is "an uneasy amalgam" that drives teaching toward the "arid business of rote learning of standards facts and theories and methods. The epistemological danger is that it makes science look like infallible received knowledge" (Millar, 1989, p. 54). Resolving the incompatibility, however, is nontrivial. Implicit beliefs and norms work in opposition to establishing scientific discourse in classrooms.

In this chapter I identify some major obstacles that I encountered and explicate why students attempted to drive classroom discourse toward more conservative outcomes. I then discuss ways in which teachers and researchers can tactically challenge beliefs implicit in school science among children—even those (e.g., lower track students) who are believed to be unable or unwilling as students to change.

CHILDREN AND THE GAME OF SCIENCE

Gee (1994) likened renegotiating science talk in the class to that of learning a new game on the schoolyard in which children must answer three central questions to successfully become a participating member: How do you win? What are the rules? Where are we in the game? Gee's articulation seemed particularly appropriate in its application to the renegotiation of science talk. Dewey (1904) criticized schools for applying one set of psychological principles to learning inside classrooms and another outside of classrooms, hence his separation between learning and "schooling." Similarly, applying Gee's sociolinguistic parameters to learning science talk illuminates differences in beliefs and processes of science learning as we compare students learning to talk science and students learning to play a new game on the playground. In addition, because playground games are used for gaining access to or status in a particular desirable group, Gee's framework allows appropriate comparisons to the increasingly central issue of equity in science. Finally, Gee's linguistic analogy fits many of the dilemmas that have previously defied characterization—I continue to apply this elegant comparison to my own struggles to reconstruct science discourse. It seemed reasonable that my students would carry their playground (out of the classroom) norms of speaking and thinking into classroom discussions as soon as they recognized that normal school rules (e.g., teacher tells/students listen) did not apply. In the following sections I explore through my students' talk ways in which these questions are resolved within the classroom community by focusing on explicit questions students posed to me during instruction.

As science educators promoting change, we are also engaged in the process of imparting a new set of values, beliefs, and dispositions to children in educational contexts rich with history and culture. There are implicit factors and pressures (often in conflict I might add) that we hope to change as we promote a new way of speaking, thinking, and acting among science students. I therefore also adopt a second educational analogy, this one described by Kohl (1984) in order to make sense of how these beliefs and practices change and how discourse can evolve within a learning community. Herbert Kohl's family was engaged in the construction business as he grew up, and he likened the teaching of children to the process of remodeling an architectural structure. It is a process of inspecting, tearing down weak supports, reframing walls and supports, and rebuilding sound structures upon which future renovation will be both safe and fruitful. In the same way, teachers must inspect, break down obstacles, and provide new supports for the advancement of new ideas and interactions. Teachers use these kinds of processes to transform everyday classrooms into learning communities with new and powerful ways of engaging with the world. I shall use this analogy as well to discuss the transformation of science discourse among lower track science students.

How Do You Win in Science?

Many factors work against teachers as they try to set new academic goals for students. Teachers (including me) operate in a school culture that is often transparent. It is a culture, however, that can mask lowered science students' aspirations working against teachers desiring change in classroom discourse. Take for example traditional classrooms that presume the teacher is the sole proprietor of correct answers—whose job, in fact perceived and self-imposed responsibility, is to disseminate these factual answers at appropriate times. In such an instructional context, those students who can receive and retain the greatest number of factual answers or skills for correct problem solving often win the game of science. It is a reasonable arrangement for both teachers and students to operate in this way but the implicit messages which directly contradict many attributes of science talk never have the opportunity to surface. When I worked to change discourse in my classroom, often the clash of assumptions caused discomfort and subsequently revealed stark contrasts between my students' and my own expectations. I begin each of the following sections with a student quote that has most poignantly captured my struggles to change classroom science discourse.

"Is this correct? Why don't you give us the answers?"

At times this question originates as a vehicle to drive the conversation toward a minimalist outcome. It can be a truncation in intellectual investment, a diversion from real thinking, and a ploy to rally support around the comfort that arises from the predictability of success with minimal academic engagement. As I have argued elsewhere (Yerrick, 1998, 1999; Yerrick, Pedersen, & Arnason, 1998), both students and teachers often find themselves entrenched in a negotiation process for minimal standards that reflects of the larger educational context of schools themselves (Cusick, 1983; Goodlad, 1984). In such cases, it is essential that teachers have a firm foundation for choosing against-the-grain teaching strategies (Hogan & Clandinin, 1993), or they will find themselves compromising away their values to simply survive. Like a builder without any architectural plans, a teacher could get lost in the morass of details inasmuch as the building process for both requires a long-term plan and understanding of the basic foundation upon which they hope to build.

When teachers understand clearly why their new practices are more likely to result in science discourse, fielding student challenges and negotiating obstacles will be less hazardous. For example, there are times when legitimate questions arise and must be dealt with as the students have a right to know what is expected of them. The question of *correctness of answers* often originates in earnest as legitimate confusion arises when years of student socialization are confronted. During such challenges there is a definite need to engage students in talk about the talk, or meta-discourse, to renegotiate the winning of the game of science. Take for example this first vignette, in which earth-science students were asked to propose solutions to a variety of real-world problems. Schools enculturate students to expect closure over every science topic in 50 minutes or less. In this vignette, students are asked to work though an application of a scientific concept to a real-world event, and they expect that my role as the teacher is to provide specific and correct answers at every turn.

In the face of a potential evacuation from a threatening Hurricane Bertha, small groups of students were asked to make predictions and gather data to support their hypotheses to questions like

- If there is an evacuation ordered, could you outrun this coming hurricane?
- What will happen to the air pressure in our school when it arrives?
- What aspect of the storm will be the most damaging?
- What determines the path of this oncoming storm?

Students were given access to current hurricane data, charting tools, data collection devices, and natural disaster information from the National Oceanic and Atmospheric Association and other sources. However, at first students were at a loss as to how to begin work on their problems. Hurricane season arrives early in the school year, and norms for collective problem solving and classroom discourse had not yet been established. Students indicated their distrust in the teacher's assignment of a real-world problem—inserting their school histories into the context.

Yerrick:	Okay, so do you all [think] that I know the answers to your problems?
Shawn:	Hmm?
Yerrick:	Do you think that I know the answer to your problem?
Sly:	I think that you have some idea for the answer, you know. I don't think, me personally, I don't think you really like know the definite answer.
Yerrick:	D'Nisha, do you think I know the answer to your question?
D'Nisha:	Mmm. [3-second pause] Maybe mine, yeah. Maybe the problem that we have, yeah.
Yerrick:	Okay. Um, if I know the answer, what then can be my role with your group? What should I do with your group if I know the answer?
D'Nisha:	[Shakes head indicating no]
Jerrell:	[Quietly] Give her the answer.
Yerrick:	[Amused] "Give me the answer." That's what Jerrell said, "Give it to her."
D'Nisha:	Mmm, well is it to lead us in the right direction? So that we can try to halfway understand what we're supposed to do with it all and stuff.
Yerrick:	[Smiles and nods]
Sly:	I didn't understand at first either, until I looked—until I, I had to look at it myself. Then I just started playing with it, and—it was like, "Cool."
Yerrick:	So you had barometer set up [Sly nods yes] and a pressure sensor set up with yours? Um, did you have to explain to Marcus and Xiao what you were doing?
Sly:	I was like, "Well, the reason I done this is because this right here works like this." So I just explained it to him.
Yerrick:	Tomorrow we can set up your MBL to go over the weekend or even measure the barometric pressure over 45 days.
Sly:	All right.

Students were at an impasse in being able to approach the problem that I had posed. Getting out of the impasse required that my agenda was made more explicit. Students could not grasp the problem, as they thought a correct factual answer had to be found by the end of the class period. Although students had been engaged in hands-on problems with real-world phenomena and technology, there was a lingering question about what the point of the activity was as it did not fit their experiences from other science classes.

When I recalled that these students had been selected for this class on the basis of their failure in prior science classes, it became understandable why they would question a teacher's encouragement to develop their own ideas independent of correct answers. This pattern is especially true for the students who had failed this very same earth science class the year before with another teacher. Building a learning community among lower track science students is especially challenging in early stages of the process. As blueprints provide a reference for an architect, the knowledge of how daily materials and activities connect to the foundation can assist teachers in not succumbing to the easier path of simply providing factual answers when pressured by their students, peers, or superiors.

"Is this science class or what?"

Of course, decisions about correct answers are part of a larger set of issues connected with one's epistemic orientation toward science (Yerrick et al., 1998). As teachers we must carefully consider why and how we use questions as a primary vehicle for teaching. If questions are meant to guide a predetermined teacher patter (as characterized by Lemke's [1990] description of a monologue masquerading as a dialogue), would it not be better just to tell them the correct answer than ask them to read your mind? Several authors have argued that the very nature of teachers' questions should be given careful consideration before student tasks are assigned (Gallas, 1995; Hogan, 1999; Roth & Bowen, 1995). Duschl (1990) contended that the role of teachers should be honest and open sharing of all their expertise with students mutually assisting students in exploring and debating current scientific theories and their applications.

Such was my approach as I struggled to teach predominantly Black lower track science students in a teacher education project situated in a rural North Carolina high school. Most of my high-school students had long histories of academic failure, and "doing what you are told" carried a particularly negative connotation to these students when they interpreted science teachers' instructions. After a long renegotiating process, the expected science discourse of the class revolved around placing personal theories on the table, gathering evidence to confirm or refute those theories, and openly debating personal theories with peers (Yerrick & Hoving, 2003). Such a discourse community was not typical at this school but rather reflected Anyon's (1981) and Fordham's (1996) descriptions in which students of lower socioeconomic status or specific ethnicities were given access to different kinds of knowledge.

In the following excerpt, students were invited to discuss the differences between our science class and other science classes. Winning the game of science had been redefined over several months with these students in small-group and large-group discussion settings. In contrast to their past science experiences, these students were asked now to engage in making decisions based on their own research, connecting ideas to the real world. Marcus described how the course met his needs as a learner better than the year prior.

Yerrick: Marcus. You were in other science classes before you were in this one. What is something that counts for science in our class that wouldn't have counted for science in another class?

Marcus: Um, the way we learn in here is a lot different from the way we learned in other classes. In other classes, most of the time it's bookwork or busy work. But in here we learn different ways by talking you know, by talking about different situations and what we studied. We do work or whatever, but it's hardly ever we use the book. We usually relate everything to real-life situations.

Jackson: That's the basically what I said.

Yerrick: And that wouldn't count in the other classes? Is that what you're saying?

Marcus: Yup. That's what I'm saying.

Jackson: Nope. It wouldn't count.

Yerrick: Why wouldn't they count?

Jackson: Because the teachers....

Marcus: Because, I guess—the other classes—we.... Most of the time we're working out of the book and we basically just do, what the teacher tells us to. Well, we do what *you* say too. [We do] what you tell us to do in here, but, in the other classes, we do things that the teacher tells us to do, and it's just like, you know, "Do this work here," without really talking about it like we do in here. Then so it's easier for us in here to, you know, to learn.

Yerrick: Karmen, do you understand what he's saying?

Male: I do. I feel it.

Yerrick: Mercedes, can you explain it to me? Explain what Marcus is telling me.

Mercedes: He's saying that the way he, the way that he learn in this class is more hands-on and all activities and stuff like that. So it's like we don't have to go home and study in a book. Just from my own observations—we all observe and, you know, the resources and research and stuff that we be doing here. Things like that.

Yerrick: Okay. Karmen, is that what you heard him say?

Karmen: Uh-huh. Pretty much.

Yerrick: Sly, do you have something to add to that?

Sly: Uh, they pretty much summed it up very nicely.

The students described the contrast between their regular schedule, filled with book learning throughout the day as well as simple, non-thought-provoking directives, and our earth-science class, in which they were required to make connections with real-world phenomena and to talk through their thinking publicly with their peers. These lower track Black students self-reported that negotiating a new science discourse was helpful to their learning science despite recommendations of other teachers in their school—teachers who believed that these same students were unwilling or unable to be successful science students (Gilbert & Yerrick, 2001). It is a testament to these students' versatility that they could carry out a significantly different kind of discourse for learning for this one hour of their day.

"Can I get my C moved up to a B for that?"

In order to shift the focus away from book work and factual answers, the classroom science discourse must require students themselves to be responsible for generating models for debate, modes of proof, and means for deciding the merits of any proposed ideas. Teachers must at times withdraw themselves from the position of authority over all scientific answers and other classroom content contributions. Teachers cannot always maintain the role of convincing students of the authority of canonical scientifically accepted ideas. Rather, teachers must at times become facilitators, even co-contributors to the process of constructing knowledge. Research on students' thinking processes (Ausubel, 1968; Driver, 1991; Posner, Strike, Hewson, & Gertzog, 1982; Roth & Lucas, 1997) and students' prior knowledge (Anderson & Smith, 1987; Driver, 1990; Osborne & Freyberg, 1985) has established that the insertion of *scientifically proven* knowledge claims will make little difference in children's actual beliefs about the world—that is unless students themselves are required to participate in ways that emphasize their roles as listeners, interpreters, evidence generators, and critical examiners of knowledge claims.

For this reason it is important to carefully craft the questions and activities put before students so as not to reinforce the false notion that science (or any other discipline) is about the gathering of correct answers (Ball, 1988; Lampert, 1990; Roth, 1998; Russell, 1983). In the next vignette, I was working diligently to deflect questions of the "correctness" of students' interpretations as it had so often truncated rich discussion. More students engaged in discussions and students more often referred to evidence in their reasoning when the students themselves offered evidence and reasoning. In this case, though I understood the reason light bulbs "burn out," Dimitri had found a way to resolve the class debate by citing evidence of the properties of burnt magnesium. Dimitri surmised in this renegotiated science discourse that only a chemical change accounts for the evidence of burnt magnesium and light bulbs and eloquently devised a thought experiment that even his adversaries understood. Moreover, Dimitri did this (as he always did) out of a motivation to win in the game of science.

Yerrick:	Brent had a reason. Do you remember Brent's reason for saying it was chemical?
Samuel:	Cause you can't bring it back.
Yerrick:	That's very good Samuel that's very good.
Dimitri:	I know how to tell. . . .
Erin:	Which one can't you bring back?
Yerrick:	He said you can't bring [the burned magnesium] back.
Erin:	You can't bring physical things back?
Dimitri:	I think I know how to test it. Put a current through it.
Yerrick:	Why? Tell us why that test would work?
Karlton:	Okay go get it!
Dimitri:	Because electricity would flow through the magnesium, but if you put it on the ash it wouldn't flow through it after it was burnt up.
Yerrick:	That's a good test. . . . That is a good test.
Tasha:	Good Dimitri!
Yerrick:	What do you all think about that?
Samuel:	Well, that's going to be real hard unless you have the ash on . . . you're going to have the ash like perfect laid out before and after you burn it.
Erin:	What did you say Dimitri?
Dimitri:	You put electricity on it when you burn it. And then see if it goes through. It'll tell ya'.
Yerrick:	We could certainly try that. Even if it's a small current. . . .
Erin:	That'll tell you if it's chemical or physical change right?
Dimitri:	If it's physical then [the electricity] gonna' go through it, but if it's chemical it won't go through it.
Erin:	O-o-o-o-h!
Yerrick:	What do you think?
Erin:	That's a good experiment!
Yerrick:	It is.
Karlton:	A way in which you could check if it goes through it is to just check it with and Ohm meter for resistance.

8. PLAYING THE GAME: OVERCOMING OBSTACLES

Yerrick: Resistance. Kind of like a volt meter. We could do that. Okay. Erin what do you think about that?

Dimitri: I think I should get my C moved up to a B after that.

After many experiments and much discussion, Dimitri had found a way to resolve the class debate about whether burning out a light bulb involves a chemical or a physical change. Further, by deflecting the role of content expert, I had raised students' expectations in this lower track classroom to be critiquers of knowledge claims. More importantly, Dimitri's contribution was significant as it was evident that he had embraced the game of science and won—not only because he generated the correct answer but because he found a way to resolve the classroom argument through a single public experiment. Though this group of lower track students began the academic year expecting to engage as "receivers" of knowledge (Yerrick, 1998), Dimitri's negotiation for a grade change reveals that he understood classroom discourse had been renegotiated into a context of proof and refutation of knowledge claims (Popper, 1963).

Along with these changes in discourse come new challenges in planning and assessing students' learning. The previous vignette may support Duckworth's (1987) claim that children's misconceptions can be fruitful venues for scientific discourse as children explore the validity of their own knowledge claims. Teachers must have at their disposal various kinds of knowledge to judge the scientific appropriateness of questioning techniques, problems to solve, beliefs and interactions of students, and ways to bring them all together. In similar ways, builders must have an immense knowledge of mechanical, electrical, and architectural knowledge to build a cogent, inhabitable, functional structure. Planning for instruction goes beyond presenting concepts and their applications to students to encompass creating an entire learning community that values what children see and understand about the world, finding ways to perturb commonsense thinking that might lead to common pitfalls. Thinking about assessing such interactions also challenges teachers to reach beyond canonical knowledge to evaluate the significance of student contributions to the classroom knowledge community. Duckworth asked, "Isn't it enough that this application of chemical change is new to them?" Acting scientifically does not require that students generate knowledge new to the field, nor does it lock students into replicating canonically accepted knowledge to demonstrate their expertise. Clearly, Dimitri brought several students' ideas and past experiences together to spontaneously generate a test, a thought experiment of sorts, that resolved the issue for some students before the experiment was even conducted—a feat I was unable to accomplish. I honored his request for a grade change that semester because of his continued efforts to win this game of science.

WHAT ARE THE RULES OF SCIENCE?

There is a substantive challenge in shifting classroom discourse from that requiring students to passively receive knowledge to a discourse embracing the constructing of rational arguments. However, this challenge is surpassed by the call of reform to give all students equitable access to scientific knowledge. If resolving gender and ethnic equity issues in science classrooms were simply a matter of presenting scientific knowledge to all students, the issue could long ago have been put to rest. Rather, scholars of equity issues have argued that power resides in understanding implicit rules and their appropriate use in context. If this is true, we are faced with the decision of which rules we are going to teach students about science and who decides which rules best exemplify scientific thinking.

To answer this question we must more closely examine how science has been presented by its membership to the public. Scientific discourse communities do not operate as objective, open, and cooperative according to some insiders (Kuhn, 1970; Mayr, 1982; Traweek, 1988); their and some norms of discourse run contrary to our accepted notion for democratic schooling (Ballenger, 1997; Tobin, 2000; Yerrick, 1998). Until recent decades, the exclusivity of the scientific community left the wider public to judge them by only the reports and final research products that had been cleaned up for public view (King & Brownell, 1966). Claims of scientific objectivity buttressing the production of facts and theories have since undergone close scrutiny by sociolinguists and ethnographers of scientific settings (including the indigenous discourse of laboratory settings). The discourse reported in these studies contrasts starkly with the final products, facts, and methods by which scientists have historically been judged (Latour & Woolgar, 1986).

Whose account is accurate regarding the rules guiding scientific knowledge production? Lemke (1990), a sociolinguist and former nuclear physicist, gave credence to these anthropological accounts and argued that students should learn the rules of collaborative inquiry and critical analysis as well as rational argumentation. If science is to be played as a game to win, then are we to believe that scientists rationally design experiments and calmly and objectively analyze and construct arguments with their rivals? Is this a reasonable and accurate lens for us to view our students' interactions? Lemke suggested not and challenged our preconceptions about the rules of science as an insider:

> How does scientific theory become established? ... Historically, a new theory always begins as somebody's way of talking about a topic or problem. They argue for their theory, or someone else does, and convince others. A faction appears that lobbies for the theory in many ways: by research papers,

8. PLAYING THE GAME: OVERCOMING OBSTACLES

> by experimental tests of predictions, by talks at scientific meetings, by writing books and textbooks, and teaching students, and so on. In the end, a community of people . . . the most powerful people within that community, determine which of the theories get published most, used most, taught most. (p. 125)

Lemke focused on the daily grind of the oral and spontaneous nature of science, including the members' treatment of ideas. Similar to Kline's (1970) portrayal of mathematics, Lemke's science was not strictly composed of arguments in clean final versions. Instead, influential individuals, groups, and even paradigms shaped the social construction of ideas. Certainly history has born out that scientists like to win, and in recent decades insiders' accounts have revealed that scientists are not always immune from bias, personal attacks, or politics in their fierce competition for answers. This process, described by philosophers and social scientists, leaves communities susceptible to social coercion, which reinforces existing power structures within classrooms.

When we teachers consider such a view of scientific work, our expectations are shifted for students working together scientifically. We should anticipate students using social prowess to advance a particular theory, expect bias to sometimes interfere with open discussion and explanation, and associate authentic engagement in scientific ideas with debate that may not always reach consensus. Classroom communities—just like scientific communities—are social contexts, and they will not be immune from similar influences. In fact, it can be argued that if students are passively going about the normal business of classroom discourse, they are likely not engaged in science.

However, not all disagreement and debate is conducive of continued inquiry and exploration. In fact, if left to their own devices students can evoke forms of classroom politics that exclude ideas of certain community members and thwart critical thinking. Students can influence classroom discourse through their engagement and maintain their control over the direction of inquiry despite the teacher's efforts to include a diversity of contributors to science talk.

Like the builder who takes into account a variety of perspectives, the teacher alone must often decide on what benefits the building of this learning community as a whole. A builder would not sacrifice the soundness of the building's structure to respond to the aesthetic whim of a designer who desires an entire external wall of glass for a better view. Likewise, teachers must balance the desires, perspectives, and contributions of the whole and decide what design best represents all vested parties' interests. Teachers need to find ways to level the playing field for all members through the establishment and enforcement of rules to assure equitable access for students who are underrepresented in science. Gee's second notion of *rules for participation* can be useful in establishing more democratic settings. Additionally, Delpit (1988) argued that when students are taught the rules of power and access specifically, they are more likely to be empowered within the discourse community.

"I talk like this in the hallway. Why can't I say that now?"

One of the most essential rules of renegotiating students' classroom talk is the establishment of every member's right to contribute and disagree without fear of personal attack. In schools students can be shunned into silence through the pressure of peers (Cobb, Yackel, & Wood, 1991). I am not implying that every idea students contribute is of equal explanatory power. Every idea that is contributed, including those of the teacher, should be viewed with healthy skepticism. However, when sarcasm and personal attacks invade classroom science talk, the results hinder everyone's progress. Consider Erin's plight as she makes a guess at the onset of a friction debate:

Yerrick: Okay. We're all in agreement about that part. Now where or what gives the force to slow down the cart? Friction, gravity, what?
(Several hands go up)
Erin: The rotation of the earth's rotation?
(Class laughter)
Micah: (Joking to Samuel) Don't rip on people!
(Referring to class rules written on the wall)
Samuel: (Still laughing with Micah) You were laughing too!
Yerrick: Why are you laughing at her?
Samuel: I don't know.
Yerrick: (to Erin) Doesn't that make you feel like "Gosh, it must have been a stupid answer," when someone laughs like that?
Erin: Yes.
Tasha: Number five of the class rules.
Yerrick: You shouldn't laugh at people.
Samuel: I'm s-s-s-orry. If you [teacher] wouldn't have made a big deal of it she wouldn't have done nothin'. She would have ... (makes a shrugging gesture)
Yerrick: I know. I just want to set a precedent that regardless of what the answer people share, we shouldn't laugh.
Samuel: All right.

8. PLAYING THE GAME: OVERCOMING OBSTACLES   245

While it would seem that this violation of class rules had been resolved, one cannot overemphasize the effect upon the contributor of such guesses in the class. There are all kinds of gender and ethnic equity issues wrapped up in moments like these that need to be balanced when teachers promote a discourse in science, which has historically been represented as Anglo and male dominated. Despite efforts to re-engage Erin, her idea was suppressed and lost.

Erin:	The earth is rotating well . . . (mumbles something, eyes downcast, becoming more quiet). Never mind.
Yerrick:	So tell me why that would exert more force on the car that is heavier?
Erin:	I don't know.
Yerrick:	Okay, We'll leave that as a possible answer. We haven't shown you right or wrong. But there might be a better explanation. There might be.
Erin:	Yeah.
Samuel:	(in a sarcastic tone) Oh I don't know. I don't think there is. That's a pretty good one Mr. Yerrick.
Yerrick:	Samuel.
Erin:	(recognizing Samuel's tone) Don't cut on me!
Samuel:	(laughing) I'm not!
Yerrick:	Okay here's how we could do this [disagree] differently.
Tasha:	Maybe it's. . . .
Yerrick:	Now someone might. Now this is how we might do this [disagree] in a different way. Someone who can support their explanation like gravity or friction and give a reason might be listened to a little more often.
Yerrick:	(acknowledging Tasha's raised hand) Yeah go ahead.
Tasha:	It just might be cause it's lower to the ground.
Samuel:	(snickering and lowering his head to regain his composure)
Yerrick:	Samuel.
Samuel:	I'm not.
(class laughter.)	
Samuel:	I'm not!

Cobb et al. (1991) described how such emotional acts of confusion, withdrawal, and embarrassment are learned specifically in the context of regular classroom discourse. What is worse is that Erin's embarrassment came as a result of her deliberately taking a risk to engage in the science talk. Erin's feelings of being further marginalized as a lower track science student were directly precipitated by her attempts to engage in new kinds of scientific classroom discourse. What is the cost? Through oppositional discourse patterns Samuel and other student leaders undermined my attempts to bring in other less vocal members to science talk. Through these socially constructed norms they evoked feelings of embarrassment in others and showed intolerance for wrong answers, which consistently prevented open discussions from occurring.

In order to widen the participation of females and other students historically excluded by confrontational White males, I had to demonstrate that no student's idea was inherently less valuable and that disagreements need not take the form of hallway brawls or shouting matches. Samuel was trying to convince me that he was just talking normally—the way students engage with one another outside of school. However, I could not allow the way he treated the tentative conjectures of females in the class. Erin and Samuel had not invited me to tell them how they could talk to each other outside of class. However, negative talk like ridicule that was acceptable outside science class needed to be curtailed for science talk to develop. Although promoting debate and sustaining public criticism of ideas, a teacher's efforts can be dangerously intermixed with students' attempts to gain peer recognition and status (Yerrick, 1998).

"Why can't we just combine these ideas or take a vote?"

It would be naïve to think that allowing children of any age to simply talk and debate would spontaneously generate powerful ideas such as photosynthesis, inertia, and explanations for the seasons. Literature clearly documents in each of these subject areas that children construct their own disparate meanings that are consistent with their prior beliefs—ignoring opportunities or evidence that does not fit their prior schemas (Osborne & Freyberg, 1985; Shapiro, 1994). In these instances it would be irresponsible for teachers to allow students to blindly agree with socially powerful students and equally inappropriate for teachers to simply insert the correct answer and thwart all student debate. In order to consistently follow Duckworth's (1987) notion of "wonderful ideas," teachers must learn to balance the origin and orchestration of ideas to honor student contributions. It would be disingenuous to allow students to propose and follow a course of investigation only to be corrected at the end with the teacher's authoritative answer. Students would accurately sense the futility of the task and inquire, "Why didn't you just tell us in the first place?"

Expert teachers therefore arrange for discrepant data or disagreements to expose weaknesses in children's thinking (Ball, McDiarmid, & Anderson, 1989; Roth, 1998). Within the debate around such data teachers also build specific rules for engagement in problem solving, such as the exclu-

8. PLAYING THE GAME: OVERCOMING OBSTACLES   247

sion of dichotomous solutions that are only apparent to the expert. For example, a teacher should separate the notions of the earth's tilt and its proximity to the sun when debating the cause of the earth's seasons. Although novices may want to merge or synthesize these ideas, differences are borne out by data and careful application. In the following event a student resists the boundaries of such a problem and, hence, engagement with the core ideas implicit in it. The problem was specifically framed to directly challenge the widely accepted Aristotelian notion that objects naturally slow down without the aid of external frictional forces. Four alternative student designs were proposed for a magnetic monorail, and students were asked to choose a defensible design to present to the rest of the class. Within the problem framework, if students misunderstood the inertia concept, they were unable to agree with another group's design. Thus, if students believed that objects remain in motion indefinitely without frictional forces, some groups would have to argue against the need for certain craft features (e.g., emergency wheel). Students agreed to the rule that they were not allowed to change the problem in order to debate the core ideas describing objects' motion.

Despite the instructions, Cindey attempted to change the problem by combining all the features so the craft would have all the special tools of both designers, thus removing the need to argue. Unfortunately, she was not talking about the scientific problem—she wanted to have all the features and be on the winning team. Understanding the conflict in her approach, Cindey's group members wanted to give a separate answer, leaving Cindey to fend for herself as she confounded the task.

Cindey: Can I give my own answer?
Yerrick: Actually, just ask Samuel some questions about his model. Why don't you agree [with Samuel]?
Cindey: Well, the reason is because number four you could put an emergency wheel on it in case it slowed down. . . .
Samuel: Yeah, but there's not an emergency wheel on it.
Cindey: The teacher said you could put one on it.
Micah: He said pick one or the other.
Cindey: Right, but then if you have to stop, the wheel's gonna start it right up.
Samuel: Okay, put an emergency wheel on it. . . . Go on.
Brent: Why would you want a spring on it?
Cindey: But, what I'm saying. . . .
Mark: Why would you want a spring on it if you already had a wheel?
Cindey: Excuse me, I'm not done!
Kristy: But, like you said. . . .

Samuel:	Okay, then let's say the spring has a wheel and a spring. That's what you're doing, you're changin' em! You can't just change the whole thing!
Cindey:	No I'm not changing em, but the little spring is gonna push....
Tasha:	Cindey is actually.... You're trying to say....
Micah:	I think....
Mark:	You ought to be a politician.
Samuel:	She's trying to fix it so she's ... //
Mark:	//Politician ...
Samuel:	// ... trying to put something on that can't be there.
Brent:	She's just trying to....
Samuel:	She's changing the problem!
Brent:	Yeah.
Samuel:	You can't do that Cindey!
Yerrick:	Right.

From the outset the class had agreed to a set of rules in order to play the game and debate the problem at hand. Each small group accepted the rule to not combine attributes of different models, defending their choice or refuting other group's choices. Because she tried to change the problem, Cindey was told she could not play the game anymore. Students recognized Cindey's actions as maneuvering around the rigor of debate in order to win by agreeing with all student groups. Even her own group recognized her actions as more political than scientific, exposed her rule bending, and uninvited her to continue playing. These students recognized the rules of public debate and when students violated the nature of the problem to fit their individual social goals. In doing so, students demonstrated their understanding of the rules of power (Delpit, 1988, p. 281) even more than if they had simply accomplished the task successfully.

Throughout their pushes and pulls I was forced to re-examine my assumptions of classroom talk. Science discourse does not always operate democratically and consensus on a singular answer is not always the goal. Cindey wanted to pacify all students by combining incoherent perspectives—thus avoiding legitimate cognitive conflict. I could not allow such circumvention of refining their collective thinking about the evidence, even if other students in the class may have voted for her idea. But how do you establish fairness amidst a context in which members want to win? Lessons learned about overcoming obstacles come through successes and failures, trials and errors. As argued above, teachers must remember that they are likely the only community member who understands the larger goals for scientific classroom discourse, much like a builder understands the whole picture whereas contractors focus only on their journeyman skills. Teachers need to stage specific talks to bring to light what the real goal of science talk is. Teachers must teach the rules for how scientists win and which tactics help the group progress on both an individual as well as a collective level.

WHERE ARE WE IN THE GAME?

Of the three questions Gee likened to the process of learning scientific discourse, his third categorization was the aspect to which I had paid the least attention as a beginning teacher. Although I had considered different rules and ways to reward student scientific participation, helping students through talk about the talk was something I had given little thought. However, researchers (Delpit, 1995; Hogan, 1999; Michaels & O'Connor, 1990) have stressed the importance of metadiscourse in learning new forms of discourse inasmuch as classrooms are places that rarely provide reasoning or explicit guidance for the ways students are expected to think, speak, and act. Consequently, when teachers attempt to make fundamental shifts in the normal classroom discourse, they should expect students to inappropriately insert other forms of talk to artifacts and events staged by the teacher for science talks. Students can misjudge the opportunities teachers create for debate, conjecture, and exploration. Sustaining regular metadiscourse surrounding commonly shared learning events allows the teacher opportunities to provide students assistance for interpreting cues, find alternative ways to transcend intellectual obstacles, and better orchestrate discussions in which resolving diverse student opinions and observations can be a challenge.

"What do you mean we're not done yet?"

College students can misjudge their place in the game as often as high-school students since the tactics they have used to master school science discourse and succeed have also misled them in their understanding of the nature of science discourse. College students who have been successful in memorizing facts and taking exams often resort to a tactic of labeling phenomena with fancy terminology. This is likely due to their extended socialization in science courses as they prepared themselves for secondary science teaching careers.

The following excerpt is the only example not taken from a lower track high-school context. Rather, it is extracted from a larger set of data collected in a senior-level college science course for preservice teachers. It serves, in part, to demonstrate how students can inappropriately apply conventions and measures for prematurely concluding well-designed science tasks. The students here who are caught at an impasse and are unaware of how to judge their place in the game are all too willing to label a phenomenon and move on without distilling the essence of the problem posed to them. We join Matt, Shawn, Jeff, and Frank in the midst of a small-group experiment in which their uncertainty of where they are in the game of science causes uncertainty in their reliance on consensus to resolve disagreements.

Shawn: That's what I was thinking at first too, but we came to the consensus. . . .
Frank: If it wasn't used up then the battery would never die.
Shawn: The battery is not used up just the only thing is its ability to produce chemically. . . .
Mike: Yeah it gives it the chemical charge because that gets used up.
Shawn: Well it won't produce any more electrons so the pump has stopped.
Frank: But ya see do we have to come to the same opinion?
Shawn: No.
Frank: Because this is where we get behind I mean . . . whatever.
Jeff: So what are you saying Frank. . . . What is your point again?
Frank: As far as electricity or moving on?
Shawn: No, but you've got a good point because in asking that I'm starting to think . . . no I don't have to agree it's good not to agree. But if we don't agree somewhere one of us is gonna go into the next step with the wrong analogy . . . then we've got old baggage you've got to unlearn all those things and it's harder to unlearn it. . . . I'm thinking that the wire has thrown off our results totally because the other wire was this thin.

(holds up a small piece of thin wire)

Jeff: So why would that make a difference?
Mike: It can hold more heat maybe. . . .
Shawn: More electric[ity].
Frank: The larger size wire has more resistance.
Jeff: So how does that explain what is going on?
Mike: Less resistance though would make it hotter.
Frank: It's Bernoulli's Principle; the smaller the channel the faster it goes . . . and the hotter it gets.
Mike: Yeah, that makes sense.

(All members nod, write Bernoulli's principle in their journals, and discussion ends)

All four members of this small group had been experiencing frustration as they conducted experiments, constructed mental models, and proposed solutions to questions posed in the *Physics By Inquiry* curriculum (McDermott, Shaffer, & Rosenquist, 1996). Finally, one of the group members spoke aloud regarding the frustration he had been journaling about for days. Shawn truncated the discussion of whether charge gets "used up" in the circuit because of his discomfort in not knowing the correct answer. Unable to put all the available data into their current model, the students agreed to throw out their data in favor of an entirely inappropriate analogy of Bernoulli's Principle. The proposed explanation likely suited the members of this group because they were more comfortable labeling an event with scientific terminology than exploring their thinking further. More importantly, the adoption of this analogy did not lead to further application or explanation. It simply ended the engagement of all group members on the problem at hand.

These students were in need of assistance in understanding the larger purpose of their explorations. Because they were unable to maneuver through the mire of disagreements and find the greater value in sorting through their available data, they were lost in the ways of playing the game of science. Students can easily grow accustomed to the processes of attaching nomenclature to the biological world and using formulae and computing answers to chemical and physical theoretical problems. Unfortunately, such processes do not entirely capture the work of biologists and physicists despite their usefulness in passing college science courses. Naming a phenomenon with a scientific term and selecting an event that conformed to their prior beliefs allowed them to believe they had completed the task. However, what they left the conversation with was a conclusion that will later confound their efforts to develop a consistent theory describing electric current. In the interest of "not having to unlearn" misinformation and the pressures of "getting behind," the group has unwittingly fallen into older patterns of science learning. Engaging in metadiscourse can assist students in their construction of order from apparent chaos and give them tools to reach beyond the obstructions next time instead of embracing a commonsense notion that directly contradicts the available evidence—as further exploration likely would have revealed.

"I don't understand what this has to do with our examples"

Even when students are coached in the recognition of stages in the development of scientific arguments, there is no assurance there will be an easy transfer of discourse norms and thinking into the problem setting. In the following excerpt, students are discussing their difficulties encountered during the day's classroom instruction. As we talked about the classroom talk I engaged these rural, Black high-school students in a conversation in order to work through why their attempts to work independently in small groups had been unsuccessful.

Yerrick: I think that's part of what scientists do. I don't think that they, that someone gives them a homework problem, they go to their lab, and then they work all by themselves, get the answer, and then come back and show it. I mean, it's. . . . I don't think that's what scientists do. So what I want to help you to try to do is, as a group of people, work more like scientists do and then help you to do it. But this is what happened. I let people pick their problems. And after they selected their problems they went to their groups. And then, and I had to chuckle to myself. Yeah. [D'Nisha laughs] Everyone just kind of looks at each other, and then they raise their hand and wait for me to come help them get started again.

D'Nisha: Well, the questions seemed like, like the ones that we had, we thought it was going to be easy. But when we got into the small groups . . . it was like, "Gosh! What are we supposed to be doing? I mean, this is not helping us at all. We don't understand it." [Shaking her head, she lowers her voice] I didn't get it. [Speaking normally again] I thought it was going to be easy, but I didn't realize. . . .

Yerrick: So D'Nisha says one of the problems that she had was that she underestimated how hard the problem was.

D'Nisha: It's not like I didn't understand it on the board, but when it came down to actually doing it, I was like. . . .

Jerrell: Yeah, me too.

D'Nisha: It didn't seem like the same question that was on the board.

Yerrick: Even though it was the same question. It just didn't *seem* like the same question. Okay, Jerrell, what was it like for you?

Jerrell: Yeah, . . . I think it was really fun. I didn't know whether it looked good. I didn't know nothin' about hurricanes and what you want me to tell you about it—or them all [the class].

Yerrick: So, so the problem that you had about hurricanes, you didn't even know where to find the information.

Jerrell: I thought we were just going to have to like—plus when he just gives that sheet, with the protractor or something like that. I could have had that. But I don't know. . . .

Yerrick: That's certainly a, a first step. . . . Tony, tell me about your, your experience when you first looked at that problem and what to do on it. Why is it hard?

Antoine: It wasn't that hard.

Yerrick: No? . . . I mean if it's not hard then you probably are getting close to the answer.

Antoine: I done forgot. But I'm glad I got it written down on the paper. [Nods.]

These students were engaged in a discussion about the talk pertaining to an event they had experienced as a group earlier that day. Through this metadiscourse not only were students able to hear how and why the structure of their task was designed to be more scientific, I also learned as a teacher about how and why such tasks regularly broke down. Without such interactions I could have modified my pedagogical approach but missed the true reason students were struggling. Renegotiating the rules and goals of science is a difficult challenge, especially against the backdrop most science teachers face of only one contact hour per day. Inasmuch as schools are unlikely to implement sweeping and profound changes in organization or pedagogy in every class, these subtle classroom discourse changes cannot occur without the cooperation of students. Getting students in a position where they both understand and assist in the process of change requires that they understand the rules, the ways to win, and their place in the game. Without such steps, struggles to appropriate new norms of discourse will likely be met with insurmountable resistance aimed at a return to classroom speaking, thinking, and acting that is more familiar to students.

CLASSROOM IMPLICATIONS FOR CHANGING THE GAME OF SCIENCE

I have argued in this chapter that students become comfortable with a kind of classroom talk that contrasts scientific discourse and that engagement in school science can lead to obstacles in engaging in ways of talking that are more scientific. The implicit and explicit messages conveyed about science pervade students' interactions with teachers and the content. The are so deeply rooted that any teacher seeking to open up the discourse and invite the construction of arguments using evidence is likely to encounter unforeseen obstacles. It is therefore essential for teachers to understand the nature of the science talk they intend to promote and to bring to bear many strategies for guiding students into the new discourse in equitable ways.

There are two central assumptions of this chapter: (a) Teaching science in ways consistent with recent reforms requires fundamental shifts in the roles, rules, and responsibilities of all members of a classroom scientific community and (b) any fundamental shifts in well-established norms of discourse will promote uncertainty and often obstacles to change. One common strand throughout each critique is that the discourse we seek to promote in order to truly reform the way science is taught must depart from normal classroom discourse. This is a great challenge because students and teachers seek predictability, comfort, and closure—even if it is artificial. Science discourse as it is conducted in laboratories and even in sterile lecture contexts does not always seek such ends, and we must therefore continually be vigilant in our assessment of classroom talk and what outcomes are consistent with our philosophical interpretation of science is.

The obstacles teachers face as they seek to promote scientific discourse have diverse origins and take on many faces. Students make seek closure and simple factual answers when further study and closer examination are needed instead. Teachers may feel compelled to move ahead quickly and cover more content because of the pressures of mandated testing and the perception of imposed curricular constraints. Students and teachers together may revert to normal classroom discourse as such rules, patterns, and rewards are far more certain and predictable. Students may not want to play by new rules. Sometimes it is even the students who have been most successful in the old science classroom discourse who will most strongly resist new ways of learning and evaluating science talk.

The ultimate reality is that both students and teachers alike have the power and often the propensity to distort classroom discourse away from anything scientific (Millar, 1989; Roth & Lucas, 1997; Russell, 1983; Smith, 1999). For this reason it is important to consider the teacher's role through Kohl's (1984) analogy of a builder. Teachers can cover over imperfections and weaknesses in students' knowledge of science just as a builder can place drywall and plaster over rotted support beams. Such would be the case if a teacher recognized that learning science is not best portrayed by receiving facts but continued to promote such a discourse anyway. It is as Schwab (1962) described, "a corruption . . . and we will know that it is" (p. 229). The results are found in the fatigue and failure of that imperfect knowledge, just as fatigue contributes to the collapse of a building. I believe Kohl's analogy goes a long way in describing the inspection, planning, and reparations required to successfully transform classroom discourse. Whether builders are inspecting buildings or science teachers are observing learning communities, both share the common vision that what they will create will be a lasting monument built on a strong foundation.

For students to reach beyond the obstacles of normal classroom discourse in which there is a single correct answer to be received from the infallible teacher, children must be encouraged to view their personal theories in meta-reflection against those of others. Teachers must also become masters not only of their content but also of building a learning community in which each student can safely but critically engage ideas. Debating personal theories then becomes an opportunity for teachers to pose problems with specific restrictions so as to ferret out children's beliefs through public discourse. Opening up classroom discourse in such ways often results in questions with no immediate answer. Consider questions from a 6-year-old like, "Why don't spiders spin cocoons like butterflies?" and

"Why does the earth spin?" Typically teachers may respond with, "That's a good question. Let me look that up and get back to you," or they may assign the question as an extra-credit project to be reported on at a later date. Such responses are spawned by the hidden curriculum of school science whose implicit messages about teaching is, "I believe that my job as a science teacher is to tell you what I have figured out. Your job is to receive it. Hold on and *I'll* go think about your answer. It's not your job to engage in experimentation or knowledge construction." Regardless of their origin or nature, obstacles can and must be negotiated since, as teachers, we are ultimately responsible for the representation of science that our students experience.

We need to actively renegotiate the discourse of classrooms so that students will think and speak about their world in inquiring ways. We need to model and scaffold science discourse with our students such that, when we ask children to think about a scientific question, they know without reservation that we really mean it. When we teach them how and invite them to try, children will do just that—construct instead of receive knowledge.

REFERENCES

American Association for the Advancement of Science. (1989). *Science for all Americans: Project 2061*. Washington, DC: Author.
Anderson, C. W., & Smith, E. (1987). Teaching science. In V. Richardson-Koehler (Ed.), *Educators' handbook: A research perspective* (pp. 84–111). New York: Longman.
Anyon, J. (1981). Social class and school knowledge. *Curriculum Inquiry, 11*(1), 3–41.
Ausubel, D. (1968). *Educational psychology: A cognitive view*. New York: Holt.
Ball, D. L. (1988). Unlearning to teach mathematics. *For the Learning of Mathematics, 8*(1), 40–48.
Ball, D., McDiarmid, G. W., & Anderson, C. W. (1989). *Why staying one chapter ahead doesn't really work: Subject specific pedagogy* (Issue Paper 88-6). East Lansing, MI: National Center for Research on Teacher Education. (ERIC Document Reproduction Service No. ED 305 330)
Ballenger, C. (1997). Social identities, moral narratives, scientific argumentation: Science talk in a bilingual classroom. *Language and Education, 11*(1), 1–14.
Cobb, P., Wood, T., & Yackel, E. (1991). Analogies from philosophy and sociology of science for understanding classroom life. *Science Education, 75*, 23–45.
Cusick, P. A. (1983). *The egalitarian ideal and the American high school studies of three schools*. New York: Longman.
DeBoer, G. (1991). *History of ideas in science education: Implications for practice*. New York: Teachers College Press.
Delpit, L. (1988). The silenced dialogue: Power and pedagogy in educating other people's children. *Harvard Educational Review, 58*, 280–298.
Delpit, L. (1995). *Other people's children: Cultural conflict in the classroom*. New York: New Press.
Dewey, J. (1904). The child and the curriculum. In R. D. Archambault (Ed.), *John Dewey on education: Selected writings (1964)* (pp. 339–358). Chicago: University of Chicago Press.

Driver, R. (1990). *Children's ideas in science.* Philadelphia: Open University Press.
Driver, R. (1991). *Pupil as scientist?* Philadelphia: Open University Press.
Duckworth, E. (1987). *The having of wonderful ideas and other essays on teaching and learning.* New York: Teachers College Press.
Duschl, R. (1990). *Restructuring science education: The importance of theories and their development.* New York: Teachers College Press.
Fordham, S. (1996). *Blacked out: Dilemmas of race, identity, and success at Capital High School.* Chicago: University of Chicago Press.
Gallas, K. (1995). *Talking their way into science: Hearing children's questions and theories, responding with curricula.* New York: Teachers College Press.
Gee, J. (1994). *Science talk: How do you start to do what you don't yet know how to do?* Paper presented at the annual meeting of the American Educational Research Association, New Orleans, LA.
Gilbert, A., & Yerrick, R. (2001). Same school, separate worlds: A sociocultural study of identity, resistance, and negotiation in a rural, lower track science classroom. *Journal of Research in Science Teaching, 38,* 574–598.
Goodlad, J. (1984). Access to knowledge. In J. Goodlad, *A place called school.* New York: McGraw-Hill.
Hogan, K. (1999). Thinking aloud together: A test of an intervention to foster students' collaborative reasoning. *Journal of Research in Science Teaching, 36,* 1085–1109.
Hogan, P., & Clandinin, D. J. (1993). Living the story of received knowing: Constructing a story of connected knowing. In D. J. Clandinin (Ed.), *Learning to teach, teaching to learn: Stories of collaboration in teacher education* (pp. 193–199). New York: Teachers College Press.
King, A. R., Jr., & Brownell, J. A. (1966). The curriculum and the disciplines of knowledge a theory of curriculum practice. *The disciplines as communities of discourse* (pp. 67–97). New York: Wiley.
Kline, M. (1970). Logic versus pedagogy. *American Mathematical Monthly, 77,* 264–282.
Kohl, H. (1984). *Growing minds: On becoming a teacher.* New York: Harper & Row.
Kuhn, T. S. (1970). *The structure of scientific revolutions* (2nd ed.). Chicago: University of Chicago Press.
Lampert, M. (1990). When the problem is not the question and the solution is not the answer: Mathematical knowing and teaching. *American Educational Research Journal, 27,* 29–63.
Latour, B., & Woolgar, S. (1986). *Laboratory life: The construction of scientific facts.* Beverly Hills, CA: Sage.
Layton, D. (1973). *Science for the people.* London: George Allen and Unwin.
Lemke, J. (1990). *Talking science: Language, learning and values.* Norwood, NJ: Ablex.
Lortie, D. C. (1975). *Schoolteacher: A sociological study.* Chicago: University of Chicago Press.
Mayr, E. (1982). *The growth of biological thought.* Cambridge, MA: Belknap Press.
McDermott, L. C., Shaffer, P. S., & Rosenquist, M. L. (1996). *Physics by inquiry: An introduction to physics and physical science* (Vol. 2). New York: Wiley.
Michaels, S., & O'Connor, M. C. (1990). *Literacy as reasoning within multiple discourses: Implications for policy and educational reform.* Paper presented at The Council of Chief State School Officers 1990 Summer Institute.
Millar, R. (1989). *Doing science: Images of science in science education.* London: Falmer Press.
Osborne, R., & Freyberg, P. (1985). *Learning in science: The implications of children's science.* Portsmouth, NH: Heinemann.
Popper, K. R. (1963). *Conjectures and refutations; the growth of scientific knowledge.* London: Routledge and Kegan Paul.

Posner, G., Strike, K., Hewson, P., & Gertzog, W. (1982). Accommodation of a scientific conception: Toward a theory of conceptual change. *Science Education, 66,* 211–227.

Roth, W. -M. (1998). Teacher questioning in an open-inquiry learning environment: Interactions of context, content, and student responses, *Journal of Research in Science Teaching, 33,* 709–736.

Roth, W. -M., & Bowen, G. M. (1995). Knowing and interacting: A case of culture, practices, and resources in a Grade 8 open-inquiry science classroom guided by a cognitive apprenticeship. *Cognition and Instruction, 13,* 73–128.

Roth, W. -M., & Lucas, K. B. (1997). From "truth" to "invented reality": A discourse analysis of high school physics students' talk about scientific knowledge. *Journal of Research in Science Teaching, 34,* 145–179.

Russell, T. (1983). Analyzing arguments in science classroom discourse: Can teachers' questions distort scientific authority? *Journal of Research in Science Teaching, 20,* 27–45.

Schwab, J. J. (1962). Education and the structure of the disciplines. In I. Westbury & N. J. Wilkof (Eds.), *Science, curriculum, and liberal education.* Chicago: University of Chicago Press.

Shapiro, B. (1994). *What children bring to light: A constructivist perspective on children's learning in science.* New York: Teachers College Press.

Smith, D. C. (1999). Changing our teaching: The role of pedagogical content knowledge in elementary science. In J. Gess-Newsome & N. Lederman (Eds.), *Examining pedagogical content knowledge: The construct and its implications for science education* (pp. 163–198). Dordrecht, Netherlands: Kluwer.

Tobin, K. (2000). Becoming an urban science educator. *Research in Science Education, 30,* 89–106.

Traweek, S. (1988). *Beamtimes and lifetimes: The world of high energy physicists.* Cambridge, MA: Harvard University Press.

Yerrick, R. (1998). Reconstructing classroom facts: Transforming lower track science classrooms. *Journal of Science Teacher Education, 9,* 241–270.

Yerrick, R. (1999). Re-negotiating the discourse of lower track high school students. *Research in Science Education, 9,* 269–293.

Yerrick, R. (2000). Lower track science students' argumentation and open inquiry instruction. *Journal of Research in Science Teaching, 37,* 807–838.

Yerrick, R. (2004). Seeing *IT* in the lives of children: Strategies for promoting equitable practices among tomorrow's science teachers. In A. J. Rodriguez & R. S. Kitchen (Eds.), *Preparing mathematics and science teachers for diverse classrooms.* Mahwah, NJ: Lawrence Erlbaum Associates.

Yerrick, R., & Hoving, T. (2003). One foot on the dock and one foot on the boat: Differences among preservice science teachers' interpretations of field-based science methods in culturally diverse contexts. *Science Education, 87,* 390–418.

Yerrick, R., Pedersen, J., & Arnason, J. (1998). "We're just spectators": An interaction of science epistemologies and teaching. *Science Education, 82,* 619–648.

CHAPTER EIGHT METALOGUE

Emergent Nature of Classroom Talk and the Game of Science

Cynthia Ballenger
James Paul Gee
Randy Yerrick
Wolff-Michael Roth

Cindy: I am talking as a teacher–researcher in this response. As such, I am especially struck by Randy's account of the uncertainty of teaching. The chapter makes this point early and then the examples illustrate it numerous times. It is a great value to make this clear and in my opinion cannot be over emphasized. In my own teaching and research I have found that I cannot accurately predict or account for the student ideas and accounts they will draw upon and co-construct in our "science talks." This is where research on children's misconceptions is of limited use for me. I may understand from conceptual change research what ideas children commonly use for describing specified science content, but I cannot predict how the collective discourse will be affected by children's culture, history, language or other parameters of real students. I think it is important to note that the difficulties and uncertainties of teaching, however well you have planned, are often the most interesting and significant things that happen. Planning is one thing, and the responding to what happens is another.

Michael: Both of you are making highly relevant remarks about the relationship between plans and situated actions (Suchman, 1987). Although we now know that situated actions are not determined by the environment, many science teachers and science educators still seem to believe that teacher actions should bring about, that is, determine, specific student activities, conversations, and learning. Randy, do you have any suggestions for teachers in preparation for such uncertain paths?

Randy: I hesitate offering any pat answers for prescriptive pedagogy. Yet I find in Brice-Heath's (1983) "slave-master" a good analogy for a teacher of other peoples' children, at least for beginning a science discourse that students can engage in successfully. She has described using a single tape with her micro-cassette recorder that requires her to transcribe and listen more carefully to student talk before her next teaching episode. I have often placed myself in contexts in which I do not know the culture of my students, and it forces me to listen more carefully to others' descriptions of science and the world and helps make me more alert or sensitive in finding instances for meaningful science connections. I have used focus groups as well to help me understand how the context of school itself shapes their talk and goals for science achievement. It helps me to also know what students want to know.

Michael: It is evident that you value listening to your students, to learn the structure and topics of their discourses.

Randy: It is quite a challenge for me to listen carefully and devise problems and strategies that actually bring students' ideas in contact with one another as a problem for students to solve. Once I think I understand their ideas and experiences surrounding a particular event, and I figure out how scientists resolved a similar debate, I need to consider which artifacts were available to scientists at the time. The question for me then becomes, "How can I make *my* problem, *their* problem?" For example, I recall students beginning to an interesting discussion about what caused a recent eclipse. The way I made my problem their problem was to ask, "Which travels faster across the sky, the sun or the moon?"

Michael: Do you have any evidence that you were successful in this move?

Randy: I was sure I had been successful because it led one faction of the room to ponder aloud, "But wait, if one travels faster than the other, wouldn't they eventually catch up? And that can't happen because I've never seen the moon out during the day." Of course, the best part was that it brought the whole class to a vote over how many high school students had seen the moon during the day. Two raised their hands and the rest described such an event as impossible. Within twenty minutes we were standing behind the school looking up at the moon's first crescent at 8:45 am. I heard about that startling event from my students for weeks. Without a doubt, hearing children's ideas and devising ways to make them intersect is the best advice I can give.

Michael: Randy used the metaphor of the game, which struck me a bit as odd; I am not sure I liked it all too much.

Cindy: I was not so taken with the metaphor of the game myself. Clearly it has some value, but I think if you start there, you see some things about what students are doing and their relationship to science, and not others. I notice among my Haitian students, for example, that they seem to wonder in science discussions, "How can I enjoy this?" This is a legitimate question I think. They could joke and fool around—sometimes they do. But among the things they've found to do is a practice where, whatever the topic, many of them seek a way to relate it to when they were in Haiti.

Michael: This practice seems to be consonant with the role of lifeworld discourse that Jim and I write in our own chapters; this lifeworld discourse is intrinsically connected to a physical world shot through with meaning.

Cindy: I agree. The children's stories appear to be very meaningful to them, stories of gardening with their mothers before their fathers were able to bring them here when we are studying plants, or watching the sun go down in Haiti where you can really see it on the horizon when we are exploring astronomy. These stories make up some of the data on the basis of which they argue out their claims, plan their experiments, and so on. I'm not against competition in science, I haven't thought about it in this sense, really. However, I think the game metaphor biases what we see. Even the middle-class boys, who in my experience ask of any activity first of all, "How can I win?" have other capacities, and those who really engage in science are doing something more than just competing.

Randy: I agree that such a perspective biases what I see. I wonder how much of my affinity to this metaphor, however, was built into the context I often operate. My research often focuses on the individual and collective student accomplishments, observing their fierce competition for status in lower track high-school science classes. American schools often foster a microculture of school generated from many levels of tracking in which students are sifted and sorted routinely on the basis of what they are told they *cannot achieve*. In North Carolina, this gap and competition for status was exacerbated by the desire to not "act White" (Fordham, 1996), which placed me in an even more precarious position as a White, male, teacher-researcher. Admittedly, there are consequences and costs for this group as well as others by focusing on the individual achievements as the most pronounced or productive of events.

Michael: But there are domains in everyday life of the students where they do not compete or feel that they have to do so—going to a dance or playing in a band. Participation in such events may be more appropriate metaphors for doing and learning science in schools. Penny Eckert (1989) writes about the differences in interpersonal relations between middle-class and working-class student cultures.

Randy: I think, however, that others have described competition and bravado as one of the main forces that shapes the context of high-energy physics (Traweek, 1988), so it is certainly an aspect of science (whether positive or negative). Although I will not tout winning as the only representation of science and the ultimate goal, I do believe that many attributes of science do get reflected in our American schools early and cultural forces in American high schools drive the classroom science discourse toward the direction of personal gain. Perhaps there is a need to focus on the variety of ways in winning both collectively and individually and how they specifically function in American schools—in particular those with alienated high-school students.

Jim: I would agree that there is no singular way to capture the endeavor of scientific speaking, thinking, and acting. Learning to compete and win within a game can never exactly parallel the processes of science, but the analogy was meant to assist and in some ways helped science educators to understand to process of acquiring or appropriating a discourse more than my analogies of foreign languages. But the issues remain the same. How does the "novice" get started—initially get up and running—in a language/discourse she does not yet know? The way you have inserted your students' authentic questions and scenarios does further our understanding of how to help students get started in science discourse. You don't get started in acquiring French by speaking English, and you don't get started in your acquisition of discourse X (e.g., science) by practicing discourse Y (e.g., a lifeworld discourse). Whether in competition or collaboration, science students require a way to achieve such convergence of meaning.

Michael: How do you think that such a convergence can be achieved?

Jim: Not just by technical terms, stipulations, and definitions, though these all play a role, but also, I would argue, by ensuring that members of its discourse share what I will call *situated meanings* by having similar experiences. Without competition and/or collaboration children cannot operate in *any* scientific discourse. People do not operate within discourses based solely on their own individual mental and bodily powers, but rather on the resources distributed throughout the other people, expressions, objects, technologies, and settings whose coordination (whose "being in synch") constitutes the discourse. A fourth grader, given a gun, can harm people well beyond the "intrinsic" powers of her body. Similarly, discourses offer "tools" with which people perform well beyond their "individual" capacities. Is there implicit harm to others in using these powers in school? I imagine that there are.

Cindy: I only meant to highlight that the metaphor allows one to see certain components of children's talk and that there are undoubtedly things it makes less visible. I think it may be Jim's approach to new situations actually, but it doesn't encompass all the capacities we might and do bring to bear on figuring out how to participate in activities. So I didn't mean anything except to bring to your attention how I think it functions. I thought you made a good point about the question "Where are we in the game?" That is an interesting question. Far more interesting, open, and useful for me than perhaps, "How can I win?"

Randy: I think you are right, and I am glad that you explicate for the reader the inherent bias that my perspective can impose on the way we view science talk between and among children. I need to continually remind myself that any effort for my lower track students to engage in science, whether to compete or collaborate, is always in direct response to who they believe they are as students and where their place is in the larger context of school and society.

Michael: So do you think you need the competitive game structure?

Randy: There are a variety of games which can be played which do not emphasize individual attention, although if you look at high-school athletics, it is a rare instance, and certainly not the norm. For example, a kind of game that I encourage with elementary children is one where the entire group wins and shares in victory when we keep a ball batted up into the air more than fifty times. If we all achieve our mark of fifty, we share in the victory and complement everyone without the existence of a loser. It is amazing how much more children encourage and offer support to one another without ever blaming or praising the individual. This game is played in stark contrast to the competitive games we encourage in high school—those that celebrate the winning shot at the buzzer or penalty shot past the goalie that do, by their very nature, raise up heroes and heroines in competition. My students relate to these games far more and likely take more quickly to competitive kinds of science talk. There is, of course, the issue of fair play and real social costs to the losers that speaks to the equity issue of aiding those who already have a leg up.

Michael: In this, I think Randy again provides evidence for his own sensitivity to students' needs, concerning the structure of the learning environment and the content, appropriating school science discourses. And he articulates the contradictions and paradoxes inherent in such an approach.

Jim: I appreciate Randy's use of real student questions to illustrate the obstacles teachers face in changing classroom discourse. I believe classrooms are full of contractions and paradoxes, and these questions taken directly from classroom talk help both teachers and researchers to focus on the interactivity and co-construction of norms discourse in science. Though I am not in any position to judge the curriculum or pedagogy invoked, I can take the vignettes offered and comment on children's allowance to "have a voice" and speak to issues germane to "specialized" ways of using language. My main concern is one of the well-known paradox about our educational system: It works best with children who already know, and worse with those who need to learn; schools are good places to confirm and practice what you have already acquired, poor places to acquire it in the first place. I wonder if someone would like to speak to the issue of how overcoming such obstacles can somehow dull the sharpness of this paradox, at least in science class?

REFERENCES

Brice-Heath, S. (1983). *Ways with words: Language, life, and work in communities and classrooms*. New York: Cambridge University Press.

Eckert, P. (1989). *Jocks and burnouts: Social categories and identity in the high school*. New York: Teachers College Press.

Fordham, S. (1996). *Blacked out: Dilemmas of race, identity, and success at Capital High School*. Chicago: The University of Chicago Press.

Suchman, L. A. (1987). *Plans and situated actions: The problem of human-machine communication*. Cambridge, England: Cambridge University Press.

Traweek, S. (1988). *Beamtimes and lifetimes: The world of high energy physicists*. Cambridge, MA: Harvard University Press.

CHAPTER NINE

Teaching Science in Urban High Schools: When the Rubber Hits the Road

Kenneth Tobin
City University of New York

There is a problem with the very idea of national standards (National Research Council, 1996) and goal setting that does not involve active formulation of those who are to accomplish them. The problem does not reside with the intentions of those who worked hard to arrive at consensus or with the idea that students should know and be able to do science as a means of improving the quality of their social lives. The problem involves the representation of standards as decontextualized, abstract texts and their remoteness from the practices and interests of those required to change. Exhortations for all Americans to accomplish specific standards have a democratic ring, an altruism that is difficult to critique because the associated rhetoric is so often positioned as above and beyond substantive analysis. When I argue about the pros and cons of standards, others implicitly position me as not advocating high standards, as expecting less from different segments of society, and as advocating continued oppression of underrepresented social and cultural groups by denying them access to a power discourse with the potential to change their social and economic lives. To be explicit on these points: I do not advocate low standards nor do I expect urban youth to benefit less from science education. Here I explore the historical, cultural, and social factors associated with teaching and learning practices (Bourdieu, 1990) when science curricula are enacted, especially in urban high schools.

Recently the problems of science education in urban high schools have received the belated attention of the research community in science education (Barton & Tobin, 2001). Even within fiscally challenged urban school districts there is a significant range in the types of schools available to students and the quality of the curricula they provide. For example, within the city of Philadelphia there are private and public schools. The public domain is stratified. Magnet schools offer programs likely to appeal to students with special talents and interests, charter schools operate according to an espoused charter and therefore are quite diverse (e.g., schools for dropouts, schools for performing arts), and neighborhood schools provide a comprehensive curriculum to students who live close by. To complicate matters, some schools have characteristics of magnet and comprehensive neighborhood schools in that they have small learning communities (SLCs), that is, schools within a school; some SLCs draw students from across the district while the remainder caters to students from a defined region proximate to the school.

There is a tendency for neighborhood schools to be the lowest achieving and least diverse according to indicators of social class and ethnicity. For example, City High, the neighborhood school involved in this study, has a student body that is almost entirely African American, most from homes where economic hardship is a way of life. Within neighborhood schools there is significant variation in the ways that teachers, students, and other resources are organized to enact curricula. In reference to science, structural characteristics that make a difference include the presence of science departments, specified curriculum frameworks, special purpose rooms in which science is scheduled, certified teachers, tracking, block scheduling, time at which science is scheduled, proximity of a science class to lunch, core studies in science, budgets for science equipment and materials, laboratory assistance, up-to-date textbooks, computers, and whether there is a statewide testing program in science. The nature of structures such as these is often overlooked in recommendations for reform, and even within one school district there can be significant variations in every structural item in the foregoing list of examples.

In this chapter I examine structures in science education from a historical context situated in my science teaching in the decade beginning in the early 1960s in Western Australia, first in a rural junior high and then in a large suburban comprehensive high school. I show how certain structures associated with the school and community were affordances for the development of a particular inquiry-oriented approach to science teaching, not unlike the approach advocated in most reform documents of the past 30 years. Then I examine the challenges I experienced when I taught science at City High, an inner-city school with students who were greatly different than me in terms of their social and cultural histories.

LEARNING TO TEACH BY TEACHING

When I began to teach science my understanding of theory was almost totally implicit. I graduated with a 2-year certificate to teach in primary schools (i.e., Grades 1–7) in Western Australia. Despite a lack of relevant qualifications I applied to teach science and mathematics and was assigned to Rural Junior High, situated in a town several hundred miles north of the main population center in the state. On my first visit to the school I observed that the classroom in which I would teach science and mathematics was an ideal multipurpose room. There were two fronts. One had chalkboards and a teacher's desk in the corner furthest from the door. Movable tables, each of which seated two students, faced this chalkboard. At the opposite end of the classroom (i.e., the science end) was a demonstration bench, behind which there were chalkboards and pinup boards to support teaching directed from the science end. Several rows of lab benches were oriented toward this end of the classroom. Each sidewall contained resources for displays, storage, and work on long-term projects. Water, gas, and electricity were available at the ends of each row of lab benches and at regular intervals along each side bench.

As a 19-year-old beginning teacher I was comfortable with the youth culture of Australians, and I could make sense of the practices of my students and anticipate how they would respond to my interactions with them. The students were mostly of European descent although a small proportion was of indigenous, Aboriginal Australian origin. There was a broad spectrum of social classes with caregivers ranging from wealthy farmers at one end to unemployed individuals at the other. Perhaps I could cope because the number of students was relatively small. Intuitively I knew what to do, and the students responded positively to my assertive approach. I was energetic and confident that I would make a positive difference to their lives, and, despite my relatively poor subject matter preparation, it never occurred to me that I would be ineffective. The students accepted me as their science teacher and gave me the respect I expected from them.

I was the only teacher of science and mathematics in the school, and my approaches to teaching and learning were shaped by a need to prepare students for external examinations, the resources available to support enacted science curricula, and my experiences of science as a student at high school and college. The science teacher I replaced had ordered the equipment and supplies I needed to enact the science curriculum. A storeroom, located at the science end of the classroom, was well stocked with chemicals and equipment that was too bulky, fragile, expensive, or dangerous to store in the cabinets and cupboards in the classroom. Straddling the chalkboards and pinup boards were open shelves on which essential solutions and chemicals were located. The value of these artifacts was more than to establish a context for the teaching and learning of science. Their accessibility while teaching was a critical factor for me to develop a teaching style that included demonstrations as an essential component. As I taught I relied on being able to reach for what I needed to illustrate particular points so that students could see and experience specific examples of the phenomena being discussed.

The equipment, materials, and supplies available to support the enacted science curricula were ordered once a year from lists provided by the State Department of Education to ensure uniform levels of support in schools across an area of about 1 million square miles. The availability and use of these lists did not allow my relative inexperience as a science teacher to limit the resources available to support the enacted curriculum. Each year I had to order equipment and materials from the list and submit the order for review by the State Department of Education. An inventory of stock on hand also was required, and personnel in the central office considered what was ordered in relation to what was already available in the school. Toward the end of each year a senior administrator from the central office evaluated all science teachers; this evaluation included observations of teaching and the manner in which classrooms, equipment, and supplies were maintained and used. Hence, there was a system of rules and practices that supported particular ways of teaching and learning, and efforts were made to maintain a level of consistency across an entire state.

As the only science teacher in the school I had to figure out how to teach and improve my teaching based on my own instincts and ingenuity. Since the nearest high school was 30 miles away I had little support from in-field colleagues. However, each month the principal reviewed my long-term planning for each subject, and every day he visited my classroom to review my daily preparation. Although his field was not science the principal made clear his expectation that I would be thoroughly prepared. Because the school was small I had opportunities to coplan and coordinate the curriculum with other high-school teachers, each of whom was a beginning teacher. The teachers learned from each other in a generic sense, but in this small rural high school with the principal as the only senior teacher, my teaching was the main resource for me to learn to teach science and to learn science.

LEARNING TO TEACH SCIENCE WITH OTHERS

After 2 years of teaching at Rural Junior High, the person responsible for science education in the state declared that my teaching was promising and that I should focus on learning more science and teaching advanced courses to older students. I was transferred to a suburban high school in a city of almost a million people, a move that allowed me to improve my knowledge of science by studying part time at a nearby university. By Australian standards the comprehensive high school was large, consisting of about 1,300 students and 100 teachers. Most of the students were Caucasian from a wealthy suburban community, and a diverse minority consisted of immigrants to Australia and students from a working-class neighborhood. The high school had departments for physical science and biological science. The departmental structure brought me into daily contact with junior and senior science teaching colleagues who were resources for my continued learning of science and how to teach it. Senior colleagues undertook formal and informal reviews of the enacted curricula in my science classrooms. I had only one junior science teacher colleague. He was an invaluable resource for testing my ideas in terms of my understanding of science, how to teach various concepts and labs, and how to deal with challenging students.

The senior faculty communicated a clear expectation that the science curriculum would include laboratory activities, demonstrations, lectures, small-group discussions, individual assignments, regular homework, and a strong emphasis on learning the big ideas of science, facts to support them, process skills, and competence at solving quantitative problems. To assist me in preparing my personal teaching notes for each course my head of department provided a syllabus and his teaching notes from the previous year. The department also provided other structural support for my teaching and learning to teach. In addition to department heads having a responsibility to maintain adequate stocks of equipment and supplies there was a qualified lab technician and an assistant to provide support for laboratories and demonstrations. Procedures were in place for requesting equipment and materials to support the enacted curriculum, clearing away and re-storing equipment, preparing reagents, replenishing supplies, and helping out with demonstrations. As was the case in the rural school, each science room contained a basic toolkit of chemicals, solutions, equipment, and gas, electricity, and water.

Although the curriculum emphasized conceptual learning, laboratory activities were regarded as core to all enacted science curricula. This trend was supported by the design of classrooms in which science was taught, the supply of equipment and materials, evaluation and supervision of teaching, and in larger schools by technical support. My teaching and learning practices and associated dispositions developed and adapted to fit this structure. The explicit theories used as a foundation for the curriculum were psychological, especially Piaget's developmental theories and their applications to science curricula in other parts of Australia, England, and the United States. Cognitive learning hierarchies also were emphasized in conceptualizing a curriculum in terms of the major and the minor concepts needed to support the learning of big ideas, especially through the uses of basic and integrated process skills. The result was advocacy for a learner-centered curriculum that prioritized active involvement of learners and a set of teaching roles intended to mediate meaningful learning. A key role of the teacher was to mediate the students' efforts at sense making by connecting knowledge from labs (concrete experiences with materials), models and diagrams, and symbolic representations such as equations and formulae. Monitoring of student participation and learning were regarded as essential parts of effective teaching.

IT JUST DOESN'T WORK HERE

I became a science teacher educator and researcher of teaching and learning science in 1974. Following extensive experience with science teacher education in Australia and the United States I was drawn back into the classroom by an unexpected problem. Student teachers from the University of Pennsylvania were experiencing extreme difficulties while teaching in urban high schools in Philadelphia. They seemed unable to successfully enact my suggestions and openly questioned the applicability of theory and research to urban high schools. I was perplexed at their relative lack of success and concerned that what we knew from theory and research did not seem to transfer well into practice, especially in urban high schools like those in Philadelphia. Accordingly, I decided to gain first-hand experience of teaching science at City High School, a neighborhood high school in West Philadelphia.

Departments Are Out

City High school has more than 2,000 students; about 99% are African American, and approximately 90% are from home conditions of economic hardship. Historical accounts show that since the school's opening in 1972 (e.g., Salley, 1975) violent conflict between rival gangs, assaults on teachers, and disruptive students creating dysfunctional learning environments had been common occurrences. Research and conventional wisdom of the period suggested that interdisciplinary SLCs could provide safer and more productive learning environments for these students (e.g., Fine, 1994). A former principal at City High wrote

> At least 10% of these students have had involvement with the criminal justice system; of those about 75 "return from incarceration" each year. Additionally, at least 200 of the female students are teen mothers and another

200 will give birth this year. Historically, fewer than 40% of our entering ninth graders graduate. (Lytle, 1998, p. 1)

When he was appointed to the school the principal had to reduce the budget by 10%. He decided to restructure the school, cut the budget by 15%, and use the additional 5% to support innovative plans for programs and priorities. Lytle noted

We also agreed to sub-divide the school into six (now ten) small learning communities, or schools-within-a-school, in which students take all but their physical education classes. In the process we shifted the school's core organizing construct from subject/department to caring/support/personalization. (p. 2)

The demise of departments occurred with the stroke of a pen, and since the fall of 1995 small learning communities (SLCs) have evolved, as the traces of departments have all but vanished. However, in dismantling departments many of the structures needed to support science education disappeared, often to the detriment of teachers and students.

Although City High had one laboratory assistant for its 14 science teachers there was no science department and no person designated as in charge of science. A rule structure no longer existed for assigning work to the lab assistant and coordinating access to the equipment and supplies needed to support the teaching and learning of science. Instead the laboratory technician was situated on the third floor of the building, close to labs that were no longer used. Similarly, most of what remained of the science equipment and materials was stored on the third floor. The extent to which teachers were supported by the lab technician depended on their proximity within the school, the social capital they enjoyed with her, and whether the equipment and supplies they requested were accessible.[1] Since there was no longer a science department there was no formal process to order, organize, maintain, disseminate, and retrieve equipment and supplies. An assistant principal, formerly chair of a science department at another school, supervised the lab technician. However, the assistant principal was too busy with school-wide matters to ensure that her roles and associated practices adapted to the changing structure of the school and the needs of science teachers. Budgets for science equipment and supplies were now distributed across 10 SLCs, and each science teacher in

---

[1] I located equipment scattered throughout the third floor in locked cupboards and storerooms. The equipment ranged from being in good working order, to being in sealed unopened boxes dating back 30 years, and to being in total disrepair and no longer usable. Dirty glassware, accumulations of used chemicals, and aged chemical reagents (e.g., unstable ethers) also were widespread and evidence of neglect.

a SLC had the responsibility to plan ahead to obtain the resources needed to sustain a science program of required courses and electives. There was no formal process for ordering large items of equipment that might be used school wide and no equipment lists from which to order, record what was in stock, and coordinate the use of science equipment and materials. Over time equipment acquired prior to the change to SLCs was either squandered for the use of one or two teachers or available resources were no longer used because of disrepair and lack of interest or expertise of teachers.

A Community of Disruptive Misfits

Incentive was an SLC created for students who were unable to settle into high-school life. Reasons for students being assigned to *Incentive* included disrupting the learning of others, low achievement, repeated absence, and problems with the criminal justice system. Teaching and learning in *Incentive* was difficult for teachers and students. There were few suitable textbooks to use, scarce equipment and materials to build a program around, and a group of students who were "disruptive misfits," removed from other parts of the school so that the majority could learn. Unsurprisingly, most students did not fit too well in the *Incentive* SLC either, and enacted curricula did not seem conducive to learning. Absences were frequent, and the motivation of students to learn was low. Most of the teachers were older females, and the only male was a young African-American graduate from City High, uncertified to teach math and without a college degree.

As a beginning teacher in the fall of 1997 Spiegel was assigned to teach science in the *Incentive* SLC. Spiegel grew up as a working-class youth, and anyone could see that he "pumped iron." He was underqualified to teach science but opted to teach it because of a relative shortage of science teachers. Biology, his area of certification, was not his college major. Spiegel preferred an approach to teaching and learning that centered on laboratory activities and, within the limits of what he knew and could afford, he purchased simple materials from the hardware store to use in labs he had done and enjoyed in college or high school. Also, he arranged to use a laboratory on the third floor every other week, taking advantage of the lab assistant to set up equipment and materials as long as she had sufficient prior notice. Otherwise the science program in *Incentive* was characterized as covering subject matter and writing notes on a chalkboard for students to copy into their science folders.

The day-to-day enactment and long-term planning of the science curriculum in *Incentive* was constrained by the lack of a system for ordering, storing, and maintaining equipment, materials, and other supplies. The decision to limit students to one part of the school also constrained the enacted curriculum. *Incentive* was located on the ground floor where non-teaching assistants and school police could easily monitor the hallways and classrooms to ensure that there were no outbreaks of disruptive behavior. Spiegel taught science in what was originally an art room that was spacious, contained a trough with running water, but otherwise had few resources to support science teaching and learning.

Incentive, like all SLCs, ran on a block schedule in which students satisfied a graduation requirement of four science courses by taking science in only one semester a year. Spiegel, the only science teacher in *Incentive*, taught three class periods a day and was responsible for teaching all science courses. Accordingly, almost 75% of his teaching assignments were out-of-field. However, Spiegel worked around this by teaching topics about life science, even in physical science courses, mainly because he was most comfortable with the subject matter and also because he knew of activities to enact, thereby increasing the chances that students would find them interesting and therefore be more ready to learn what he wanted them to learn. Although outsiders like myself were critical of this practice, it made sense to Spiegel. He could better connect with the students' interests and more competently mediate their learning of science when he was teaching in his field. From this perspective, what might be regarded as irresponsible and unprofessional conduct can be considered an important pedagogical move.

Because Spiegel was the only science teacher in the SLC he had no science colleagues with whom to interact on a regular basis in the way I had during my early years of teaching. The other teachers in the SLC focused on their own teaching, and the SLC coordinator was completely occupied in dealing with student problems and her role in professional development; reviews of the curriculum were virtually nonexistent. Consequently, Spiegel had to figure out on his own how to teach the students in the *Incentive* SLC without the assistance of colleagues. He often remained in his classroom at lunchtime and watched TV with his students and talked to them, thereby getting to know students and building social capital.

WHEN THE RUBBER HIT THE ROAD

My teaching of students in *Incentive* was guided by my history of science teaching and perceptions of the necessity to enact a curriculum that connected what was to be learned to what the students knew, could do, and were interested in learning about. As a 54-year-old beginning to teach inner-city youth in a resource-starved urban high school I knew very little about the culture of my students and could not predict how they would respond to my efforts to interact with them. I felt "culturally other" as I walked down the hallway and weaved among the students. At the door to the science classroom students, seated on the ledge of an empty display cabinet, ate their breakfast and jousted verbally with one another. To a person they ignored my daily greeting to them, and efforts to engage them in conversation were disregarded or dealt with summarily. Their verbal fluency in an unfamiliar dialect made it virtually impossible for me to readily communicate with them. I felt uneasy about the extent to which I would succeed in my efforts to teach science to these students at this school. I had been in education long enough to know that I had to earn the right to teach them and showing students respect and building rapport were keys to eventual success.

My first lesson involved chromatography, and the students participated willingly and seemed to enjoy their experience with a lab that was relatively unstructured, in which they separated the different colors in inks and dyes from marker and writing pens. I left the class optimistic that I would soon feel more comfortable and that the students would respond to a form of science that was quite foreign to them; one in which they participated actively in labs. I looked forward to the follow-up lab and headed toward the grocery store to purchase the candy needed as a source of color. My plan required students to measure the dispersion of colors from candy such as M&Ms in a variety of solvents that included water, saline solution, and alcohol. The laboratory was a means to an end. I wanted them to do science and in so doing learn to talk about chromatography.

The second lesson was a struggle. About half of the 18 students in attendance on the first day were absent on the second. Those who were there for the first lesson were resistant to participate in a follow-up lab that was so similar to the first. "We did that arready Tobin!" I was unprepared for their unwillingness to do the lab and unsure of how to gain their participation. Those who were not there for the first time, about 9 students, had to be shown what was accomplished the day before. I had not anticipated the necessity to be prepared for sporadic attendance. Also I was unsuccessful in my endeavors to have those present the day before teach those who missed the first lesson. I struggled through the lab, and at the end of the lesson I was dejected and unsure of how to proceed. Throughout that lesson, and in all subsequent lessons, I planned thoroughly and was optimistic that the students would learn. Although there were rare moments when some students were cooperative, the hardest nuts to crack were beyond me and refused me the right to be their teacher. It seemed as if the more I thought about my teaching, especially while teaching, the less possible it was for me to enact practices that minimized disruption and created learning opportunities for most students.

There were many approaches to teaching that I would not consider. For example, I was opposed to enacting activities in which students copied from out-of-date textbooks. Also, I refused to write on the chalkboard for students to copy for long periods of time. I knew such approaches would keep students occupied, and it is ironic that most would have preferred that I conduct the class that way. Unless safety was an issue I would not eject students from my class for disciplinary reasons. Unlike resident teachers in the school I would not write up students for misbehavior thereby getting them an automatic suspension. It seemed hegemonic to create productive learning environments by excluding those students who most needed education. My unwillingness to suspend students was seen by many as a sign of weakness and further justification to disrespect me. The perception of most students was that effective teachers threw out troublemakers and set work to be copied from the textbook and chalkboard. Contrary to this view, I was determined to enact a curriculum to involve students in the doing and learning of science such that what was learned was potentially transformational.

During my teaching at City High I continuously accessed research and theory on urban education and African-American psychology and consulted with numerous colleagues, especially African Americans. I also spoke with other teachers in the school, including the principal, about my struggles and determination to succeed. I was learning a great deal, but none of it was making much of a difference to my teaching. My analyses of the videotapes of my teaching depict a serious teacher who is uncertain about what to do. At times my teaching appeared to be unfolding in slow motion and I looked uncertain. At a micro level, when I undertook a frame-by-frame analysis of my teaching, it was apparent that the students were not coparticipating with me and that frequently their actions were not synchronous with mine. For example, when I approached a group to discuss possible solutions to a problem the students turned away and continued to interact as I endeavored to speak to them. Often their verbal remarks were insensitive and intended to insult me. "Back off man. You gonna get hurt!" "Oooo! Yo' breath stinks man. Get outta ma face!" Comments such as these occurred frequently and left me wondering how to proceed.

My teaching habitus (Bourdieu, 1986, 1990) was in constant breakdown. Resources I needed to support my teaching were apparent to me, such as equipment and materials, lab technicians, and rules that were enforceable. Without such structures my teaching practices had no chance of success, and I could only teach with a conscious awareness of the rapidly unfolding events. Frequently the students breached my efforts to teach normally. My dispositions to deal with the unfolding events of the classroom were inappropriate for these students in this school at this time, generating practices that became resources for the students to pursue their own goals. I did not anticipate how students would respond to my teaching, and my actions became slower, deliberative, and reactionary. I attempted to remedy some of my earlier mistakes, but the damage was done and the way to repair it was not apparent. The more I tried to teach the students the greater were the opportunities for them to show their disrespect for me. There was the potential for a downward spiral as more students regarded me and my practices as targets of disrespect. A major part of my difficulties was associated with my lack of understanding of the culture of my students and my inability to build social capital with them.

It is not just a teacher who brings habitus to a science classroom. Students do so too. The youth who attend City High develop culture in their homes and the institutions associated with lives at home, in the streets as children and adolescents, and at school. My reading led me to *Streetwise* (Anderson, 1990), a book based on Philadelphia and the practices needed to safely navigate the streets. I learned that safe navigation was a practice and necessitated learning from being in the streets. Even more significant to me was Anderson's (1999) later book, which identified *respect* as the currency of the streets and sensitized me to the relentless pursuit for respect in which many African-American youth participate. African Americans in Philadelphia regarded respect as a currency akin to money. It was considered most important to be respected by others, and respect had to be earned and maintained through physical or sexual prowess, and by obtaining money, dressing smartly, verbal fluency, rapping impressively, or disrespecting others, especially those in authority. Teachers make an easy target for attempts to gain respect in the peer culture by showing disrespect.

The applicability of street code to my experience of teaching at City High was immediately apparent to me. The more I endeavored to interact with students the more opportunities I provided them to disrespect me and thereby earn the respect of peers. As soon as I knew about the centrality of respect within the culture I realized that when many students came to my class they had as their principal object the maintenance of the established hierarchies of respect among their peers that would obligate some students to be disrespectful to me and to flagrantly break class and school rules. This may have been the case generally at City High but was especially the case for students from *Incentive* who were assigned to that SLC because of their inability to succeed in high school. Until I understood fully the implications of *respect* as the central activity in my science classroom there was no chance for me to become an effective science teacher for these students. It was necessary for me to accept that an activity system (Cole & Engeström, 1993) with the learning of science as the object would necessarily have to align with one in which students could earn and maintain respect.

Most of my miscues were directly attributable to not understanding the students' culture. For example, holding eye contact was regarded as a sign of me having problems with the student and in some cases inviting a fight. Similarly, monitoring students as they worked individually or in groups was perceived as not trusting them to work without supervision. My inability to understand their dialect placed me in a position of having to ask students to repeat what they said and led to unwillingness for some of them to speak to me. Also, students used unfamiliar terms. For example, one student who wanted to be friendly once said to me "Heh old head." I was offended by this style of greeting and asked him to repeat what he said. He became frustrated with me and repeated "Old head. Old head!" I had heard the first time but was unaware of his intended meaning and aggressively told him to be respectful. That interaction was the beginning of a downward spiral in our relations and an indication that I needed to learn the students' language. I listened to sports talk radio to increase my understanding of sports and dialect. But I made no effort to learn about their main interest, music, especially rap. With hindsight I realized that two sources of media I could have accessed for lessons on the culture of my students were a radio station listened to predominantly by African-American youth and a newspaper read mainly by the working class. In hindsight a better result might have been achieved if I had asked them in which sports they were interested, which radio and television stations they accessed, and even how best to learn their dialect. However, my experience at the time was traumatic, and my concerns were with safety, finding ways to interact so that students acknowledged my existence, and eliciting civil responses from students. The necessity for me to create social capital was compelling.

I focused on enacting a curriculum around the doing of science and identifying topics that were interesting and that could connect with the canon of science. Hence, when we studied Newton's laws we explored issues such as air bags in motor vehicle accidents, how safety belts work, why helmets should be worn while riding a bicycle or playing hockey, and how padded gloves reduce injuries in boxing. Similarly, when we studied music I brought in glass containers and drinking glasses so that they could experience resonance and understand the armonica invented by Benjamin Franklin. I secured a videotape of an Australian football grand final at which an aboriginal rock band Yotu Yindi played a didgeridoo during the halftime show, and I brought in a didgeridoo for them to play. As was the case with the chromatography lab, the only students to get involved were those with whom I had built sufficient social capital and who knew that I respected them. Colleagues continued to remind me of the importance to connect the curriculum initially to the interests of students, not to my interests. Rather than build an enacted curriculum around the interests of students my disposition was to make curricular decisions, to select activities I considered to be of interest to students. Contrary to this approach is the potential of developing high-school science curricula from the interests of students (Seiler, 2001).

LEARNING FROM STUDENTS

The students enacted practices that made it difficult for me to teach them, and I was unable to mediate those practices to create productive learning environments for them (Tobin, 2000; Tobin, Seiler, & Walls, 1999). I had to learn the culture of inner-city African-American youth, and my learning had to include deeper insights into my own teaching, particularly about dispositions about which I was unaware. If I were to succeed as a teacher of these students at City High I knew it would take more than researching and theorizing. I needed practice, a coach with direct experience of the enacted curriculum, and critical dialogue that would bring forth the perspectives of all the key stakeholders (e.g., learners and teachers). Also, I knew that my lack of social capital in this classroom was a key problem.

I believed that the best coach would be one of those students who were most difficult to teach. With these insights I hired Tyrone, a student from my class who was often in trouble with authority, had a quick temper, and whose aggression was a problem for adults and students in the school. Tyrone was regularly absent from school, and he believed he was suspended as often as he was because of his thug-like appearance. Tyrone repeated Grade 9 for 3 years and later dropped out of school. He was smart and felt that school did not meet his needs. He found it repetitive and felt that the curriculum was well beneath his intellectual capacity and irrelevant to what he felt he needed to know. In the following section Tyrone's advice to science teachers is provided as an edited transcript from a conversation he had with me about my teaching of his science class.

"They'll Come to You"

See me, if I was a teacher, I would tell my students in the beginning of the year, I'm not going to make you do nothing you don't want to do. If you don't want to work don't even come to school because I don't want you in my class destructing my class, destructing the ones that want to work. They want to work let them work. If you don't want to just don't come to school. Stay home if you got. Do whatever you going to do. I don't care what you going to do. I'm trying to teach you all. If you don't want to learn you don't have to be here. You can leave whenever you want. I'm not going to stop you because I'm trying to teach.

You cannot give them leisure. You can give them room to breathe, but you can't give them too much room where they going to run over you. See, once they see that they can run over you they going to keep doing it regardless of what you do or how you try to do it.

When you come to Ms. Jackson's class, you sit down. It's real hot in her class, but, hey, you not coming there to feel the heat. You're not coming there to feel cold. You're coming there to work. She'll tell you sit down. You can complain all you want about being hot. She don't want to hear it. All she wants to hear is if she asks you answer a question, or asking a question when you come to her class. Ain't none of that running around, playing around, no. You come to her class, you sit down, you do your work. And when you look up, and say you get out 12:29. You look at the clock. You see 12:29. Ain't no getting up to walk out the door. You sit there 'til she tell you that you can walk out that door. If she feels that as though you're not going to do it she'll give you a detention after school, before school, or Saturday detail.

Never put a group of students together that you know is not going to do the work. If you really want everybody to learn, find the smartest students in the class an' put at least one in each group so that whoever can't understand will get help from one of his peers. To make sure that each person in the group will participate, give each student a different part of the lesson to do, and tell them that they will get a group grade. That way his peers will make sure they do what they have to do.

When the class starts to get out of control, the teacher should separate the students to a more focused surrounding. If that do not work pull the students that are making the most noise out of class and tell them, that if they are not going to work just put their heads down on their desk and not to disturb the others, and if that do not work give them a Saturday work detail.

So you just wait for them. You wait 'til they think they ready to work. They'll come to you. You don't have to come to them. They'll come to you.

Doing What You Have Got to Do

Tyrone's advice was of extreme value to me as a teacher because he was able to see when students were being disrespectful to me and within that context provide advice on how to create productive learning environments. At first sight his advice not to teach students who do not want to be taught seems at odds with the goals of schooling. The very idea that a teacher would stop trying to teach unmotivated students seems abhorrent. Yet Tyrone had seen many teachers endeavoring to unsuccessfully engage unmotivated students while those who wanted to learn were left to wait. Tyrone's strategy was to design a curriculum around those who wanted to learn and achieve at a high level. In good time the others would see the error of their ways and come forward for teaching. Efforts to teach those who did not want to learn were bound to end up in failure, and with these students in particular, the efforts would be opportunities for rebuttal and disrespect.

One day, in a moment of frustration, I implemented Tyrone's advice abruptly. I was using a diagram on the chalkboard to explain the trajectory of a basketball. Hardly anybody was paying attention. Of the 14 students in the class 3 had their heads down and others were talking. I was annoyed and admonished the class, and, instead of demanding their attention, I calmly invited those who wanted to understand the issue to pull up a chair at the front of the room. Four students, including Tyrone, sat in a small semicircle, and I spoke quietly to them. The protests were immediate. "Heh man. You s'posed to be teachin' us!" I ignored them, and just as Tyrone predicted, one by one they came to me, until most of the students sat in the expanding semicircle.

Most of Tyrone's advice suggests a clear and firm rule structure with inconvenient consequences for students who do not comply. Unsurprisingly, Tyrone is an advocate of holding students responsible for their own learning, attendance, and adherence to rules. If they want to learn and yet break the rules then there ought to be consequences firmly applied. He does not want to see any time spent in class disciplining students. He was constantly reminding me of the time wasted at school and wanted teachers to use all the time available to teach.

This suggestion is easier said than done. First of all it goes against the grain to allow students to sleep in class or allow them to talk to one another while I am teaching. Yet Tyrone has a point when disruptions relentlessly unfold. There were times when I simply could not keep up with the inappropriate behavior and when I felt that students were intentionally giving me the runaround. At other times I felt an obligation to maintain an orderly and safe environment. Without being dramatic, there were many times when I did not feel safe while teaching the class.

So, how can this be done? The answer may lie in Tyrone's vignette. Set up an activity with the object of earning respect and then do what you have got to do to consistently apply the rules so that all students participate. If that necessitates having students copy from the chalkboard then do what has to be done. As you earn respect as a teacher it is possible that more productive forms of activity will become possible.

My chief concern with the advice of Tyrone and the many student researchers with whom we have worked at City High is that their suggestions about good teaching and learning are within their experiential realm, enacted curricula that have been characterized by the relative academic failure of urban high schools and African-American students. In learning from their advice it is important not to reproduce forms of teaching that are exemplary but propagate forms of practice that reproduce oppression and disadvantage. Rather than agree with their renderings of what is and should be it is important to present and discuss alternative ways of enacting curricula. Through cogenerative dialogues with students it seems desirable to create an awareness of structures that shape schooling that may be hegemonic, inducing forms of practice that feel normal but are in fact oppressive to them. For example, in regards to science education the relentless quest for higher test scores on the very tests that are demonstrative of the failure of urban schools is a practice that is potentially hegemonic. The presence of the tests focuses enacted curricula in ways that produce learning environments that resemble those from my class, some students participating, some with their heads down, and high levels of absence from school. One focus for discussion with urban youth might involve what they would like to learn and accomplish from science education.

MORE THAN GOING THROUGH THE PACES

My experiences in teaching students in the *Incentive* SLC provided me with insights into what to do and what not to do in teaching science to African-American youth from inner-city schools. Some of what I learned is easy to say and less easy to enact. Meet the students at the door as they enter the room and speak with each of them about some aspect of their lives that is not related to class. This necessitates that you know their names and details of their lifeworlds. Knowing about the students' interests and hobbies is a good way to have something substantive to speak about while showing respect for them. It is also essential to understand their language and use it without projecting an image of trying to be cool. Using the students' language at a place such as the door to the classroom is a sign that you respect their culture and do not see it as a deficiency that has no value. Elsewhere it is necessary for students to learn to code switch so that they can use the discourse of the majority culture and thereby take full advantage of their knowledge of science.

I have used the metaphor of taking difficult students to lunch as a reminder that teachers need to build social capital with the toughest of students, perhaps especially so. Taking them to lunch means more than just eating with them, but also being there when they might need assistance and reaching out to help them at such times. The most successful teachers are available to students before school, at lunchtime, and after school. For example one of our teacher researchers from City High is a science teacher who assists students with their homework, not only in science, but also in mathematics and other subject areas. Students respect him because they appreciate his genuine efforts to support them as signs of respect and care that few people are prepared to show without contingent expectations. The teacher also assists students in other ways, such as driving them home if they miss the last bus on which concession fares are available. Being an advocate for students is an essential step toward being accepted by them as their teacher. However, consistent with Tyrone's advice, it is imperative for teachers to show that they warrant respect by establishing and enforcing rules fairly and maintaining high standards.

My most critical challenge was building social capital with students and creating spaces for them to build social capital with me. Getting to know each student's family and caregivers is a priority that creates opportunities for conversations around which social capital and mutual respect can grow. Hence, if it can be arranged it is a good idea to visit the homes and meet the caregivers. Spiegel's strategy of eating his lunch and "shooting the breeze" with students while watching television seems to have significant merit as a procedure for building social capital.

In an endeavor to better understand the students' culture I asked some of them to drive with us through their neighborhoods and inform us about what can and should not be done. They showed us areas where it was and was not safe. We also experienced the contradictions of life in their neighborhoods. Whereas many neighborhoods are trash ridden, replete with crime and drugs, and dangerous for White people to be alone, the same neighborhoods are teeming with life and evidence of cross-generational caring, have many beautiful murals and gardens, and are sources of pride to the students who live there. Being with students in their 'hood is a sign to them that we are willing to see the good and build an appreciation of where they live. We have found it highly productive to allow students to borrow video equipment and produce video ethnographies of their neighborhoods. The insights they provided of their homes and neighborhoods were valuable windows into their identities and the forces that shape them.

Within an overall framework of learning science by doing science it is important to have goals that are relevant and interesting to the students. The students I have taught at City High all needed to create an identity (Roth et al., in press) that included enjoyment of science and success in doing it. Also, it is imperative that science-related attitudes develop through participation in science. These include such attitudes as interest, enjoyment, searching for alternatives, calling for data and other evidence to support knowledge claims, considering alternative evidence, persisting in completing tasks, and searching for parsimonious solutions. Many students are not well equipped to follow oral and written instructions. Although they have gone out of fashion and often are referred to as cookbook or verification labs, such activities can be used to accomplish critical goals as students follow procedures to verify aspects of science about which they have been told, have seen in a video, or have read in a book. Similarly, through the doing of labs students can build manipulative skills that are an important part of success in science. In addition, students can learn to work cooperatively with others by participating in labs and other tasks that involve collaborative work with others.

TOWARD SCIENTIFIC LITERACY

Our approach in this collaborative project to scientific literacy in urban high schools begins with the establishment of a field to support an enacted curriculum in which all participants, teacher and students, can do science and learn from one another through their activities. We recognize that structure does not fully reveal itself even when participants think carefully about it, and, when their intentions are not met, they can often identify elements of structure that inhibit and those that afford their goals. The accessibility of equipment and materials and the physical arrangement of furniture and students is an aspect of structure that is often overlooked as teachers make do with what they have, become frustrated, and feel constrained to teach in particular ways.

Within a context of learning by participating in science it is expected that participants will create a scientific habitus to support practices (i.e., patterns of action) that are important parts of scientific literacy (e.g., making sense of data in terms of science concepts). The act of doing science creates habitus, and the habitus then supports particular patterns of action that accord with the structure of the field of which the practices are part (i.e., habitus is structured and structuring). By doing science, especially while at the elbows of a person who is a central participant, students can create a habitus to support scientific practices. Just as habitus is formed by being with others in particular cultural fields, so too a scientific habitus is built by coparticipating with others (i.e., teachers and students) as they do science. The habitus is not necessarily conscious or intended, and yet it can be enduring and provide a platform from which scientific practices are launched. In this way science can gradually become part of the students' identities. Doing science and being successful are important steps in perceiving the self and science as interconnected.

An important part of doing science is conceptual, and I maintain that teachers should do what they can to have students learn the conceptual parts and thereby to access and appropriate a scientific discourse. However, if a teacher pushes too hard, students might pull back from their involvement. What seems essential is for students to get involved in the doing of science and to build an identity in which self and science are interconnected. Initially the interests, habits, and discourse will be a mix of home, street, and school, but a goal can be for this to evolve over the duration of a semester to become more driven by curiosity about science. When and how students interconnect their different ways of knowing science and add to their discursive repertoire can be mediated by teaching, especially for students who aspire to careers in science or who need a symbolic understanding of science to participate in advanced studies of science (e.g., in medicine or engineering). However, care should be taken not to reify the symbolic and recursive ways of knowing science over those practices that are enacted unconsciously and may lead to the emergence of an identity associated with success and enjoyment of science. If the chief object of science education for all is to create forms of practice that are enacted with the purpose of transforming social life then it seems clear that alternative ways of doing and knowing science might be embraced. Let those who need and want to know science in its symbolic forms enact practices to afford knowing in the appropriate ways and permit those with different goals to enact different but equally appropriate practices. A future molecular biologist might need and want to know about DNA differently than one who, as a literate citizen, must understand numerous applications of DNA testing in daily life. My experience of urban schools is that the objects of science education reside more toward the symbolic ways of knowing, and students switch off and reify their failure to succeed. Recently it has been my experience that urban students from City High like to do science and attain success. As they do so they show more enthusiasm for coming to school and getting involved. If most citizens will engage science as practices in the everyday aspects of life (as distinct from engaging as scientists or in science-related professions) then an initial goal in the core courses of urban high schools might be to focus on success, enjoyment, and active participation. Then, let students whose identities connect to science, who enjoy doing science, make choices about what form is most appropriate to their needs. As Tyrone might say, they will come to you when they want to know and be taught by you.

CODA

Intensive participation in teaching, research, and other educational activities over a period of approximately 40 years has afforded my production of extensive cultural resources that produce my teaching practices for a given field. Resonance occurs between dispositions to act in particular ways and structures of a field, producing practices without conscious awareness. In synchrony, also without conscious intent, my students developed appropriate dispositions and practices. My teaching in *Incentive* was initially characterized by failure of my teaching habitus to generate appropriate practices for the students at City High. In breakdown, I was conscious of my teaching and reactive to the unfolding events. Over time I generated habitus to support my practices in urban science classes. However, the habitus reflects the resources available and missing from the field inasmuch as it was generated through a dialectical relationship between structure and practices. Those who teach in urban schools should be alert to the potential of their teaching habitus to generate practices that are hegemonic. They can use cogenerative dialogues to describe the practices of the field and the extent to which they contribute to the transformative goals of science education.

ACKNOWLEDGMENTS

The research in this chapter is supported by the National Science Foundation under Grant No. REC-0107022. Any opinions, findings, and conclusions or recommendations expressed in this chapter are those of the author and do not necessarily reflect the views of the National Science Foundation.

REFERENCES

Anderson, E. (1990). *Streetwise: Race, class, and change in an urban community*. Chicago: University of Chicago Press.

Anderson, E. (1999). *Code of the street: Decency, violence, and the moral life of the inner city*. New York: W.W. Norton.

Barton, A. C., & Tobin, K. (2001). Preface: Urban science education. *Journal of Research in Science Teaching, 38*, 843–846.

Bourdieu, P. (1986). The forms of capital. In J. G. Richardson (Ed.), *Handbook of theory and research for the sociology of education* (pp. 241–258). New York: Greenwood Press.

Bourdieu, P. (1990). *The logic of practice*. Cambridge, England: Polity Press.

Cole, M., & Engeström, Y. (1993). A cultural historical approach to distributed cognition. In G. Salomon (Ed.), *Distributed cognitions: Psychological and educational considerations* (pp. 1–46). Cambridge, England: Cambridge University Press.

Fine, M. (1994). *Chartering urban school reform: Reflections on public high schools in the midst of change*. New York: Teachers College Press.

Lytle, J. H. (1998, April). *Using chaos and complexity theory to inform high school redirection*. Paper presented at the annual meeting of the American Educational Research Association, San Diego, CA.

National Research Council. (1996). *National science education standards*. Washington, DC: National Academy Press.

Roth, W. -M., Tobin, K., Elmesky, R., Carambo, C., McKnight, Y., & Beers, J. (in press). Re/making identities in the praxis of urban schooling: A cultural historical perspective. *Mind, Culture, & Activity.*

Salley, C. (1975). The individualized study program at University City High School: A case study in the dynamics of innovative ineffectiveness (Doctoral dissertation, University of Pennsylvania, 1974). *Dissertation Abstracts International, 35,* 4943A.

Seiler, G. (2001). Reversing the standard direction: Science emerging from the lives of African American students. *Journal of Research in Science Teaching, 38,* 1000–1014.

Tobin, K. (2000). Becoming an urban science educator. *Research in Science Education, 30,* 89–106.

Tobin, K., Seiler, G., & Walls, E. (1999). Reproduction of social class in the teaching and learning of science in urban high schools. *Research in Science Education, 29,* 171–187.

CHAPTER NINE METALOGUE

Expanding Agency and Changing Social Structures

Kenneth Tobin
Randy Yerrick
Wolff-Michael Roth

Randy: You have argued that science education reform rhetoric tends to be "positioned above and beyond substantive analysis." Michael wrote in an earlier metalogue that he doesn't understand American educators' embrace of standards and their contribution to them. It's something of a "wag the dog" kind of phenomenon if I understood him correctly. This would suggest that teachers and even teacher educators are deferring their own knowledge and expertise to visions that are poorly articulated and often inadequately assembled or thought through.

Ken: Those who participate in standards writing are our colleagues and they are well intentioned. There are some dilemmas that are not easily overcome for those who see standards as having a role in reform. For example, some people in our field assume that there is such a thing as standards that can be right for an entire community that is as socially and culturally diverse as the K–12 children of the United States. Creating a consensus among participants and an activity system to support articulated standards extends considerably beyond writing, saying, and proclaiming.

Michael: It is interesting to me that a culture that emphasizes individuality so much at the same time attempts to make everyone the same in some respect. That is, structures such as standards and curricula are being put in place that have yet to be shown to lead to an enhancement of agency and better life for individuals.

Ken: There needs to be a better understanding of the agency-structure dialect that operates in social life, in this case in science. As agents we are not free to do whatever we want. I do not accept the national science education standards formulated in the United States as optimal for all Americans, yet I am not free to disregard them and it may appear to others that I endorse them. Those with power create rules that become part of the structure of my social life and through necessity my agency coexists dialectically with that structure. I can change the structure as I act in the world; but my power to do so is truncated by the actions of others (many who are more powerful), and their actions also are part of the structure that shapes my agency. Teacher and teacher educators cannot act as if they are free of structure; nor can they be aware of all parts of the structure that constrains their power to act in the world.

Michael: But they could also be more active in changing their professional and personal life conditions rather than taking these social structures as given and unchangeable. Thus, I see national standards as only temporary formulations that we abandon if they are shown to lead to the same disadvantages among particular social groups—the poor, women, African American and First Nations—as other, past curricular frameworks that have been abandoned.

Randy: I think I understand (at least in part) the intent and message of the National Standards that, as you aptly pointed out, are at least partially informed by our well-intended and competent colleagues. Yet it is the local and state standards that tend to wield the most power among teachers in my experience. Teachers are often put in the position of devaluing their own knowledge and experience, compromising their values under administrative and assessor pressure.

Michael: Such a view of the individual as being determined in his or her action by a specific set of structures, here the National Standards, returns us to behaviorism. Ken wants to move the discussion and have us look at the dialectical relation of agency (e.g., teacher) and structure (e.g., National Standards).

Ken: I tend to view a person's actions in social life as multifaceted. We always participate in multiple activity systems, and sometimes the objects of those activities align and at other times they conflict and we encounter contradictions. To complicate matters even more, much of social life occurs at a level that is beyond conscious awareness and is unintended. The ideology that is part of the macro structure of the United States saturates projects like the science education standards and the efforts of a community to enact them. As I have noted previously our agency or power to act in the world coexists dialectically with structure as a whole and this includes an ideology that reproduces stratifications of various types and oppresses groups such as those I write about in urban schools in the United States.

Michael: And that is exactly where I think we have a lot of work to do, articulating and analyzing the ideology and then use our agency to change the structures for the ultimate purpose of improving the life conditions for and with those currently oppressed.

CHAPTER 9 METALOGUE: EXPANDING AGENCY

Ken: I can accept that position in part, but the dialectical relationship between agency and structure will necessarily truncate agency in cases where the stakes are considered high by powerful others. To not teach to the standards in some schools in most school districts would cost a teacher his or her job. Decisions to act in ways that go against the grain may carry costs.

Randy: How would you categorize the knowledge you gained during the initiation you experienced as an older, White, male teacher?

Ken: I was aware of my cultural otherness to an extent that I had not previously experienced as an educator. Perhaps more salient was the symbolic violence I experienced when my cultural capital did not afford my objects. Of course as I began to flounder the students were opportunistic and used me as an object for earning respect of their peers by dissing me.

Michael: And thereby they taught you a lesson in the sociology of agency and structure, and how the reach of structure can be truncated through appropriate agency.

Ken: I *did* experience intentional sociological violence as well as symbolic violence as efforts were thwarted to create social capital by building relationships with others and enact cultural capital.

Michael: Perhaps the violence you experienced is similar to the violence that the students experience when you are "successful" in making them do, perhaps voluntarily, what you want them to do? At the same time, these experiences, the resistance at the heart of your personal agency-structure dialectic allowed you to change and become a better teacher for this situation.

Ken: Looking back on it, I had a lot with which to contend. The creation of a new identity as teacher of these students at this school in a field in which I did not have the social and cultural tools needed to meet my goals.

Randy: "Dis"ing a teacher is the quickest and most rewarding of behaviors for these students to earn status among their peers. This is why I don't think that you were the source of the problem. Others though, outside of your situation, who have never stepped into this kind of role, might offer quaint advice or quick fixes with snappy lessons or disciplinary models.

Ken: That is precisely why I did not encourage anybody to come and observe. I only allowed coteachers into my classroom. Even the research team had to engage in coteaching. Viewing from the side was not permitted because the responsibility for promoting the learning of students is no longer shared. I do not regard the division of labor as appropriate if there are problems and everyone in the space is not endeavoring to overcome those problems. There is no room for second-guessing of outsiders and that includes experts on the side.

Michael: At the same time, working with and at others' elbows and subsequently engaging in cogenerative dialogues, which also included students, allowed the identification of common goals for dealing with structure and enhancing the agency of teachers and students alike.

Randy: What if there are seemingly no common interests in your teaching and their life goals?

Ken: Common interests can evolve, and it seems to me that earning respect and social capital are key initial steps. Of course none of this can be linear, but I do see the respect-rapport dialectic as a fundamental that must happen in a community and that has got to be a goal for all participants, not just a teacher.

Michael: And therefore, your approach is also expanding individual agency to a certain level of collective agency.

Randy: What if your students themselves tell you in a focus group that 80% of the class is with you but you're just going to have to expel the other 20% because they're too militant? Can we give up on them if the advice of even their peers is that there is no hope?

Ken: I hate to see any action as giving up. Re-assigning some individuals to another place is inevitable if they insist on civil violence. I cannot tolerate any one individual disrupting an activity for which there is a consensus among the majority. As an interim move they may have to be assigned to another place. I was naive to think I could maintain a productive learning environment even though some students wanted to disrupt my efforts to teach others. It is one thing to opt out personally and another to disrupt the agenda of others. As Tyrone might say, "let them come to you when they want to learn."

Randy: Tyrone's advice does tell us something about these kids. They are reacting not to you but to anyone who presumes to influence them that they assume is an "other." I, too, have experienced this and have no glowing recommendations. What I will say is that I understand helplessness you felt and that convincing them that you are "on their side" is a major challenge—albeit an important one. I agree that we need to turn inward to the community of students—the microcosm we hope to influence. So the questions become "Where do you look for such knowledge" (e.g., churches, community leaders, gang members, parents)?

Ken: We have been quite successful in getting students like Tyrone interested to participate as researchers; insiders who ever so gradually learn to become very solid critical ethnographers. We are beginning to include parents in much the same way. When students, parents and other stakeholders have a voice in the research all sorts of change seem possible.

Michael: Again, the level of agency that comes from your collective effort is much more appropriate to the complexity of the problem and for dealing with the hard questions surrounding existing social structures. However, it is not only collective agency that is enhanced,

CHAPTER 9 METALOGUE: EXPANDING AGENCY

within the group, you also afford the expansion of individual agency of the participants, teachers, students, and researchers.

Ken: What we learn in our research we have managed to use in teacher education where students are now an integral part of "learning to teach kids like me." I do think we need to proceed slowly in moving into the community even further. We should avoid gimmicks because risks can be very high. There were many times when I did not feel safe and I could not expose new teachers to risks such as these. The way forward does not reside with the efforts of heroic individuals. As Michael says, there is a great deal of affordance that comes from collective efforts. The challenge for me as I look back is to build a consensus on the objects and associated activities.

Michael: And consensus concerns above all the shared responsibility all stakeholders need to take for what happens in the classroom.

Ken: When there is a shared responsibility for attaining outcomes then cogenerative dialogues can occur in which local theory is cogenerated among participants who have shared in the experiences of the class. I see great hope for science education in urban schools as long as we can elicit the aid of all stakeholders in building consensus and shared responsibility for attaining it. Of course this can never occur if efforts are only local. Agency, even collective agency, always coexists dialectically with structure and much of what happens, even at a collective level, occurs at a level that is beyond conscious awareness.

CHAPTER TEN

When the Classroom Isn't in School: The Construction of Scientific Knowledge in an After-School Setting

Margaret A. Gallego
San Diego State University

Noah D. Finkelstein
University of California, San Diego

> *The [National] Commission [on Mathematics and Science Teaching for the 21st Century (NCMST)] is convinced that the future well-being of our nation and people depends not just on how well we educate our children generally, but on how well we educate them in mathematics and science specifically.*
>
> —NCMST (2000, p. 4)

Recent calls for improving school-aged children's achievement in science have stirred debate and public discussion at every level of the educational process. These discussions have been prompted by the recognition of our nation's failure to adequately prepare students. Whether tested in the international (National Center for Education Statistics [NCES], 1995) or national (NCES, 2000) arenas, our children are not mastering mathematics or science sufficiently. According to a National Assessment of Educational Progress study (NCES, 2000) less than one third of students achieve a "proficient" level of science achievement—that is, only a small fraction of the students tested demonstrated competency in challenging subject matter. At 12th grade, almost half the students do not attain "basic" levels of achievement, meaning that they lack "partial mastery of the knowledge and skills that are fundamental for proficient work at a given grade" (NCES, 2002, p. 1). As a result, a broad range of calls for improving student achievement has been issued (e.g., Department of Education, 2002; NCMST, 2000).

In many ways, these calls echo the national emphasis on science education that was present in the 1940s to 1960s (Lopez & Schultz, 2001). In the aftermath of World War II it was recognized at the national level that science was critical in the advancement of society (Bush, 1990). The Cold War and Space Race of the 1950s and 1960s added urgency to the development of science and a new generation of scientists. In the early 1960s our need (however distant—literally and figuratively) for science and science understanding became a national obsession. This movement influenced politics and national affairs, prompting the development of rigorous curricula that would ensure students' understanding of science and therefore reclaim our collective status as the world's leading nation (French, 1986; Swartz, 1991). The "race to the stars" played out our national interests on a world stage and focused efforts on the cultivation of scientists with a very specific goal/mission.

The new science crisis is a high-stakes endeavor of a different sort than that described earlier. A critical difference between calls for reform of the mid-20th century (Bush, 1990; French, 1986; Lopez & Schultz, 2001; Swartz, 1991) and those in the last decade and a half (Rutherford & Ahlgren, 1999; Department of Education, 2002; National Science Foundation, 2002) is the explicit identification of who is targeted in the educational process. Earlier reforms focused on the development of a relatively few well-trained scientists. Although science was viewed to be in the service of all society, only a fraction of society "needed" to understand scientific content and participate in its development. Currently, we are concerned with every member of society developing a mastery of science, not in order to become scientists, but to help each individual identify how and why scientific knowledge helps each of us do the ordinary, the everyday, even the mundane. The NCMST (2000) states that, "No citizen of America can participate intelligently in his or her community or indeed conduct many mundane tasks without being familiar with how science affects his daily life and how mathematics shapes her world" (p. 14).

The National Commission further states that mathematics and science impact the public in three substantive ways. Science and mathematics are said to (a) have great explanatory power—they teach us that our world is not capricious but predictable, (b) continually shape and reshape our history and culture, and (c) provide human beings with powerful tools for understanding and reshaping the physical world itself. These features are not relegated to a few elite researchers but rather are qualities necessary for all citizens to substantively participate in our society.

Although a fair amount of discussion has centered on school-based learning, another avenue for promoting children's educational development has been in the after-school environment. Relatively recently, educational researchers have been examining the academic benefits of children's participation in after-school educational programs (Cole, 1996; Garner, Zhao, & Gillingham, 2002; Schauble & Glaser, 1996).

In this chapter, a sociocultural framework (Cole, 1996; Wertsch, 1991) is used to illustrate children's development of academic discourse/scientific knowledge while participating in a non-school learning setting. We begin with a brief description of the sociocultural perspective, in which we focus on learning in context, concept formation (everyday and scientific), and the role of language as a tool for learning. Next we provide a general description of the "Science and Tech Club," an after-school program designed as a hybrid space for promoting children's mastery of science and scientific language. Following the program description, four excerpts from video-taped recordings of the Science and Tech Club are analyzed. We focus on children's concept formation (everyday and scientific) and construction of academic discourse. Finally, we address how research conducted in non-school settings can support school-based learning and how such analysis informs teaching and learning more generally.

SOCIOCULTURAL THEORY

The central thesis of the cultural–historical school is that of the social origins of human thinking (Vygotsky, 1978). This school argues that one's thinking develops through participation in culturally mediated, practical activities. As Scribner (1990) described, the concept of the mediation of human actions (including thinking) is central to Vygotsky's theorizing, perhaps its defining characteristic. People interact with their worlds through cultural artifacts, especially language in both its oral and written forms, and these artifacts serve to mediate our interaction with the world and enable us to develop higher order cognitive skills. That is, the mediation of action plays a crucial role in the formation and development of human intellectual capacities (e.g., the learning or development of science). In this way, human thinking must be understood in interaction with social contexts—a counterpoint to the traditional American psychological orientations that view the individual as the exclusive focus of study (unit of analysis and locus of treatment/intervention).

From a sociocultural perspective, competencies are cultural phenomena—products of the individual and the social context in interaction. Therefore, the appropriate unit of analysis is the individual in interaction with others while engaged in a specific activity. From this theoretical perspective, full understanding of an individual's learning is dependent upon the understanding of the context in which learning has taken place (Rogoff, 1994). Furthermore, for Rogoff (1997), actions need to be analyzed in context by examining "how children actually participate in sociocultural activities to characterize how they contribute to those activities. The emphasis changes from trying to infer what children *can* think to interpreting what and how they *do* think" (p. 273). The social organization that influences learning is broader than the physical setting or location for learning. Here, Dewey's (1938) conception of a learning situation is useful:

> What is designated by the word "situation" is not a single object or event or set of objects and events. For we never experience nor form judgments about objects and events in isolation but only in connection with a contextual whole. This latter is what is called a "situation." (pp. 66–67)

And despite psychologists' efforts to reduce the study of objects or events in isolated or abstracted fashion, Dewey (1938) argues that

> in actual experience, there is never any such isolated singular object or event; *an* object or event is always a special part, phase, or aspect, of an environing experienced world—a situation.... There is always a field in which observation of *this* or *that* object or event occurs. (p. 67)

We believe that understanding the social organization (supportive of or resistive to learning) is useful to (science) teaching and learning in all environments, including schools and classrooms. Therefore, one central concern for science educational reform is the question, "How can we organize the social activity to maximize student learning?"

SCIENTIFIC AND EVERYDAY CONCEPTS

In educational research and pedagogy, scientific knowledge refers to the understandings of concepts within a science discipline. However, within the sociocultural theoretical framework, scientific concepts are *not* bound by discipline (science, history, art, etc.) but refer to a larger body of understanding that refers to all concepts that are formally learned through directed practices (Van der Veer & Valsiner, 1994). This broader connotation has led to a referral to scientific concepts as academic or scholarly concepts, reflecting their introduction into the school environment. Conversely, everyday concepts are spontaneously developed as one experiences the world and interacts with other persons on a daily basis. These concepts are not part of a collectively formalized system. Scientific concepts, in the Vygotskian sense, are nonspontaneous.

Individuals hold both scientific and everyday conceptions regarding the phenomena surrounding them. Mathematical or physical formalisms (e.g., a circle is the collection of all points in a plane equidistant from a given point in the plane, or $F = ma$) are scientific concepts (of mathematics and physics principles). By contrast, everyday concepts are less formalized (e.g., a circle is the round thing that everyone knows about, or when I push on an object it starts to move). Thus, one could offer two distinct descriptions of the movement of a car. One explanation is that when petroleum combusts, its chemical energy is exchanged for mechanical energy (driving a piston). The energy is transmitted through the drive train and gears to apply a torque to the wheels that begin to rotate. Given friction between the tire and the road, the car begins to accelerate forward (with drag from the air). Or the same phenomenon described in a more everyday conception is that when I push down on the gas pedal (or accelerator), the car moves forward.

The relationship between scientific and everyday concepts has been long debated and continues to hold much relevance in contemporary discussions regarding learning and development. When Vygotsky wrote on concept formation, he was concerned with addressing the two predominant views at the time. The first drew no distinction between scientific concepts and everyday concepts (Mahn & John-Steiner, 1998), whereas the second, presented primarily by Piaget, distilled one from the other. Piaget (1929/1979) proposed that the development of everyday concepts was prerequisite to the understanding of scientific concepts as foundations for new thought, but not necessarily integral to the new scientific conceptualizations. From these perspectives, Vygotsky formulated his own position. Vygotsky's central question was how children's learning changes upon entering school, that is, understanding the nature of learning and its relation to development. Through his examination of the interrelations of scientific concepts and everyday concepts, Vygotsky theorized dynamic processes: Scientific and everyday concepts are constantly unfolding and being influenced by the other. Vygotsky (1934/1987) wrote

> The development of scientific concepts begins in the domain of conscious awareness and volition. It grows downward into the domain of concrete, into the domain of the personal experience. In contrast, the development of spontaneous concepts begins in the domain of the concrete and empirical. It moves towards the higher characteristics of concepts, towards conscious awareness and volition. The link between these two lines of development reflects their true nature. This is the link of the zone of proximal and actual development. (p. 220)

As van der Veer (1998) stated, Vygotsky "argued that scientific concepts are dependent on everyday concepts in the sense that scientific concepts presuppose everyday concepts as their foundation, but that scientific concepts in their turn, are able to transform everyday ones" (p. 91). Thus, scientific and everyday concepts may be different but they are not independent of each other. A distinction between nonspontaneous (scientific) and spontaneous (everyday) concepts is useful only to the degree that division is fruitful. Far more attention has been given to Vygotsky's distinction between scientific concepts and everyday concepts than has been given to the interrelation he described between them (Mahn & John-Steiner, 1998). This dichotomization ignores Vygotsky's view of scientific and everyday concepts as aspects of a unified process of concept formation.

Understanding the relationship between scientific and everyday concepts is significant in light of current educational efforts to make school subjects (scientific concepts) like science more real-life-like (everyday concepts). For instance, schools are settings in which the majority of the curricula and academic objectives can be categorized as nonspontaneous, scientific concepts. Indeed, what school environments routinely do is "compress" real-life experiences into 6½-hour intensive sessions—schooling is a nonspontaneous (and rather unauthentic) enterprise. School accelerates learning episodes at the rate and scope not afforded in a genuine (real life) context and draws attention to phenomena (scientific concepts) that may go unnoticed. Relying exclusively on schooling to teach science can leave children with abstract, amorphous knowledge that they perceive as useless. The purpose of school becomes that of preparing for future schooling and testing. A link to real life (and its implied utility for real life) is a key element that stimulates current educational reform efforts within science education. After-school settings provide hybrid and potential alternative spaces for such learning.

In the case presented here, we examine one particular after-school site, the Science and Tech Club. By suspending the typical constraints of classrooms and school experiences (e.g., time, mandated curriculum), we can focus on children's science knowledge development and appropriation of academic discourse through their manipulation of scientific (nonspontaneous) and everyday (spontaneous) concepts. For Vygotsky (1934/1987), spontaneous concepts "develop through the child's practical activity and immediate social interaction," whereas scientific concepts develop with the child's "acquisition of a system of knowledge through instruction" (p. 168). We specifically designed the Science and Tech Club as an environment that blends the scientific and everyday worlds in dynamic manner—a hybrid space that reorganizes social interaction in order to create an activity that maximizes student learning.

THE SCIENCE AND TECH CLUB

The University of California, San Diego (UCSD) Science and Tech club is an after-school activity modeled after the Fifth Dimension Project, a longstanding program that originated at the Laboratory of Comparative Human Cognition at the University of California, San Diego. As with all Fifth Dimension sites, the UCSD Science and Tech Club is designed to improve the literacy of precollege-aged children and, in this case, focuses explicitly on improving student attitudes toward and abilities in science, mathematics, and technology. The Club borrows from the Fifth Dimension structure by explicitly creating an environment in which (a) the university and community collaborate in creating a joint activity, (b) university members promote and study learning in precollege environments, (c) child participation is voluntary, (d) activities are a purposeful mix of play and learning, and (e) the roles of the multigenerational participants (children, college students, and community staff) fluidly change (Cole, 1996; Gallego & Blanton, 2002).

The Science and Tech Club generally follows a constructionist approach (Papert, 1993) that places children as the central agents and constructors of the educational projects. Furthermore, the Club emphasizes the social and contextual nature of student learning (Scott, Cole, & Engel, 1992). That is, we recognize that it is not fruitful to separate student learning from the context in which it occurs; context is not simply a backdrop for student learning. Rather, context is intrinsic to student learning; it shapes and is in turn shaped by both the content and the student.

Science and Tech Club participants represent a wide range of age and expertise (in children and university students). Typically the Club (housed in the local Boys and Girls Club) is attended by about 10 children and staffed by 1 community staff employee and 3 to 5 university students. The children range in age from 5 to 14 years. At least one of the participating university students is a content expert, an undergraduate or graduate student in physics or engineering. These content experts are enrolled in one of two courses designed to support their interest in teaching and learning science. Each elective course (either *Physics 180* or *Electrical and Computer Engineering 198*) is a blend of content study (physics or electrical engineering) and study of student learning in these domains. In addition to meeting twice per week on the university campus, university students are required to engage in field-based teaching experiences. The university students' field-based teaching occurs in a variety of sites, after and during school, formal and informal environments. They interact with students of various ages who are enrolled in community college or high-school classes or are participants in after-school programs such as the Science and Tech Club. Additional staff support in the Science and Tech Club is provided by the community partner and by university undergraduates enrolled in other (non-science-based) field courses, such as the *Practicum in Child Development*.

Children have the opportunity to participate twice per week in the project-based work of the Science and Tech Club. Participants collectively agree upon the objects of study and engage in structured real-world activities that mix play and learning (e.g., building strobe lights or throwing water balloons). The projects vary in scope and duration from single-day activities (e.g., making light bulbs with nichrome wire) to multiweek projects (e.g., building voice-activated strobe lights) dependent upon the complexity of the topic and the extent of children's interest and commitment.

Although the university students who participate in the club change quarterly according to their course enrollment, the children's attendance in the Club is relatively consistent, though voluntary. This exchange of experienced for inexperienced university students provides children with genuine opportunities to teach novice Club participants (the university students and new peer members) the roles and responsibilities of the local Science and Tech Club culture. In addition, each course-quarter rotation provides a natural break for the shift in content. Each quarter, the Science and Tech Club members focus on one content-based theme, such as electricity and magnetism (circuit design and assembly), robotics (systems design and programming), or states and properties of matter (phases, temperature, and pressure). The data analyzed next are taken from a quarter when the club members studied mechanics (forces and motion).

For the purposes of this chapter we have selected excerpts in which the verbal and nonverbal interaction illustrates participants' (both university students and community children) use of everyday (spontaneous) and scientific (nonspontaneous) concepts in the process of learning about physics concepts (Newtonian mechanics).

The constraints inherent in the task of writing about this interaction compels us to remind the reader that we contend that it is the very combination of the verbal and the nonverbal interaction that support participants' concept formation and it is the combination of this video and audio information that informs our analysis of it. In the following, we provide the verbal text in its entirety (within the limits of present technology) and our best depiction of the physical context in which they transpired.

SCENARIO I: MECHANICS: FORCES AND MOTION—DISCUSSION

In the following, we designate adult participants by the use of the letter A. A1 is a graduate student in psychology, and A2 is an undergraduate senior in engineering. Child participants are designated by the letter C. The participating children include C1, an 11-year-old Hispanic boy, and C2, a 9-year-old Anglo boy.

On a day when the children are learning about the scientific concepts of Newtonian mechanics (force, mass, and acceleration), the activities begin with a demonstration by the university students. Children observe A1 throw an air-filled balloon to A2, and then A1 leads a discussion about what would happen if the balloon were filled with water. This group discussion is designed to introduce children to the concepts of mass and acceleration and set up the following activity, tossing water balloons back and forth.

A1: How is he [A2] going to prevent it [a water-filled balloon] from popping?
C1: Catch it softer.
A1: Catch it how? Softer.... With a softer what? His hands will be softer?
C1: It will be softer.
[...]
C2: Oooh.... He can sort of like ... he can sort of like move back and catch it ... [hand movement of catching over distance].
C1: It could be like running down the street and the speed will get lower....
A1: So what causes....
[...]
A1: The speed....
C2: No the speed going lower.

Here, A1 introduces the scientific concepts of force and acceleration by use of everyday concepts (throwing a water balloon and preventing it from breaking). Although it is unclear what C1 means by "catch it softer," the clarifying comments "move it back and catch it" refers to reducing the force applied to the balloon. (Whether over a short distance or long distance, the amount of work [force integrated over distance] required to reduce the balloon's kinetic energy to zero is constant. Hence, if the distance is greater, the force applied to the balloon is smaller and the likelihood of popping the balloon is lower.) Secondly, the statements by C1 and C2 reflect an understanding about acceleration, "speed getting lower." That is, there is some important system property other than speed, namely, the change in speed that is significant in the balloon toss. Furthermore, this seems to relate to the popping of the balloon or, in a more scientific conception, relates to the force applied to the balloon.

A few points are worthy of highlighting. First, the children tend to explain the phenomena correctly using their everyday experiences and language. Even advanced-level university students often fail to distinguish between the concepts of velocity and acceleration (diSessa, 1993; L. McDermott, 1993). Furthermore, children's use of their own words to describe physical phenomena before assigning the scientific concepts, illustrated in this example, resonates strongly with a tenet of physics education reform, "idea first, name afterward," (Aarons, 1990). The activity of throwing balloons is simultaneously everyday, familiar, and playful. Despite learning formalized physics concepts, children are doing so in a manner that is familiar and that builds on their collective experiences in the world.

SCENARIO II: MECHANICS: FORCES AND MOTION—ACTIVITY

Later that day, children are assembled in dyads outdoors. They are given both air-filled balloons and water-filled balloons. They are encouraged to mess about and discover the properties of each of the balloons. Children are prompted to throw the water balloons back and forth to one another and to keep them from popping.

> A1: C1. I want you to catch it and do what you can to reduce the force so it doesn't pop.
> [A1 tosses balloon to C1, C1 walks back as he catches it]
> [other balloons are tossed; one breaks]
> A1: C3, what are you gonna do to prevent it from popping . . . how can you reduce the acceleration? Who can help him out, how can you reduce the acceleration? . . .
> C6: [Raises hand] walk backward!

In this later episode, the adult overtly uses scientific concepts/language. Both the words *force* and *acceleration* are used in relation to the activity at-hand. Furthermore, students are problem solving in practice. Ultimately, no other balloon breaks, and students are walking backward and moving their arms as they catch to reduce the applied force to the balloons. Also evident in this scenario, the club discussion begins to include more children. Students who had not previously contributed to the class discussion now present answers that had been mentioned in the prior scenario, such as "walk backwards."

In this excerpt, adults not only use the scientific concepts of *force* and *acceleration*, but also students are responding to the adults in scientifically appropriate ways. The answer that is provided, "Walk backward!" still uses everyday language but does so in a fitting manner, one that is consistent with the physics solution. In this environment, the children are not expected to answer with the scientific conception in physics: increase the distance over which the balloon is decelerating, which reduces the net acceleration to the balloon and hence minimizes the applied force.

SCENARIO III: MECHANICS: FORCES AND MOTION—RECAP

The following week, participants revisit the water-balloons activities. Students are asked to report back on what happened. After discussing properties of what made balloons pop or not pop, the class examines the concept of *force* in more detail:

A1: The more force you get. Moving back caused what to happen? [discussion of popping balloons and what causes it . . .]
A1: . . . So force is based on acceleration and what else?
C3: Speed?
A1: Well that's part of the acceleration. . . .
C1: Mass! Mass!
A1: Mass, good.
C4: Mass is the weight!

The adult once again leads a discussion using scientific concepts. In this excerpt, the terms *force, mass,* and *acceleration* are used. In this environment students are encouraged to play with their spontaneous conceptions of physics, to guess or to mess about creatively, and to blend these with the use of scientific concepts. Scientific concepts are injected into the discourse both by the adult (in this case the adult introduces the concepts of *force* and *acceleration*) and by the students themselves (as with the case of introducing the concept of *mass* in a scientific manner). At some points the students are clearly guessing: "Speed?" In other statements, children begin to use the scientific conceptions correctly. C1 and C4 both use the scientific concept of *mass*, a term not found in their everyday lexicon. C1 appropriately answers "mass" when the adult is fishing for answers, but what level of mastery C1 possesses is unclear. C4 elaborates, relating mass to an earlier discussion about weight. Although this is not the strict scientific conception, it is an often-used approximation.

SCENARIO IV: MECHANICS: MOTION/NEWTON'S THIRD LAW—ANOTHER DISCUSSION

Later in the same day, students learn about Newton's third law: For every action, there is an equal and opposite reaction. At first the children are told explicitly that they will be learning about the law. They are told that ultimately they will build their own model hovercrafts. Then the adults provide a demonstration of the third law by connecting a balloon to a straw that has a right-angle bend in it. The straw is placed on a pin so that it will freely rotate in the plane with the right angle bend in it (see Fig. 10.1). The balloon is inflated, and then the air releases from the balloon, causing the balloon/straw apparatus to rotate about the pin. In this figure, the balloon rotates counter clockwise in the plane of the page.

A1: How is that an example of the third law?
[kids shout out dozens of words/phrases]
C: It was moving in motion. . . .

FIG. 10.1. Balloon/Straw set-up for Newton's third law activity.

C: Action! Reaction!
A1: What's the force?
C: It's the opposite reaction.
A1: What's the force?
C: I don't know.
C: The air!
C: Air!
C: Yes!
A1: Air from where?
C: The balloon!
A1: Good. The air that gets pushed into the balloon is released and then what happens?
[. . .]
[The demonstration is run a second time.]
C1: Whatever direction the straw is pointing, it's gonna go the opposite.
A1: Ok it's gonna go the opposite way, why?
C1: Because the air is coming out and it's pushing it that way. [motions left and right with his hands to show the moving air]
C3: The air inside the straw and it pushes out and [hand spinning to emphasize] I can feel the air pushing out.
[. . .]
C3: Because the straw was bended and the air sprayed out and it pushed it the opposite way.

Unlike the organization of earlier activities, the scientific concepts and terminology (Newton's third law) were presented first, followed by a demonstration designed to motivate and develop the underpinning physics concepts by using everyday language. After the demonstration, A1 asks the children to describe what happened. They are cued by the adult to explain using scientific concepts, "How is that [demonstration] an example of the third law?" It is relatively safe to assume that until this discussion the children had no association with the "third law." As if socially scripted, students begin shouting out as many scientific concepts as they can remember. Clearly little else is meant by children's use of the terms, "Action! Reaction!" C1 admits not knowing what the "force" (which moves the balloon in a circle) is after having answered with a scientific concept, "the opposite reaction." With another demonstration and some leading conversation, student responses become more detailed and address why the balloon is moving; students begin to use their everyday concepts to describe what is happening. In so doing, they demonstrate an understanding of the underpinning physical principles. C1 accurately describes what is happening in the demonstration and why, "Because the air is coming out and it's pushing it that way." His use of "pushing" and his hand gestures are everyday conceptions that demonstrate his understanding of Newton's law. C3 describes what is occurring in his own words, "The air inside the straw and it pushes out and . . . I can feel the air pushing out." Earlier in this segment each of these children stated the scientific concepts with no understanding of the physical principles. As they begin to use everyday concepts, the group collectively begins to develop an understanding of the fundamental and complex physical principle, Newton's third law.

Deviating from previous sessions, the organization of activities in the session described above resembles the social organization typical of classroom lessons. That is, science terms are presented by the adult (teacher) who then directs a demonstration followed by prompting students with questions. Previous sessions children were asked to participate in rather than witness a demonstration conducted at the outset of the session.

In addition, the question asked by the adult is formulated in a rather quiz-like manner "How is that [demonstration] an example of the third law?" that required students' to link the phenomenon to an understanding of a over arching system (scientific/nonspontaneous concepts) rather than the direct explanation of the phenomenon (everyday/spontaneous concepts), for example, "Why did the balloon move?" Although the students' participation is rather passive in comparison with other sessions, they nonetheless energetically respond to the adults prompts with a wealth of terms (some appropriate some not) eager to "show" the adults what they know.

It is unlikely that the adults (university students of electrical engineering and psychology) were aware that their verbal and physical interactions were "slipping" into interactions normative of classrooms, that is, school-like. Rather they too, were likely building upon their own everyday and scientific understandings and were undoubtedly influenced by (good and poor) models of teaching/learning from their (past and present) educational experiences.

DISCUSSION

The foregoing data and general organization of the Science and Tech Club emphasize three broad effects of discourse in creating educational environments: (a) The Development of Children: Science Discourse/Knowledge, (b) The Re-cognition of a Discipline: Science, and (c) The Development and Design of Social Organization: School Science.

The Development of Children: Science Discourse/Knowledge

The verbal play evident in participants' interactions provides evidence of the importance of language as one of the key tools with which we come to understand (Vygotsky, 1978, 1934/1987; Wertsch, 1985, 1991). In the earlier episodes, children's verbal interactions with peers and more capable adults illustrate the use of everyday concepts and scientific concepts in the process of learning. In discussion of children's development of higher order cognitive processes, Vygotsky (1978, 1934/1987) concentrated primarily on what he called *psychological tools*, the meaning-making potential of systems of signs and symbols (most significantly language) in mediating thinking. Pontecorvo (1993) summarized this Vygotskian idea of tool mediation:

> Mediation tools include the semiotic systems pertaining to different languages and to various scientific fields; these are procedures, thought methodologies, and cultural objects that have to be appropriated, practices of discourse and reasoning that have to be developed, and play or study practices that have to be exercised. (p. 191)

In the case of the Science and Tech Club, children use language to develop their understanding of physical phenomena. Children develop the concepts of force, acceleration, and mass by using scientific and everyday conceptions of the phenomena under investigation (tossing balloons). Furthermore, in so doing, students develop a language common to their community, the Science and Tech Club, which in turn reinforces the community. In no other communities within the Boys and Girls Club do adults ask children, "How is [a spinning balloon] an example of the third law?"

Vygotsky emphasized a double function of language, how it serves as a means of communication and how it comes to mediate intellectual activity. As a means of communication, language enables human beings to socially coordinate (or dis-coordinate) actions with others through meaning. The children in the Science and Tech Club use specialized language (e.g., force, acceleration, Newton's third law) to structure and coordinate ideas and activities. With the internalization of this communication, language comes to mediate intellectual activity through the discourse of inner speech. The development of this capacity for self-regulation through inner speech is what helps bring actions under the control of thought, a development to which Vygotsky assigned great importance (Wells, 1996). In the Science and Tech Club, through the use of everyday conceptions and language, children begin to develop an understanding of scientific concepts. These concepts help them learn new concepts (e.g., developing an understanding of force facilitates their understanding of Newton's third law) and regulate their own action (e.g., understanding an aspect of force allows students to talk about and shape how they engage in everyday activities, preventing balloons from popping).

The Re-Cognition of a Discipline: Science

Our analysis here focused on how children's interactions surrounding activities and science topics prompted their use of everyday (spontaneous) and scientific (nonspontaneous) concepts to develop an understanding of science content. We argue that in these moment-to-moment discoveries made by each child a reformulation, or re-cognition, of the educational discipline of science is developed.

The development of the science discipline requires a much broader lens over a substantively longer period of time than we discuss presently; yet, the excerpts provided here illustrate conceptual development of science by children. That is, the child's perception, appreciation, and wonder of and for science change. For this child, what counts as science has changed, being no longer limited to a school subject. With the arousal of each child's interest in learning "about" science is the potential for continued motivation and interest to "do" science. As educators, we must reconsider how and what we define (count) as "school science." The learning that takes place in environments such as the Science and Tech Club, illustrated here, requires us to recognize the importance and the very necessity of everyday concepts in the development of scientific (schooled) concepts. Based on the interrelationship between everyday and scientific concepts, we can support efforts to recontextualize school science to incorporate epistemological, affective, and motivational elements.

Discussions regarding the utility of everyday understandings and everyday concepts for the development of students' scientific (schooled, nonspontaneous) concepts in the discipline of science have found a new and eager audience. Educators toil to link school with the everyday *as a means* for supporting *the end*, school learning. However, we contend that the relationship between everyday and scientific concepts is not unidirectional. What we wish to underscore is that everyday and scientific concepts are interdependent. As presented above, children use everyday concepts to learn the scientific. This relates to the scientific concepts being rooted in the everyday. However, once scientific concepts are developed, they are in turn used to understand the everyday. Indeed one of the main purposes of scientific conceptions within science is to understand everyday phenomena. Scientific concepts (even the discipline of science) are challenged, encouraged, provoked, and legitimated by everyday concepts. The Science and Tech Club builds on this understanding in order to recognize the value of scientific concepts and the disciplines in which they are embedded and to use them to explore the world of the students.

The Development and Design of Social Organization: School Science

Studies in ecological psychology have documented that the physical features of a setting and how people participate in such environments are related (Gump, 1978; Gump & Good, 1976). That is, the site or setting for learning influences what and how something is taught or learned (Johnson, 1985). Clearly there are characteristics of the school setting that influence content and how teachers and students participate in these settings. Features such as compulsory attendance, standardized curriculum, mandatory assessment, and high student-to-adult ratio can be viewed as either positive/supportive or negative/nonsupportive influences to a particular curricular goal.

The participants of the Science and Tech Club acting within a nonschool setting generate the interactions described and analyzed in this chapter. Both tacit understandings and physical characteristics that collectively comprise this unique learning situation support the social organization of knowledge in this setting. Although the physical attributes of the Science and Tech Club are relatively easily identified (located in a room at the local Boys and Girls Club, tables not desks, participants of various ages, etc.), these attributes also influence the development of more tacit features regarding session involvement and Club membership. For instance, children's voluntary participation can exert both positive and negative influences on the site's social organization. One the one hand, positively, children voluntarily elect to participate based on their interest in the topic and motivation to advance their understandings of science. Such attitudes and motivation result in an atmosphere in which participants have a common investment and are therefore members of the club community. One's club membership requires a certain degree of responsibility (e.g., students are required to act safely and attentively and to participate regularly) as well as the provision of special and reserved rights (e.g., students who do not act safely or are not regular participants are not allowed to work on the more advanced and multiday projects, such as building strobe lights). On the other hand, voluntary participation has the potential to negatively influence understanding of science. Inconsistent attendance can pose particular challenges and pedagogical concerns regarding topic selection, lesson pacing, and assumptions about children's prior knowledge. Each of these instructional issues in turn influences the environment's climate and level of risk-taking, intimacy, and ambiguity the participants can tolerate while still achieving success in the setting.

CONCLUSION

It is commonly stated that schools are environments that decontextualize information in ways that make the information abstract and difficult to learn (Lave, 1988). However, various researchers (Cole, 1996; R. McDermott, 1993; Rogoff, 1994; Rueda, Gallego, & Moll, 2000; van Oers, 1998) emphasize the importance of context for all forms of meaningful concept formation (including abstract concepts). Rather than describing school learning as decontextualized, what is actually taking place is the process of recontextualization (Van Oers, 1998). Another form of recontextualization is the use of "out of school" environments for learning the type of content that may be considered abstract and school-like (scientific) concepts.

Currently, after-school learning environments are being examined with great interest for their potential in supporting school learning (Garner et al., 2002; Schauble & Glaser, 1996). Furthermore, some researchers are interested in how after-school settings can inform our educational revision, restructuring, and reform. Schauble and Glaser noted the appeal of out-of-school environments as context for educational research:

> Many classroom researchers have found it instructive to rethink the design of classrooms in light of what works in out-of-school learning environments. Because of the constraints of these environments are somewhat different

from those that operate in school, informal learning contexts can serve as laboratories for testing innovative approaches to learning. (p. 9)

As attractive as this direction for educational research is, Resnick (1991) tempers educational researchers' enthusiasm by cautioning that simply removing oneself physically from the school setting does not ensure a distinct social organization. Rather, educational activities (e.g., tutoring science camp), commonly held in a nonschool location (e.g., Boys and Girls club, parks and recreational facilities), may very well replicate in-school interactions. Without deliberate attempts to change respective roles within such arrangements, participants are likely to maintain their adult/child responsibilities as teacher and student, regardless of the physical setting. Resnick argues that change in the social organization of/for learning requires a qualitative and dramatic change in the relationship between participants.

We agree with Resnick and believe that care should be taken in advocating any educational program if its merit is based simply on its physical location. Yet, we also think that educators can build on Resnick's cautionary statement to advocate the possible use of after-school environments to restructure school practice. That is, it is possible to import, transfer, and replicate the patterns of interaction and the collaborative relationships found in some out-of-school programs (such as the Science and Tech Club). In this chapter, we have presented concept formation as the positive negotiation of both scientific (non-spontaneous) and everyday (spontaneous) concepts within a physical context.

Our review of the data were/are informed by our knowledge of science, pedagogy, data analysis, and our direct participation with the Science and Tech Club community. To the extent upon which the focus of this collective volume is on the development of scientific knowledge as evidenced by verbal discourse. Our analysis is based on students' actual verbal statements as well as our greater understanding of is meant but what is said. Rogoff (1997) reminds us "the emphasis changes from trying to infer what children *can* think to interpreting what and how they *do* think" (p. 273).

In this way, the challenge is not only to arrange learning contexts that provide student opportunities to use language as a tool for communication (overt and intended for others) as well as the use of language as instrument of thought and reason (intended for self, though at times overtly stated). Vygotsky's (1934/1998) central question, "How children's learning changes upon entering school?" presupposes that it would indeed (and necessarily) change. As a unified process, concept formation, everyday and scientific are mutually influenced by the other. Whereas the physical differences of nonschool settings may provide for variable social organization, schools and classroom can also be modified to support the academic utility of everyday and scientific concepts. Promising examples of recent science educational reforms (Klentschy, Garrison, & Amaral, in press) seek to develop children's scientific concepts and their understanding of the science domain.

Some of the modifications to classroom norms may include the increase in experimentation, demonstrations, activities, discussions, hypothesis formulations and the generation of novel questions by both teacher and students. In a complementary way, our creating opportunities for students exploration provides us with opportunities to become better listeners and interpreters of what they do and say (understanding the connection between everyday and scientific concepts), in ways that honors the process and expands what counts as science (knowledge) in classrooms/schools.

REFERENCES

Aarons, A. (1990). *A guide to introductory physics teaching*. New York: Wiley.

Bush, V. (1990). *Science—The endless frontier: A report to the President on a program for postwar scientific research*. Washington, DC: National Science Foundation.

Cole, M. (1996). *Cultural psychology: A once and future discipline*. Cambridge, MA: Belknap-Harvard University Press.

Department of Education. (2002). *No child left behind*. http://www.nochildleftbehind.gov/

Dewey, J. (1938). *Experience and education*. New York: Collier Books.

diSessa, A. (1993). Misconceptions reconceived. *American Association for Physics Teachers Announcer, 23*(4), 94.

French, A. (1986). Setting new directions in physics teaching: PSSC 30 years later. *Physics Today, 39*(9), 30.

Gallego, M. A., & Blanton, W. (2002). Cultural practices in fifth dimension sites around the world. In E. R. Garner, Y. Zhao, & M. Gillingham (Eds.), *Hanging out: Community based after-school programs for children* (pp. 137–147). Westport, CT: Bergin & Garvey.

Garner, R., Zhao, Y., & Gillingham, M. (Eds.). (2002). *Hanging out: Community-based after school programs for children*. Westport, CT: Greenwood.

Gump, P. V. (1978). School environments. In I. Altman & J. F. Wohlwill (Eds.), *Human behavior and environment: Advances in theory and research* (Vol. 3, pp. 131–169). New York: Plenum Press.

Gump, P. V., & Good, L. R. (1976). Environments operating in open space and traditionally designed schools. *Journal of Architectural Research, 5*, 20–27.

Johnson, N. B. (1985). *West Haven: Classroom culture and society in a rural elementary school*. Chapel Hill: The University of North Carolina Press.

Klentschy, M., Garrison, L., & Amaral, O. (in press). Valle Imperial Project in Science (VIPS): Four-year comparison of student achievement data 1995–1999. *Journal of Research in Science Teaching*.

Lave, J. (1988). *Cognition in practice*. Cambridge, England: Cambridge University Press.

Lopez, R., & Schultz, T. (2001). Two revolutions in K–8 science education. *Physics Today, 54*(9), 44–49.

Mahn, H., & John-Steiner, V. (1998). Introduction. *Mind, Culture, & Activity, 5*, 81–88.

McDermott, L. (1993). How we teach and how students learn—A mismatch? *American Journal of Physics, 61*(4), 295.

McDermott, R. (1993). The acquisition of a child by a learning disability. In S. Chaiklin & J. Lave (Eds.), *Understanding practice: Perspectives on activity and context* (pp. 269–305). New York: Cambridge University Press.

National Center for Education Statistics. (1995). *Pursuing excellence: A study of U.S. twelfth-grade mathematics and science achievement in international context.* http://timss.bc.edu/timss1995.html

National Center for Education Statistics. (2000). *The nation's report card: National Assessment of Educational Progress.* http://nces.ed.gov/nationsreportcard/

National Center for Education Statistics. (2002). *Setting the achievement levels.* http://nces.ed.gov/naep3/set-achievement-lvls.asp

National Commission on Mathematics and Science Teaching for the 21st Century. (2000). *Before it's too late.* http://www.ed.gov/americacounts/glenn/toc.html

National Science Foundation (2002). *Math and science partnership program solicitation* (NSF-02-061). http://www.nsf.gov/pubs/2002/nsf02061/nsf02061.html

Papert, S. (1993). *The children's machine.* New York: Basic Books.

Piaget, J. (1979). *The child's conception of the world.* New York: Harcourt Brace. (Original work published 1929)

Pontecorvo, C. (1993). Social interaction in the acquisition of knowledge. *Educational Psychology Review* [Special Issue: European educational psychology], *5*(3), 293–310.

Resnick, L. (1991). Literacy in and out of school. In S. R. Graubard (Ed.), *Literacy: An overview of fourteen experts* (pp. 169–185). New York: Noonday Press.

Rogoff, B. (1994). Developing understanding of the idea of community of learners. *Mind, Culture, & Activity, 1*, 209–229.

Rogoff, B. (1997). Evaluating development in the process of participation: Theory, methods and practice building on each other. In E. Amsel & K. A. Renninger (Eds.), *Change and development: Issues of theory, method and application* (pp. 265–285). Mahwah, NJ: Lawrence Erlbaum Associates.

Rueda, R., Gallego, M. A., & Moll, L. C. (2000). The least restrictive environment: A place or context? *Remedial and Special Education, 21*, 70–78.

Rutherford, F. J., & Ahlgren A. (1999). *Science for all Americans.* New York: Oxford University Press.

Schauble, L., & Glaser, R. (1996). *Innovations in learning: New environments for education.* Mahwah, NJ: Lawrence Erlbaum Associates.

Scott, T., Cole, M., & Engel, M. (1992). Computers and education: A cultural constructivist perspective. In G. Grant (Ed.), *Review of research in education, 18* (pp. 191–251). Washington, DC: American Educational Research Association.

Scribner, S. (1990). A sociocultural approach to the study of mind. In G. Greenberg & E. Tobach (Eds.), *Theories of the evolution of knowing* (pp. 107–120). Hillsdale, NJ: Lawrence Erlbaum Associates.

Swartz, C. (1991). The physicists intervene. *Physics Today, 44*(9), 22.

van der Veer, R. (1998). From concept attainment to knowledge formation. *Mind Culture, & Activity, 5*, 89–94.

van der Veer, R., & Valsiner, J. (Eds.). (1994). *The Vygotsky reader.* Oxford, England: Blackwell.

van Oers, B. (1998). The fallacy of decontextualization. *Mind, Culture, & Activity, 5*, 135–142.

Vygotsky, L. S. (1978). *Mind in society.* Cambridge, MA: Harvard University Press.

Vygotsky, L. S. (1987). *Thinking and speech.* New York: Plenum. (Original work published 1934)

Wells, G. (1996, September). *The zone of proximal development and its implications for learning and teaching*. Paper presented at the Second Conference for Sociocultural Research, Geneva, Switzerland.

Wertsch, J. V. (1985). *Vygotsky and the social formation of mind*. Cambridge, MA: Harvard University Press.

Wertsch, J. V. (1991). *Voices of the mind: A socio-historical approach to mediated action*. Cambridge, MA: Harvard University Press.

CHAPTER TEN METALOGUE

Scientific Versus Spontaneous Concepts at Work in Student Learning

Maria Varelas
Margaret A. Gallego
Randy Yerrick
Wolff-Michael Roth

Randy: Margaret, I interpreted your chapter to be a demonstration of how science talk is co-constructed differently in after-school programs due to certain components of such programs that are more open, conversant, and embedded in children's experiences rather than predetermined science topics. I see your chapter as a challenge for educators to think beyond the normal constraints of classroom talk and activity and look directly toward children's dispositions and abilities to better assess and plan for what children are able to do. I wonder how you would gauge the role of input of the adults and how they bring to the discussion scientific concepts such as *force*, *mass*, and *acceleration*. And how did the adults help students integrate the scientific concepts with your description of Vygotsky's "spontaneous" concepts?

Margaret: I challenge the notion that any concept children are using in the science classroom can be clearly dichotomized as *spontaneous* or *scientific*. As a hybrid space, spontaneous in a pure form is not identifiable (they are not in isolation). However, their worth is in their interdependence with the scientific. As children who go to school (consciously or unconsciously that is a factor that can be identified readily), they are always influenced in both contexts by their prior participation in the other. Teachers in the Tech Club do not systematically insert foreign scientific concepts in ways that neither blend with the experiences nor are central for solving some sort of problem at hand. As such, the context provides multiple means of "input," coordinated as potential learning formats, so beyond verbal use of *mass*, etc., demonstration, and free play.

Michael: In our respective chapters, James Gee and I make the point that the different school discourses cannot be separated from the children's root discourse. What children say using even this discourse itself, by its very nature, is never their own, because the language is always the language of the other, too. In this sense, ideas expressed in language are never entirely spontaneous. When a child asked about the relation of sun and earth responds that the former revolves around the latter, or something of the like, we do not have to impute much to the child. Everyday language is full of expressions in which the sun is the agent—"the sun comes up," "the sun rises," "the sun goes down," or "the sun hides behind the clouds" are all everyday expressions that imbue the sun with agency. We therefore have to be careful about how we employ the notion of *spontaneous concept*.

Maria: Margaret, I wonder what relationship *you* see between explanations and the two Vygotskian types of concepts (i.e., *spontaneous* and *scientific*) and how you see such a relationship played out in your data. What are *scientific explanations* and what do they necessitate? What explanations did the students offer relatively to the water-filled balloon? Is the student's answer "walk backwards" an explanation?

Margaret: It would appear from the comments I have received from other authors in this book that explanation has a certain meaning and is central to definitions of scientific literacy. I guess to answer your question I would need to understand the use of explanation in scientific discourse. Is explanation equivalent to a concept, or is it correlated to a verbal manifestation that refers to the concept? Children offer explanations all the time.

Michael: Which is just one way of being in the world; we do offer explanations without reflecting on the fact that we do this.

Margaret: Whether or not explanations are grounded in evidence or used to forward a particular position against an "other" seems to matter here. With such a definition, we could use the terms consciousness and volition within our analysis to help define when students are aware of the differences between their ideas and those presented by others (including adults) but only as they naturally play out in the text. We would like to avoid the use of more jargon. In this chapter we highlight concept formation. Just so that we may be interpreted as this way we rather focus on the deeper meanings that Vygotsky's terms always represent.

Maria: In my view, an important characteristic of scientific practice is the interplay between theory and data—between developing a network of concepts and processes that are logically linked and have explanatory power, and examining empirical evidence collected through observations and experiments (Varelas, 1996). Literacy is integral in this enterprise. In the ongoing process of scientific inquiry, written texts often serve as important cultural tools as scientists grapple with the ideas, thoughts, and reasoning of others. I think that the definition and role of explanation should be central to any discussion of children adopting linguistic registers or genres of science. Children's scientific understanding is also influenced—as Michael and Jim have described in their chapters—by the construction of new conceptual entities and the wordings to express those entities.

Michael: It may not be so much of an active, conscious construction as the emergence of new ways of doing and saying. We must take care not to overemphasize or introduce conscious effort where there is none. The root languages Jim and I wrote about embody much of the formal aspects of commonsense, namely those aspects in lived experience that are articulated.

Randy: Maria's comments remind me of Driver's distinction between commonsense explanations and scientific ones (Driver, 1991). It is a distinction that reaches beyond the concepts or events themselves and into the processes that define what counts for expertise. Part of what Driver was suggesting was, I think, a necessary connection of ideas, to evidence, and their explanatory power. Do you see her description as related to your distinction of science and everyday concepts?

Maria: This is a good connection, Randy, to work of others in the field. I recall Driver's description of commonsense explanations from some of her later work and I think perhaps Vygotsky's (1934/1987) explanation may fit well with Margaret's argument regarding spontaneous and scientific concepts. Vygotsky (1934/1987) wrote,

> The strength of the scientific concept lies in the higher characteristics of concepts, in conscious awareness and volition. In contrast, this is the weakness of the child's everyday concept. The strength of the everyday concept lies in spontaneous, situationally meaningful, concrete applications, that is in the sphere of experience and empirical. The development of scientific concepts begins in the domain of conscious awareness and volition. It grows downward into the domain of concrete, into the domain of personal experience. In contrast, the development of spontaneous concepts begins in the domain of the concrete and empirical. It moves towards the higher characteristics of concepts, toward conscious awareness and volition. The link between these two lines of development reflects their true nature. This is the link of the zone of proximal and actual development. (p. 220)

Michael: It seems to me that spontaneous concepts simply refer to everyday language that children and people use as a matter of course in everyday activity rather than to consciously reflect on the state of affairs. Researchers often conflate the way people talk as part of ongoing activity with reflective talk, where language is turned upon itself in a reflective way.

Maria: But I wonder how Margaret thinks of the issue of consciousness relative to the distinction and interplay between spontaneous and scientific concepts.

Margaret: I see interplay as one fuels the other that is interdependent. Scientific without everyday lacks significance other than for the purpose of performance (school); everyday without scientific real tasks may lack progress (doing the same over time); learning totally as independent rather than "culture" which is the culmination of many individual findings. I see this as extending beyond the discussion of just scientific concepts (as connected to the subject). All topics/subjects (a fabrication of the world systematized for the convenience of schooling) are connected to real need to know (life) and only survive within a body of people (culture) to the degree they are useful to them. In the case here, everyday concepts are probably valued (for the educational content) more in the earlier grades then phased out as one continues in school, being less and less valued when the shift is toward scientific concepts.

Randy: You have asked a very intriguing question that you do not directly answer. "Therefore, one central concern for science educational reform is, how can we organize the social activity to maximize student learning?" Do you have insights you would like to emphasize briefly in this metalogue that would give us educators some insight?

Margaret: Noah and I have indicated that there are several components of after-school programs (e.g., compulsory attendance, resources, time-frames) that should be used to inform our educational revision, restructuring, and reform. Out-of-school learning environments have fewer "physical signals" that look/feel like school and therefore may assist in children adopting a different character/role in interaction with activities (fewer cues for kids about "how they should act"). Because of the constraints of in-school environments, informal learning contexts can serve as important test-beds for innovative approaches. Traditional text-based science instruction found in the social activity of classrooms is but one kind of instruction and only when it is arranged to focus particularly on scientific concepts from both definitions—Vygotsky vs. subject matter. The Tech Club's social arrangement is necessarily physically different from classrooms because it is not held in school. However, beyond the physical difference, the opportunity to highlight and focus value on spontaneous (perhaps in this case even privilege spontaneous) over scientific could be adapted within the classroom context.

Maria: You challenge your readers to "reconsider how and what we define (count) as school science." What would be fruitful ground to consider for creating or for defining science for children? How might scientific or spontaneous explanations give educators insight for broadening the standard curricular definition of *science*?

Margaret: I think that it would be heavy handed to say there are no redeeming qualities of existing school science programs. Clearly there are. However, one must judge their success by their stated goal (e.g., good test performance, get kids excited about science). These explicit goals stated by districts, state, and federal agencies tell us a lot about their interpretations of scientific literacy and their focus (or perhaps exclusion) of explicit statements regarding scientific discourse. I think the wrong direction that many states are taking, including many parts of California, is the propensity for literacy to be taught as specific skills out of context. I think such efforts are shortsighted to cut out all content during the "literacy" instruction in classrooms. Science topics and resources can serve as a rich content area to teach literacy in a multifaceted way. As James indicated in his chapter, appropriating a discourse like science can serve as a major venue for a variety of literacy proficiencies. In this way, science in the lives of children could be more meaningfully incorporated throughout the elementary curriculum, though I would be hard-pressed to give an example of any state doing this effectively.

Randy: I agree with the assessment that elementary science instruction is all but absent from the schools I have worked in and I also have heard such comments as, "We can only fit science in after the literacy block." These comments clearly indicate to me that elementary teachers see science as outside of normal instruction and that teachers are missing a great opportunity to use science text (and nontext resources) as vehicles for teaching analysis, interpretation, and other explicitly stated goals in literacy reform. I wonder if it would be possible to accomplish such integration of instruction if it were a part of a professional-development initiative with the goal of promoting diversity and equity as well. For example, I have read of horticulture gardens in urban settings and other elementary science community projects that emphasize doing literacy and science simultaneously with some social agency. Couldn't projects like planning, funding, planting, and reporting on an urban garden or analyzing the speed limit (or other safety factors) with the intent of developing genres to effect change stand out as good examples of how out-of-school activities can demonstrate more authentic science and literacy learning?

Michael: This is just what Angie Barton and I have been arguing in a book on scientific literacy (Roth & Barton, 2004). The language students develop is not the monoglossic repertoire that they would in normal science classes, but it is the heteroglossic repertoire characteristic of talk in the community, including those that involve science, especially when it is contested (e.g., Roth et al., in press).

Randy: While these could certainly be accomplished in after school programs like Margaret has described, are there reasons these couldn't be more commonplace?

Margaret: Clearly such projects are in the right direction for supplementing text-based instruction *and* text-based assessment with more "hands on" demonstration, practice in progress. I suppose that everyday experiences will necessarily be different among individuals, and the types of differences supported by our culture (broadly understood and school/classroom cultural norms) will reveal themselves in an activity. Diversifying the means of knowledge expression will promote diversity and equity. Ultimately once alternate norms are established, changing the classroom social organization, children and teachers can begin to change, which provides the basis for children and teacher to change their thinking about what counts as science and what counts in the process of learning science. After-school arrangements have assets that schools don't (explained previously); however a drawback is the limited knowledge of some staff members (though our example drew from university students from particular content courses). I guess one could argue that resources may or may not be any worse in after-school environment than in classrooms, that is, some classroom teachers are not better or worse off than after-school volunteers in regard to science content knowledge. And in some cases, classroom teachers may be more resistant to changes in classroom norms (teacher as expert, etc.) than after-school staff. This resistance works against many educational reform efforts.

REFERENCES

Driver, R. (1991). *Pupil as scientist?* Philadelphia, PA: Open University Press.

Roth, W. -M., & Barton, A. C. (2004). *Rethinking scientific literacy*. New York: Routledge.

Roth, W. -M., Riecken, J., Pozzer, L. L., McMillan, R., Storr, B., Tait, D., Bradshaw, G., & Pauluth Penner, T. (in press). Those who get hurt aren't always being heard: Scientist-resident interactions over community water. *Science, Technology, & Human Values*.

Varelas, M. (1996). Between theory and data in a 7th grade science class. *Journal of Research in Science Teaching, 33*, 229–263.

Vygotsky, L. S. (1987). Thinking and speech. In R. W. Rieber & A. S. Carton (Eds.), *The collected works of L. S. Vygotsky (Vol. 1): Problems of general psychology* (N. Minick, Trans.). New York: Plenum Press. (Original work published 1934)

Author Index

A

Aarons, A., 301, 311
Ahlgren, A., 294, 312
Aikenhead, G. S., 2, 17, 99, 100
Alcoff, L., 82, 100
Alexander, T., 55, 70
Amaral, O., 311
American Association for the Advancement of Science [AAAS], 62, 64, 69, 81, 97, 98, 100, 118, 132, 175, 189, 232, 255
Anderson, C. W., 202, 222, 239, 246, 255
Anderson, E., 276, 285
Anyon, J., 237, 255
Apple, M. W., 97, 100
Arnason, J., 235, 237, 257
Artiles, A., 200, 202, 224
Asoko, H., 140, 162
Atkinson, D., 84, 100
Au, K., 202, 222
Ausubel, D., 239, 255

B

Bailey, F., 141, 162
Baker, S., 200, 223
Bakhtin, M. M., 30, 36, 93, 140, 141, 162
Ball, D. L., 232, 239, 246, 255
Ballenger, C., 5, 12, 14, 30, 40, 41, 175, 176, 189, 191, 193, 196, 197, 242, 255, 259
Barber, B., 161, 162
Barry, A., 11, 17
Barsalou, L. W., 25, 26, 36
Barton, A. C., 82, 100, 266, 285, 319, 320
Bastide, F., 115, 132
Bazerman, C., 32, 36, 42, 44, 84, 100
Becker, J., 140, 162
Beers, J., 63, 71, 282, 286
Begley, S., 82, 100
Bell, R. L., 81, 103
Biagioli, M., 176, 189
Bianchini, J. A., 81, 97, 100, 107, 108
Bisanz, G. L., 140, 163
Blanton, W., 299, 311
Bloome, D., 141, 142, 144, 162
Bloor, D., 82, 100
Bodwell, M. B., 33, 37
Bourdieu, P., 126, 131, 132, 265, 275, 285
Boutonné, S., 59, 69–71
Bowen, G. M., 2, 10, 17, 79, 81, 98, 104, 109, 112–116, 118, 123, 125, 126,

128, 130, 132, 133, 135, 169, 195, 197, 237, 257
Boyle, R. A., 203, 223
Bradshaw, G., 51, 67, 71, 319, 320
Brewer, W. F., 90, 101
Brice-Heath, S., 260, 264
Brickhouse, N. W., 63, 69, 84, 101
Britzman, D. P., 90, 101
Brookline Teacher Researcher Seminar, 178, 189
Brown, A. L., 25, 36
Brown, C. M., 80, 87, 88, 90, 92, 94, 100–102
Brown, D., 175, 189
Brown, J. S., 56, 69
Brown, N., 201, 224
Brownell, J. A., 242, 256
Burns, M. S., 20, 37
Bush, V., 294, 311

C

California Department of Education, 97, 101
Carambo, C., 63, 71, 282, 286
Carlsen, W. S., 81, 87, 97, 101, 102, 222
Cazden, C. B., 141, 162
Chang-Wells, G. L., 141, 164
Chen, C., 9, 17, 79, 84, 86, 87, 98, 99, 102
Chinn, C. A., 90, 101
Clandinin, D. J., 235, 256
Clement, J., 175, 189
Clinton, K., 176, 189
Cobb, P., 113, 133, 244, 245, 255
Cole, M., 16, 17, 200, 224, 276, 285, 295, 299, 309, 311, 312
Collins, A., 25, 36, 56, 69
Collins, H. M., 80, 82, 101
Collins, K. M., 11, 13, 14, 199, 200–203, 218, 222–224, 225
Conant, F., 176, 189
Cook-Gumperz, J., 96, 102
Cope, B., 175, 189
Crawford, T., 9, 17, 80, 84, 86–88, 90–92, 97–99, 101, 102
Cummins, J., 159, 162
Cunningham, C. M., 87, 97, 98, 101, 102
Cusick, P. A., 235, 255
Cutter, J., 200, 201, 224

D

Dargan, P., 112, 133
Davidson, D., 45, 46, 49, 69
DeBoer, G. E., 74, 77, 81, 88, 98, 101, 233, 255
Delpit, L., 75, 77, 175, 189, 244, 248, 249, 255
Denning, P., 112, 133
Denzin, N. K., 90, 101, 143, 162
Department of Education, 293, 294, 311
Derrida, J., 5, 45, 48, 68, 69
Dewey, J., 222, 223, 233, 255, 296, 311
DiSchino, M., 176–181, 183, 189, 193, 196
diSessa, A. A., 176, 189, 203, 224, 301, 311
Dixon, C. N., 140, 163
Driver, R., 10, 17, 140, 162, 239, 256, 317, 320
Duckworth, E., 241, 246, 256
Duguid, P., 25, 36, 56, 69
Duit, H., 47, 70
Duschl, R. A., 74, 77, 81, 85, 90, 101, 237, 256

E

Eckert, P., 55, 69, 261, 264
Edwards, D., 48, 69
Egan-Robertson, A., 141, 142, 144, 162
Ehn, P., 5, 67, 70
Einstein, A., 203, 223
Elmesky, R., 63, 71, 282, 286
Emerson, R. M., 96, 101
Engel, M., 299, 312
Engeström, Y., 25, 36, 276, 285
Englert, C. S., 200, 202, 224
Erickson, F., 204, 205, 223

F

Fairclough, N., 144, 162
Fine, M., 270, 285
Ford, D., 201, 224
Fordham, S., 237, 256, 261, 264
Forman, E., 202, 223
Foucault, M., 7, 17, 46, 70
Fourez, G., 49, 70
Fowler, A., 144, 162
Fox-Keller, E., 176, 189

AUTHOR INDEX

Fradd, S., 176, 190
French, A., 294, 311
Fretz, R. I., 96, 101
Freyberg, P., 239, 246, 256
Fujimura, J., 129, 133
Fuller, S., 80, 101

G

Gallas, K., 175, 189, 237, 256
Gallego, M. A., 16, 17, 225, 293, 299, 309, 311, 312, 315
Garfinkel, H., 42, 44, 82, 101
Garner, R., 295, 309, 311
Garrison, L., 311
Gee, J. P., 6, 7, 14, 19, 20, 24, 25, 32, 36, 37, 39, 42, 44, 45, 50, 58, 66, 70, 73, 140, 161, 162, 169, 170, 176, 189, 195, 202, 223, 233, 244, 249, 256, 259, 316
Gersten, R., 200, 223
Gertzog, W., 239, 257
Gilbert, A., 239, 256
Gillingham, M., 295, 309, 311
Glaser, B. G., 144, 163
Glaser, R., 295, 309, 312
Glenberg, A. M., 25, 26, 37
Goffman, E., 204, 223
Goldman, S. R., 140, 163
Goldstein, G., 176, 191
Gonzales, P., 43, 44, 176, 190
Good, L. R., 308, 311
Gooding, D., 176, 189
Goodlad, J., 235, 256
Goodwin, C., 90, 101, 187, 189
Green, J. L., 79, 87, 88, 99, 100, 102, 140, 162, 163
Griffin, P., 20, 37, 200, 224
Gross, P. R., 82, 102
Guba, E., 143, 163
Gump, P. V., 308, 311
Gumperz, J. J., 87, 88, 96, 102
Gutierrez, K., 141, 163

H

Halliday, M. A. K., 22, 30, 37, 98, 102, 140, 144, 163
Hamilton, R. J., 90, 101
Hammer, D., 176, 189, 190

Hanks, W. F., 28, 30, 37, 45, 63, 70
Harama, H., 46, 63, 70
Harding, P., 83, 84, 102
Hare, W., 83, 84, 102
Hasan, R., 144, 163
Heath, S. B., 159, 163, 175, 190
Heidegger, M., 49, 70, 196, 197
Helms, J. V., 98, 101
Hewson, P., 239, 257
Hicks, D., 208, 223
Hogan, K., 237, 249, 256
Hogan, P., 235, 256
Holland, D., 202, 222
Holzkamp, K., 5, 63, 70, 108
Hoving, T., 237, 257
Hudicourt-Barnes, J., 176, 189, 191
Hutchins, E., 25, 37
Hymes, D., 87, 102, 175, 190

J

Jacob, F., 176, 190
Jacoby, S., 43, 44, 176, 190
Jasanoff, S., 79, 102
Johnson, N. B., 308, 311
John-Steiner, V., 202, 223, 297, 298, 311

K

Kalantzis, M., 175, 189
Kamberelis, G., 159, 163
Keller-Cohen, D., 202, 223
Kelly, G. J., 3, 5, 8, 9, 10, 17, 39, 79–81, 84, 86–88, 90, 92, 94, 97, 99, 100–102, 105, 107, 108, 193
King, A. R., Jr., 242, 256
Klentschy, M., 311
Kline, M., 243, 256
Knorr-Cetina, K., 80, 102
Koertge, N., 82, 83, 102, 103
Kohl, H., 234, 254, 256
Kolpakowski, T., 176, 189
Kress, G., 140, 163
Kuhn, T. S., 232, 242, 256

L

Lampert, M., 232, 239, 256
Larson, J., 141, 163

Latour, B., 2, 17, 25, 80, 84, 99, 103, 110, 112, 113, 133, 176, 190, 242, 256
Lave, J., 25, 37, 108, 112, 133, 309, 311
Law, J., 117, 133
Lawless, D., 49, 70
Layton, D., 233, 256
Leach, J., 140, 162
Lederman, N. G., 81, 103
Lehrer, R., 176, 190
Lee, O., 176, 190
Lee, S., 51, 68–71, 227, 230
Lemke, J. L., 40, 44, 56, 63, 70, 87, 98, 103, 140, 141, 144, 163, 203, 209, 221, 223, 237, 242, 243, 256
Levitt, N., 82, 102
Lincoln, Y. S., 143, 162, 163
Lindfors, J. W., 159, 163
Livingston, E., 42, 44, 82, 101
Longino, H. E., 82, 103
Lopez, R., 294, 311
Lortie, D. C., 231, 256
Loving, C., 83, 84, 103
Lowmaster, N., 97, 101
Lucas, K. B., 239, 254, 257
Luster, B., 140, 164, 171, 174
Lynch, M., 42, 44, 80, 82, 84, 86, 99, 101, 103, 106, 112, 117, 133, 176, 190
Lyotard, J. F., 82, 86, 103
Lytle, J. H., 271, 285

M

Macbeth, D., 80, 86, 99, 103, 106
MacKnight, Y. -M., 63, 71
MacLean, F., 202, 223
Magnusson, S., 13, 199, 200–203, 206, 223–225
Mahn, H., 202, 223, 297, 298, 311
Marano, N. L., 200, 201, 224
Markle, G. E., 79, 102
Martin, E., 85, 103
Martin, J. R., 21, 22, 37, 98, 102, 140, 163
Martins, I., 140, 163
Masciotra, D., 195, 197
Matthews, M. R., 81, 83, 84, 103
Mayr, E., 242, 256
McDermott, L. C., 176, 190, 251, 256, 301, 312
McDermott, R., 309, 312
McDiarmid, G. W., 246, 255
McDonnell, L. M., 199, 223

McGinn, M. K., 2, 17, 59, 62, 70, 71, 79, 81, 84, 98, 104, 106, 108, 114, 115, 130, 133
McGullicuddy, K., 140, 163
McKay, S., 45, 70
McKnight, Y., 282, 286
McLaughlin, M. J., 199, 223
McMillan, R., 51, 67, 71, 319, 320
Mehan, H., 99, 103
Merton, R. K., 80, 81, 103
Michaels, S., 176, 190, 202, 214, 223, 224, 249, 256
Miettinen, R., 25, 36
Millar, R., 2, 17, 233, 254, 256
Minick, N., 202, 223
Minstrell, J., 176, 190
Mitroff, I., 80, 103
Moje, E., 222, 224
Mokros, J. R., 130, 133
Moll, L. C., 11, 142, 163, 202, 224, 309, 312
Monk, G. S., 177, 190
Monk, S., 179, 190
Morison, P., 199, 223
Mortimer, E., 140, 162
Mukerji, C., 80, 103
Mulkay, M., 80, 103

N

National Center for Education Statistics [NCES], 293, 312
National Commission on Mathematics and Science Teaching for the 21st Century [NCMST], 17, 293, 294, 312
National Reading Panel, 20, 37
National Research Council [NRC], 64, 70, 81, 97, 98, 103, 118, 133, 175, 190, 199, 224, 265, 285
National Science Foundation, 294, 312
Neale, D., 222, 224
Nemirovsky, R., 176, 190
Newman, D., 200, 224
Noble, T., 176, 191
Noë, A., 48, 70
Nystrand, M., 141, 163

O

Ochs, E., 43, 44, 176, 190

AUTHOR INDEX

O'Connor, M. C., 202, 214, 223, 224, 249, 256
Ogborn, J., 140, 163
Ogonowski, M., 176, 191
O'Regan, J. K., 48, 70
Osborne, R., 239, 246, 256
Oyler, C., 11, 17, 142, 162, 163

P

Palincsar, A., 13, 199, 200–203, 222–224, 225
Papert, S., 299, 312
Pappas, C. C., 11, 17, 139, 140–142, 159, 161–164, 171, 172, 174
Pauluth Penner, T., 51, 67, 71, 319, 320
Pedersen, J., 235, 237, 257
Penick, J. E., 205, 224
Petersen, J. C., 79, 102
Pfundt, H., 47, 70
Piaget, J., 269, 297, 312
Pickering, A., 49, 70
Pinch, T., 79, 102, 106, 176, 189
Pineda, E., 140, 164
Pontecorvo, C., 306, 312
Popper, K. R., 5, 232, 241, 256
Posner, G. J., 90, 104, 239, 257
Pothier, S., 176, 191
Potter, E., 67, 82, 100
Potter, J. T., 63, 69
Pozzer, L. L., 51, 71, 319, 320
Prothero, W., 87, 102
Punamaki, R. -L., 25, 36
Puttick, G. M., 33, 37

R

Resnick, L., 310, 312
Rheinberger, H., 176, 190
Ricœur, P., 63, 70
Riecken, J., 51, 67, 71, 319, 320
Rife, A., 17, 139, 143
Roberts, R. S., 205, 224
Robertson, D. A., 25, 37
Rodriguez, A. J., 3, 17, 97, 103
Rogoff, B., 201, 224, 295, 309, 310, 312
Root-Bernstein, R. S., 176, 190
Rorty, R., 49, 70–81, 85, 86, 100, 103
Roschelle, J., 203, 224

Rosebery, A. S., 30, 33, 37, 176, 177, 189–191
Rosenquist, M. L., 176, 190, 251, 256
Roth, W. -M., 1, 2, 5, 7, 8, 17, 39, 45, 46, 49–51, 54, 55, 59, 62, 63, 67, 68, 70, 71, 73, 76, 77, 79, 81, 84, 98, 104–106, 108, 111–116, 118, 123, 125, 126, 130, 132, 133, 135, 169, 193, 195, 197, 225, 227, 230, 239, 246, 254, 257, 259, 282, 286, 287, 315, 319, 320
Roychoudhury, A., 54, 55, 71
Rueda, R., 309, 312
Russell, T., 239, 254, 257
Rutherford, F. J., 294, 312
Rymes, B., 141, 163

S

Säljö, R., 48, 71
Salley, C., 270, 286
Schaffer, S., 176, 189
Schauble, L., 176, 190, 295, 309, 312
Schoultz, J., 48, 71
Schultz, T., 294, 311
Schwab, J. J., 5, 9, 13, 14, 18, 77, 81, 104, 202, 203, 205, 222, 224, 231, 232, 254, 257
Scott, P., 140, 162
Scott, T., 299, 312
Scribner, S., 295, 312
Seiler, G., 278, 286
Shaffer, P. S., 251, 256
Shapiro, B., 246, 257
Shaw, L. L., 96, 101
Sherin, B., 176, 189
Sleeter, C., 202, 224
Smith, D. C., 93, 94, 100, 104, 222, 224, 254, 257
Smith, E., 239, 255
Smith, J. P., 203, 224
Snow, C. E., 20, 37
Sohmer, R., 176, 190
Stone, A., 202, 223
Storr, B., 51, 67, 71, 319, 320
Strauss, A. L., 144, 163
Strike, K. A., 80, 81, 85, 90, 97, 104, 239, 257
Suchman, L. A., 119, 134, 259, 264
Sutton, C., 87, 104, 140, 164
Suzuki, S., 40, 44

Swartz, C., 294, 312
Szymanski, M. H., 96, 102

T

Tait, D., 71, 51, 67, 319, 320
Templin, M., 201, 203, 223
Third International Mathematics and Science Study [TIMSS], 1, 18
Tierney, C., 176, 190
Tinker, R. F., 130, 133
Tobin, K. G., 14, 15, 63, 71, 76, 77, 242, 257, 265, 266, 278, 285–287
Tomasello, M., 28–30, 37
Traweek, S., 2, 18, 80, 104, 111, 134, 136, 138, 242, 257, 261, 264
Trent, S., 200, 202, 224
Turnbull, D., 2, 18, 99, 104

V

Valencia, R., 200, 224
Valsiner, J., 296, 312
van der Veer, R., 296, 297, 312
van Oers, B., 309, 312
van Zee, E., 176, 190
Varelas, M., 11, 17, 139, 140, 142, 161, 162, 164, 169, 171, 172, 174, 315, 317, 320
Viennot, L., 176, 191
Vygotsky, L. S., 5, 13, 16, 18, 30, 140, 164, 170, 202, 204, 295, 297, 298, 306, 307, 310, 312, 315, 316–318, 320

W

Wacquant, L. J. D., 126, 132
Wade, P. D., 81, 103
Wallat, C., 88, 99, 102
Walls, E., 278, 286
Warren, B., 30, 176, 189, 191
Watson-Verran, H., 2, 18, 99, 104
Wells, G., 140, 141, 164, 307, 313
Wenger, E., 25, 37, 112, 133, 134
Wenzel, S., 140, 164, 171, 174
Wertsch, J. V., 5, 16, 18, 140, 164, 202, 224, 295, 306, 313
Winer, L., 56
Wittgenstein, L., 49, 71, 85, 86, 88, 104
Wolcott, H. F., 143, 164
Wollman-Bonilla, J. E., 159, 164
Wong, S. -L., 45, 70
Wood, T., 244, 245, 255
Woolgar, S., 2, 17, 84, 103, 110, 112, 113, 133, 176, 190, 242, 256
Woszczyna, C., 59, 70, 71
Wright, T., 176, 190
Wyndhamn, J., 48, 71

Y

Yackel, E., 244, 245, 255
Yerrick, R., 1, 14, 39, 46, 55, 73, 105, 135, 169, 193, 225, 231, 235, 237, 239, 241, 242, 246, 256, 257, 259, 287, 315

Z

Zecker, L. B., 140, 141, 162, 164
Zeitsman, A., 175, 176, 189
Zhao, Y., 295, 309, 311

Subject Index

A, B

Academic language
  acquisition of, 24, 39–42, 298
  defined, 20, 21, 187
After-school programs, *see* Out-of-school programs
Alphabetic code, 20
Assessment, 2, 4, 5, 40, 65, 74, 80, 81, 98, 107, 118, 170, 229, 241, 254, 267–269, 281, 288, 293, 298, 308, 319, 320
Authentic science, 10, 105, 106, 111–113, 135, 137, 319
Bilingualism, 8, 12, 56–59, 61–63, 188, 200

C

California standards, *see* Standards-based teaching approach
Chéche Konnen Center (CKC), 12, 177, 178, 194
Child-centered instruction, *see* Student-centered instruction
Classroom management, *see* Discipline in classrooms
Code switch, 45, 49, 281
Communities of practice, 2, 6, 7, 9, 10, 13, 112, 130, 132, 175, 211, 222
Comparing scientific and everyday language, 67, 82, 87, 111, 149, 159, 161, 169, 175, 186, 231, 296–298, 301, 302, 305–310, 317, 318
Comparing scientific and school communities, 10, 14, 16, 43, 74, 75, 99, chapter 4, 169, 173, 201, 216, 217, 228, 232, 242, 243, 253, 260
Concept formation (Vygotskian), 16, 17, 295, 297–300, 310, 316
Concept mapping, 52–55
Conceptual profile, 230
Constructivism, 83, 88, 140, 142, 175, 200, 202
Cognitive psychology, 26
Contextualization of language, 50, 51, 66, 88, 97, 158, 183, 186, 187, 265
Cookbook activities, 113, 114, 116, 118, 123, 283
Cultural capital, 7, 11, 108, 162, 265, 271, 273, 276–278, 281, 282, 289, 290
Cultural knowledge, 12, 13, 46, 50, 55, 84, 172, 194, 196, 234, 259, 260, 266, 274, 277–282, 290

327

Cultural reproduction, 7, 8, 46, 50, 62, 68, 76, 106, 108, 196, 235, 243, 281, 288

D

Deficit approach, 11, 13, 161, 200, 225, 229
Descriptive studies, 9, 10, chapter 3
 defined, 80
 norms and, 80, 98, 100, 106
 of scientific practice, *see* Science studies
Dialogic inquiry, 11, 139–143, 152, 156, 159, 208, 216
Disabilities, students with, *see* Special needs, students with
Discipline in classrooms, 2, 273–282, 289, 290
Discourse, defined, 45, 237
Discourse acquisition, 7, 11, 262, *see also* Language acquisition

E

Epistemology, 3, 9, 35, 36, 80, 82–84, 86, 99, 106, 118, 127, 137, 230, 233, 237, 308
Equity issues, 3, 43, 75, 194, 199, 233, 242–245, 263, 266, 319, 320
Everyday language, 32, 33, 43, 50, 59, 61, 161, 175, 196, 297, 301, 302, 305–311, 316, 317, *see also* Lifeworld language
Expanded language, 33, 36

G

Gap gazing, 3
Guided inquiry, 13, 200, 201, 204–208, 211, 215, 216, 221, 222, *see also* Inquiry
Guided Inquiry supporting Multiple Literacies (GIsML), 13, 201, 202, 205–208, 216, 226

H, I

Habitus, 15, 275, 276, 283, 285

Heteroglossic language, 51, 64, 67, 69, 319
Identity issues, *see* Social language
Initiation/response/evaluation (IRE), 141, 159
Inquiry, 92, 93, 106, 115, 126, 130, 137–140, 159, 175, 177, 186, 188, 194, chapter 7, 226, 232, 242, 243, 251, 266, 317
Inscriptions, 113–116, 118, 121–123, 125, 129, 130
Intersubjectivity, 214, 227
Intertextuality, 11, chapter 5, 169–172

L

Language
 acquisition, 19, 22–24, 28, 30, 33, 36, 39–42, 46, 48, 64, 73, 139, 170, 186, *see also* Discourse acquisition
 defined, 45, 51, 63, 68
 emergence of, 49, 54, 61, 64
 evolution of, 48, 49, 51, 54, 55, 127
 language-in-use, 47
 uses of, 85–87, 99, 175, 184–187, 194, 196, 203, 214, 295, 306, 307, 310
Learning disabilities, *see* Special needs, students with
Learning from children, *see* Professional development
Lifeworld language, 22–24, 27, 30, 40–42, 66, 169–171, 195, 196, 261, 262, 277, 281
Lower track students, 14, 74, 233, 234, 237, 239, 241, 245, 261, 263, 266

M

Marginalized students, 14, 43, 225, 245
Meaning-in-use, 28
Meaning making, *see* Sense making
Mediation, 200, 226, 295, 306
Messy science, 2, 110, 125, 126, 129, 174, 243
Metadiscourse, 14, 235, 249, 251, 253
Misconceptions, 47, 229, 230, 241, 247, 251, 259
Monoglossic language, 51, 64, 67, 74, 319
Multiculturalism, 68, 195

SUBJECT INDEX

Multiple literacies, 13, 201, 202, 227, *see also* GIsML

N

Nature of science, 1, 2, 4, 9, 10, 16, 79, 81, 83, 85–88, 99, 105–107, 110, 131, 249, 253, 255, 261, 262
Norms, 89, 90, 99, 183, 186, 216, 226, 228, 230, 233, 235, 242, 246, 251, 253, 263, 306, 320

O, P

Out-of-school programs, 16, 17, 294, 295, 298, 299, 308–310, 315, 318–320
Perspective taking, 28–30
Postmodernism, 82, 83, 86, 90
Professional development, 4, 273, 319
  learning from children, 14, 177, 178, 188, 193–196, 259, 260, 278, 291
  teacher knowledge, 3, 106, 126, 222, 267, 320
  teacher training, 12, 105, 106, 138, 177, 195, 268, 269, 273, 299
Puzzling children, 12, 176–180, 187, 188

R

Reform, science education
  historical and current foci, 2, 3, 39, 81, 111, 118, 225, 232, 253, 266, 287, 293, 294, 298, 301, 311, 318–320
  equity issues in, 2, 3, 6, 12–15, 194, 225, 242, 266, 294
Registers (linguistic), 8, 73, 75, 87, 140, 147–149, 151, 155, 157, 159, 172, 173, 317
Relativism, 82–84
Revoicing, 147, 156, 213–217, 220, 221, 226

S

Sabir, 49, 51, 52, 59, 64, 67–69, 75
Scaffolding, 14, 33, 36, 59, 161, 173, 213, 214, 222, 226, 227, 255
Science for All, 3, 5, 13, 62, 68, 75, 81, 194, 199, 201, 205, 208, 221, 232, 265, 284, 288, 294

Science-in-the-making, 9, 99, 110, 113, 127, 131
Science studies, 9
  criticisms of, 82, 83
  defined, 79
  norms and, 80, 81
  purpose of, 84–87, 98, 99, 106
Scientific knowledge, *see also* Nature of science
  building, 126, 130, 137, 143, 151, 158, 159, chapter 7, 228–230, 237, 239, 246, 251, 255, 270, 283, chapter 10
  classroom, 111, 114, 118, 141–143, 149, 151, 159, 200, 233, 237, 241, 254, 270, 306, 315
  scientists' views of, 110, 125, 136, 228, 242, 243, 252
  subjectivity of, 110, 242, *see also* Messy science
  what counts as, 110, 111, 131, 141–143, 149, 176, 202, 222, 238–242, 296, 297, 307, 311, 318–320
Scientific language, *see also* Academic language
  acquisition of, 46, 48, 49, 59, 62, 74, 140, 149, 161, 171, 174, 233, 262, 263, 295, 302–306
  defined, 50, 68, 203
Scientific literacy, 11, 113–116, 131, 140, 141, 149, 154, 157, 159–162, 169–173, 283, 316, 319
  as focus of reform, 2, 81, 98, 199, 232, 319
  defined, 171, 172
  school-based literacy, 6, 16
Scientific method, 2, 110
Scientistic attitudes, 131
Second language, *see* Bilingualism
Sense making, 2, 11, 26, 32, 35, 42, 56, 94–96, 110, 114, 116, 129, 139–141, 149, 154, 157–162, 169–171, 176, 202, 205, 270, 283, 306
Situated activity, 20, 86, 87, 200, 202, 259
Situated cognition, 25, 90
Situated meanings, 8, 23–28, 36, 40, 41, 88, 96, 262
Small learning communities (SLC), 266, 270–273, 276, 281
Social language
  acquisition of, 22, 23, 28, 30, 33, 36, 40–42, 75
  defined, 20, 21

Social language *(cont.)*
  identity issues and, 40, 41, 45, 46, 51, 62, 63, 142, 152, 263, 283, 284
  situated meanings and, 24–28, 36, 40, 41
Socioculturalism, 11, 16, 48, 140, 142, 161, 200, 226, 295, 296
Sociolinguistics, 3, 87, 89, 96, 99, 233, 242
Special needs, students with, 13, 59–61, 199–201, 205, 208, 210, 218, 221, 225, 226, 229
Spontaneous concepts, 170, 243, 296–298, 300, 303, 305, 307, 310, 315–319
Standards-based teaching, 15, 74, 97, 98, 107, 108, 195, 199, 225, 226, 229, 265, 287–289
Student-centered instruction, 195, 270, 273, 277, 278, 282
Substantive structures, 13, 14, 110, 202–205, 208, 212, 214, 221, 222, 229
Syntactic structures, 13, 14, 110, 202–205, 208–222, 228, 229

T

Teacher knowledge, *see* Professional development

Teacher role, 55, 56, 69, 74, 76, 91–93, 98, 105, 107, 127, 132, 135, 136, 140, 142, 147, 149, 153, 154, 158–161, 173, 177, 183, 186, 195, 196, 201, 205, 209, 211–217, 222, 226, 228, chapter 8, 259, 260, 267, 270, 274–281, 288, 289, 305
Teacher training, *see* Professional development
Tools, 26, 52, 76, 113, 124–126, 129–132, 179, 182, 186, 208–214, 218–222, 226, 247, 251, 294
  conceptual, 209–211, 221, 228
  cultural, 140, 289, 317
  discursive, 218, 262
  language as, 16, 41, 54, 140, 208, 295, 306, 310
  scientific, 10, 131, 132, 172, 175, 202, 235, 269
  using, 10, 112, 124, 131, 132, 172, 210, 214, 220, 226

Z

Zone of proximal development, 226, 227, 297, 317